Student Book

To support the Linked Pair pilot scheme

NEW GCSE MATHS LINKED PAIR
AQA GCSE in
Methods in Mathematics & Applications of Mathematics

Matches the AQA GCSE Specification for 2011 onwards

Keith Gordon • Trevor Senior

William Collins' dream of knowledge for all began with the publication of his first book in 1819. A self-educated mill worker, he not only enriched millions of lives, but also founded a flourishing publishing house. Today, staying true to this spirit, Collins books are packed with inspiration, innovation and practical expertise. They place you at the centre of a world of possibility and give you exactly what you need to explore it.

Collins. Freedom to teach.

Published by Collins
An imprint of HarperCollins*Publishers*
77–85 Fulham Palace Road
Hammersmith
London
W6 8JB

10 9 8 7 6 5 4 3 2

ISBN-13 978-0-00-741005-7

British Library Cataloguing in Publication Data
A Catalogue record for this publication is available from the British Library.

Commissioned by Katie Sergeant
Project managed by Emma Braithwaite
Edited by Joan Miller
Proofread by Joan Miller and Marie Taylor
Answers checked by Joan Miller and Marie Taylor
Exam board specification checked by Keith Gordon and Trevor Senior
Design and typesetting by Graham Brasnett
Concept design by Nigel Jordan
Illustrations by Ann Paganuzzi
Index by Michael Forder
Cover design by Angela English and Julie Martin
Production by Kerry Howie
Printed and bound by Printing Express, Hong Kong

Browse the complete Collins catalogue at: www.collinseducation.com

Acknowledgements

The publishers wish to thank the following for permission to reproduce photographs. Every effort has been made to trace copyright holders and to obtain their permission for the use of copyright material. The publishers will gladly receive any information enabling them to rectify any error or omission at the first opportunity.

Pretzels p.6 © zoranm/istockphoto.com; Talking Heads p.6 © René Mansi/istockphoto.com;
The Gherkin p.20 © kmiragaya/shutterstock.com; Student p.38 © kate_sept2004/istockphoto.com;
Tropical beach with palms p.62 © Kushch Dmitry/shutterstock.com; Cargo ship p.90 © Rafael Ramirez Lee/shutterstock.com;
Traffic jam p.106 © McCaig/istockphoto.com; Horse race p.130 © sylvaine Thomas/shutterstock.com;
Man on mobile phone p.130 © Phil Date/shutterstock.com; Sunflower p.152 © vladacanon/istockphoto.com;
Floor pattern p.170 © apdesign/shutterstock.com; Brick wall p.170 © syzx/shutterstock.com;
Aeroplane p.186 © Fernando Jose Vasconcelos Soares/shutterstock.com; Railway tracks p.186 © Péter Gudella/Shutterstock.com;
BP oil spill p.204 © Keystone USA-ZUMA/Rex Features.

With special thanks to Mark Burley (Cape Cornwall School), Andrea Frame (Walton High), Richard Furniss (Retford Oaks High School), Debbie Guy (St Aldhelm's Academy), Marek Milejski (The King David High School), Karen Moss (Wellgrove School) and Peter Roberts (The Crypt School).

INTRODUCTION

This book should be used alongside **Collins New GCSE Maths AQA Modular course** and provides all the extra Foundation and Higher content that you need to study for the Linked Pair exams. The content is divided into the four main sections of your course. These sections are colour-coded:

M1 – Methods in Mathematics

A1 – Applications of Mathematics

M2 – Methods in Mathematics

A2 – Applications of Mathematics

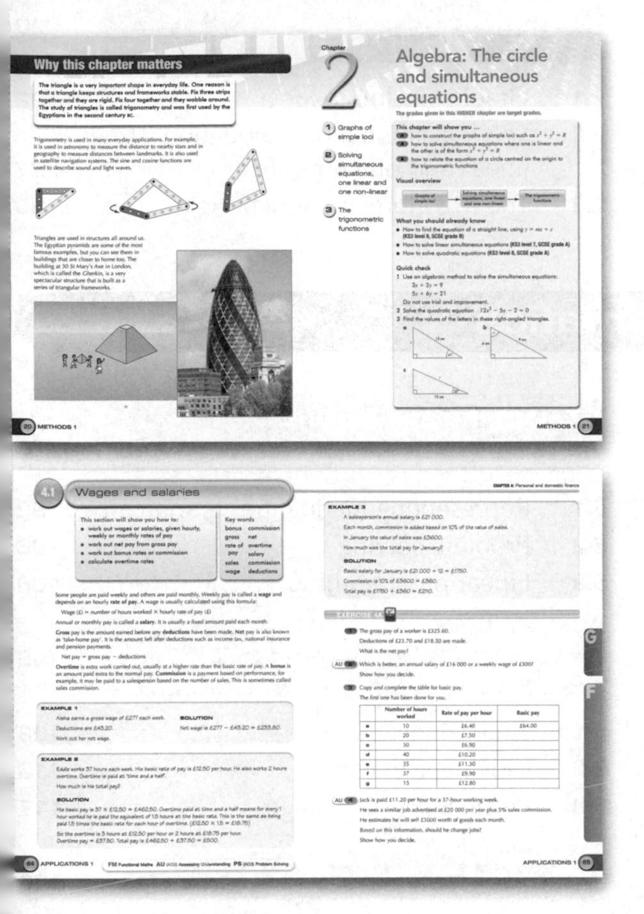

Why this chapter matters

Find out why each chapter is important and how maths affects the world around us. See how maths links to other subjects and is used in the real world.

Chapter overviews

Look ahead to see what maths you will be doing and how you can build on what you already know.

Colour-coded grades

Know what target grade you are working at and track your progress with the colour-coded grade panels at the side of the page in each exercise.

Use of calculators

Look for the icon to see where you must or could use a calculator to do exercises.

Grade booster

Review what you have learnt and how to improve your grade with the Grade Booster panel at the end of each chapter.

Worked examples

Before you start an exercise, take a look at the examples in the blue boxes. These take you through questions step by step and help you understand the topic.

Functional maths

Discover how people use maths in everyday life and practise your functional maths skills by looking out for questions marked **FM**.

New Assessment Objectives

Practise new parts of the curriculum (Assessment Objectives AO2 and AO3) with questions that assess your understanding, marked **AU**, and questions that test if you can solve problems, marked **PS**. There are also plenty of questions that test your basic maths skills (AO1).

Exam practice

Prepare for your exams with exam-style questions and detailed worked exam questions. Find extra hints and tips to help you score maximum marks.

CONTENTS

Unit M2 - Methods in Mathematics

Unit A2 - Applications of Mathematics

Endmatter

Why this chapter matters

Mathematicians like to be organised. One way of being organised is by sorting things into sets that have properties in common or by using patterns.

A set can contain anything that we choose. For example, it may be a set of numbers, a set of objects, a set of people, a set of chemicals.

Sets can be large or small.

Sets can be used to help to solve problems.

For example:

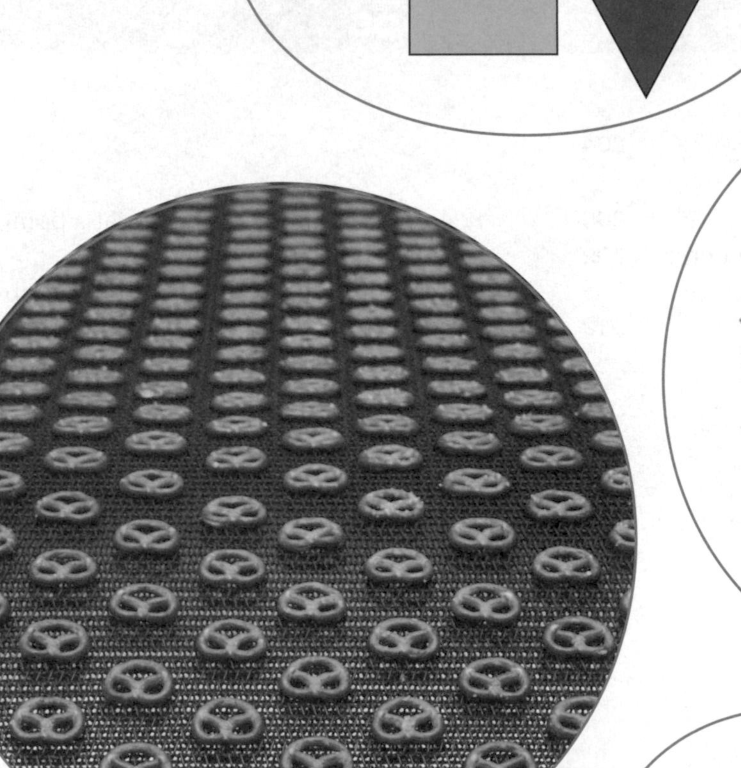

3 9 1 11 7 5

Chapter 1

Probability: Set notation and Venn diagrams

The grades given in this FOUNDATION and HIGHER chapter are target grades.

This chapter will show you …

- **D** how to understand and use set notation to describe events
- **C** how to use Venn diagrams to solve probability problems
- **B** how to shade Venn diagrams to show unions (∪) and intersections (∩) of sets

Visual overview

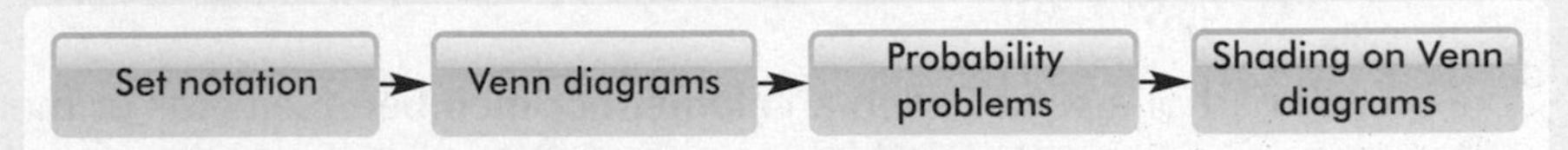

What you should already know

- How to add and subtract decimals between 0 and 1 **(KS3 level 4, GCSE Grade G)**
- How to list all the outcomes of an event in a systematic manner **(KS3 level 5, GCSE grade E)**
- How to calculate simple probabilities **(KS3 level 5, GCSE grade E)**
- That the total probability of all outcomes in a particular situation is 1 **(KS3 level 6, GCSE grade D)**

Quick check

1 Students can choose to take two languages from French, German and Spanish.
List all the possible choices.

2 There are three red, four blue and five white counters in a bag.
A counter is chosen at random.
Write down the probability of choosing:

a a red counter

b a white or a red counter

c a counter that is not red.

3 An ordinary dice is rolled.
What is the probability of its landing on:

a an even number

b a number greater than 1

c 5 or 6?

1.1 Set notation

This section will show you how to:

- understand probability notation
- use set notation with probabilities

Key words
complement
element
empty set (∅)
member
probability (P)
set
universal set (ξ)

Capital letters are often used to represent a **set**. For example, the set of odd numbers less than 10 could be represented by A.

$A = \{1, 3, 5, 7, 9\}$

Each part of a set is called an **element** or **member**. Elements or members of a set are usually written inside curly brackets.

The **probability** of event A happening is written **P(*A*)**.

Suppose event A does not happen. This is written as A'. This is called the **complement** of A but is read as 'A dash'.

The probability of event A **not** happening is written **P(*A′*)**.

The **universal set** is a set that contains all elements used and is written ξ.

The **empty set** is a set containing **no** elements and is written $\varnothing$.

EXAMPLE 1

You are given that $P(A) = 0.6$

Work out $P(A')$.

SOLUTION

$P(A') = 1 - P(A) = 1 - 0.6 = 0.4$

EXAMPLE 2

$\xi = \{1, 2, 3, 4, 5, 6, 7, 8, 9, 10\}$ $A = \{1, 3, 5, 7, 9\}$ $B = \{2, 3, 5, 7)$

A number is chosen at random from the universal set. Write down:

a $P(A)$ **b** $P(B')$

SOLUTION

a There are five elements to choose from out of 10 so $P(A) = \frac{1}{2}$

b There are four elements in B and six elements that are **not** in B, so $P(B')$ is $\frac{6}{10} = \frac{3}{5}$

FM Functional Maths **AU** (AO2) Assessing Understanding **PS** (AO3) Problem Solving

EXAMPLE 3

A = {red, green, yellow, blue, black, white}

B = {purple, green, yellow}

What is the probability that a colour chosen at random from A is also in B?

SOLUTION

Only green and yellow are in both sets so, from A:

P(green or yellow) is $\frac{2}{6} = \frac{1}{3}$

EXERCISE 1A

D

1 $\xi = \{1, 2, 3, 4, 5, 6, 7, 8, 9, 10, 11, 12\}$

$A = \{1, 5, 8, 10, 12\}$

$B = \{2, 3, 4, 6, 8, 10\}$

An element is chosen at random from ξ. Write down in its simplest form:

a $P(A)$ **b** $P(A')$

c $P(B)$ **d** $P(B')$

AU **2** $\xi = \{1, 2, 3, 4, 5, 6, 7, 8, 9, 10\}$

$A = \{1, 3, 5, 7, 9\}$

$B = \{2, 4, 6, 8, 10\}$

a What do you notice about A and B'?

b What do you notice about A' and B?

3 $\xi = \{1, 2, 3, 4, 5, 6, 7, 8, 9, 10\}$ A = {multiples of 3} B = {prime numbers}

a List the elements of A.

b Write down $P(A)$.

c Write down $P(A')$.

d List the elements of B.

e Write down $P(B)$.

f Write down $P(B')$.

4 $P(A) = 0.1$ and $P(B) = 0.3$. Write down:

a $P(A')$ **b** $P(B')$

5 $P(A) = 0.25$ and $P(B) = 0.55$. Write down:

a $P(A')$ **b** $P(B')$

1.2 Venn diagrams and probability

This section will show you how to:
- use set notation with probabilities
- understand the meaning of union (∪) and intersection (∩)

Key words
intersection (∩)
union (∪)
Venn diagram

Diagrams that represent the connection between different sets are called **Venn diagrams** and are named after John Venn who introduced them in about 1880.

Below is a Venn diagram for two sets.

Let A represent the number of students who study French and B represent the number of students who study Spanish.

12 students study French only.
15 students study Spanish only.
4 students study both French and Spanish.

16 students study French.
19 students study Spanish.
All students study at least one language.

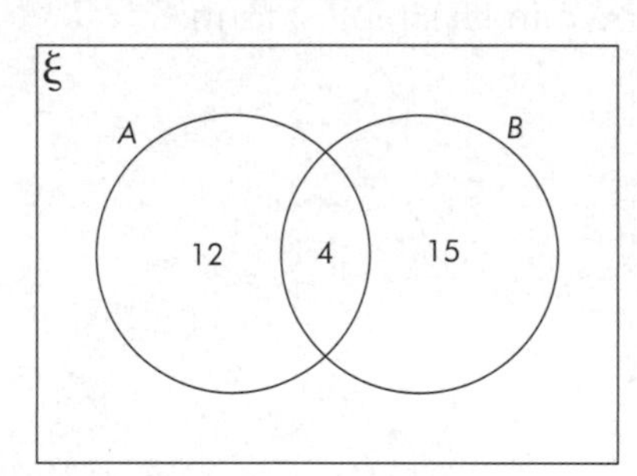

The overlapping part of the sets is called the **intersection** and is written as $A \cap B$.

The combined set that contains all of A and all of B is called the **union** and is written $A \cup B$.

The universal set, ξ, is shown as a rectangle.

EXAMPLE 4

ξ = {1, 2, 3, 4, 5, 6, 7, 8, 9, 10}

A = {2, 3, 4, 5}

B = {4, 5, 6, 8}

a Show this information in a Venn diagram.

b Use the Venn diagram to work out:

i P(A) **ii** $P(A \cup B)$ **iii** $P(A \cap B)$

SOLUTION

a

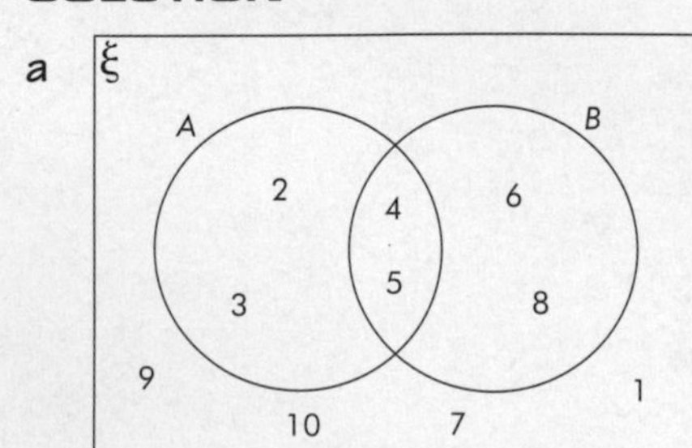

b **i** $P(A) = \frac{4}{10} = \frac{2}{5}$

ii $P(A \cup B) = \frac{6}{10} = \frac{3}{5}$

iii $P(A \cap B) = \frac{2}{10} = \frac{1}{5}$

EXAMPLE 5

The Venn diagram shows the numbers of people with fair hair (A) and the number of people with blue eyes (B).

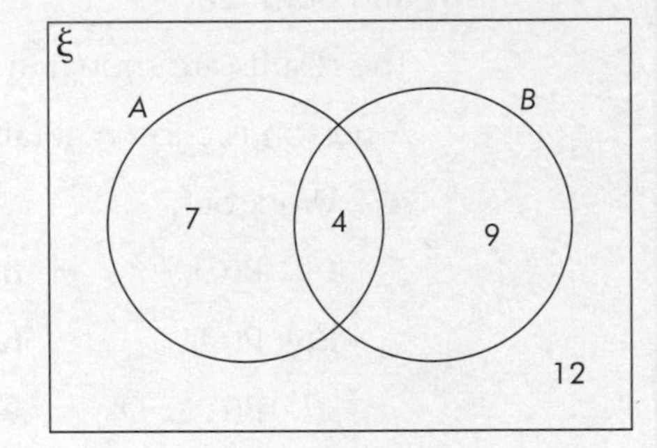

a How many people are there altogether?

b What is the probability that a person chosen at random has blue eyes?

c Work out $P(A')$

d Work out $P(A \cap B)$

e Work out the probability that a person chosen at random has fair hair but does not have blue eyes.

SOLUTION

a $9 + 4 + 7 + 12 = 32$

b There are $9 + 4 = 13$ with blue eyes, so $P(B) = \frac{13}{32}$

c A' means not in A or does not have fair hair.

$9 + 12 = 21$ do not have fair hair, so $P(A') = \frac{21}{32}$

d $A \cap B$ means has fair hair **and** has blue eyes. There are only four of these, so $P(A \cap B)$ is $\frac{4}{32} = \frac{1}{8}$

e P(fair hair but **not** blue eyes) $= P(A \cap B') = \frac{7}{32}$

EXERCISE 1B

1 $\xi = \{1, 2, 3, 4, 5, 6, 7, 8, 9, 10\}$ $A = \{1, 2, 4, 8\}$ $B = \{1, 3, 4, 9, 10\}$

a Show this information in a Venn diagram.

b Use the Venn diagram to work out:

i $P(A)$ **ii** $P(A')$ **iii** $P(B)$

iv $P(B')$ **v** $P(A \cup B)$ **vi** $P(A \cap B)$

AU 2 The Venn diagram shows the numbers of students who walk to school (S) and the number of students who walk home from school (H).

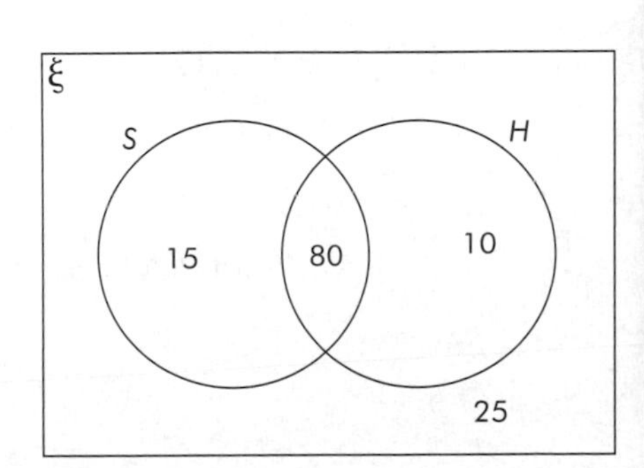

a How many students are there altogether?

b **i** Work out $P(S \cap H)$.

ii Describe in words what $P(S \cap H)$ represents.

c A student is chosen at random.

What is the probability that the student only walks one way, either to or from school?

C

3 A survey asked 100 people if they liked cats (C) and dogs (D).

The results are shown in the Venn diagram.

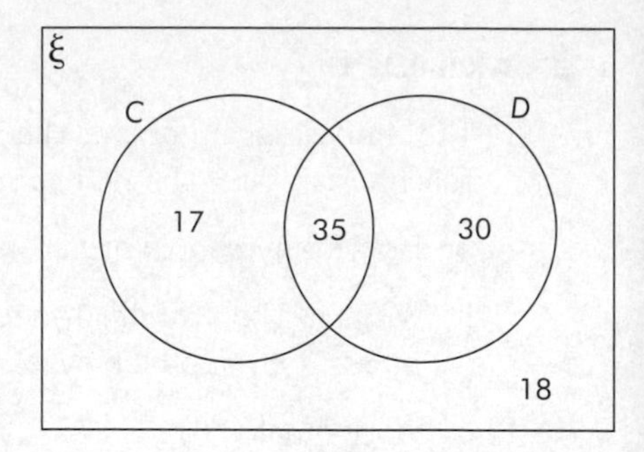

A person is chosen at random.

a Work out:

i P(C) **ii** P(C')

iii P(D) **iv** P(D')

v P($C \cup D$) **vi** P($C \cap D$).

b Work out the probability that a person likes dogs but does not like cats.

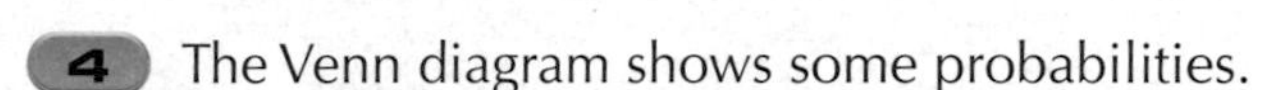

4 The Venn diagram shows some probabilities.

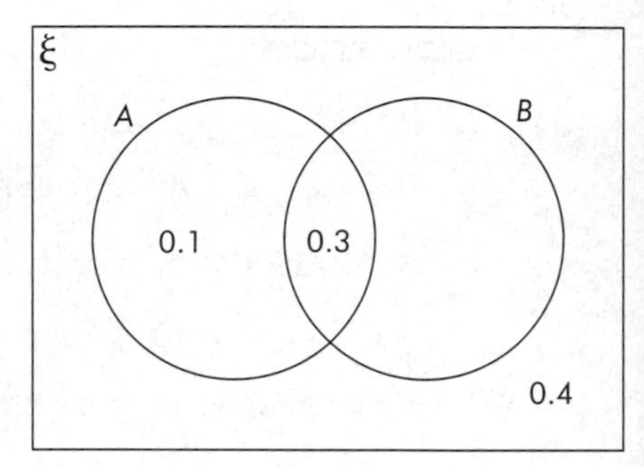

a Copy and complete the Venn diagram.

b Work out:

i P(B)

ii P($A \cup B$)

iii P($A \cap B$).

5 The Venn diagram shows some probabilities.

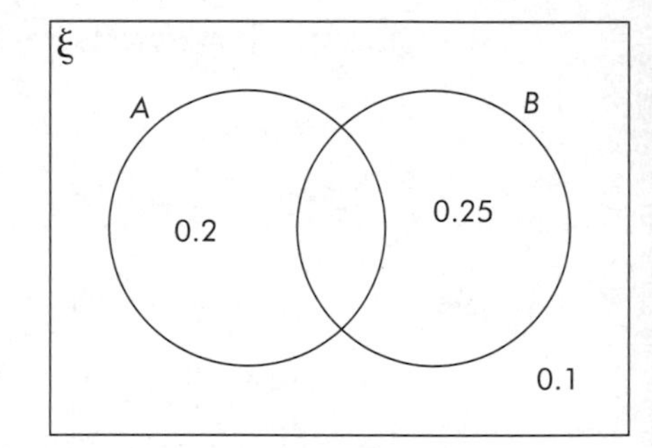

a Copy and complete the Venn diagram.

b Work out:

i P(A)

ii P(B)

iii P($A \cup B$)

iv P($A \cap B$).

PS 6 P(A) = 0.7

P(B) = 0.6

P($A \cup B$) = 0.9

Work out P($A \cap B$).

7 P(A) = 0.12

P(B) = 0.45

P($A \cap B$) = 0.07

Work out P($A \cup B$).

1.3 Shading on Venn diagrams

This section will show you how to:

- shade areas of Venn diagrams to show unions (∪) and intersections (∩)

Key words
complement
element
intersection (∩)
member
union (∪)

You have used **union** (∪) and **intersection** (∩) for two sets, for example, $A \cup B$ and $A \cap B$.

These can be represented by shading on a Venn diagram. Diagrams for the first four shadings are given on the formula sheet for M1, but for this unit you also need to know how to shade all other possible areas.

All **elements** (or **members**) of A will be in the shaded area.

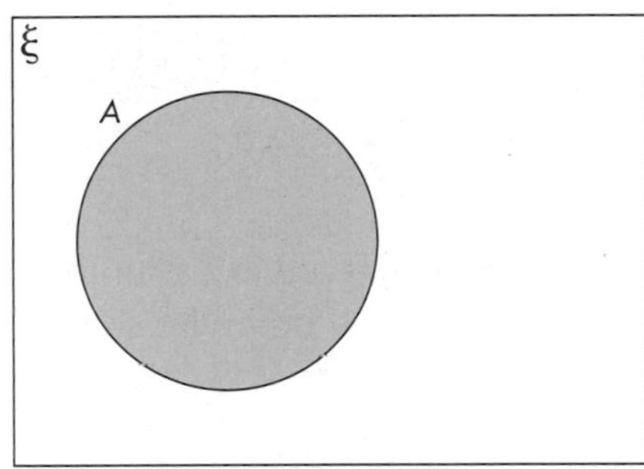

A is shaded

All elements that are in the **complement** of A, so they are **not** in A, will be in the shaded area.

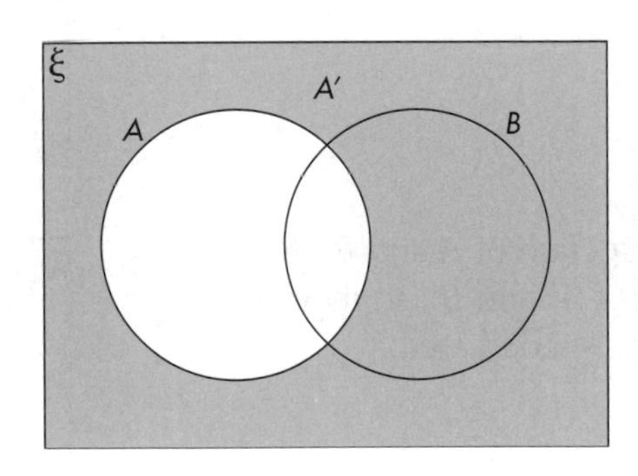

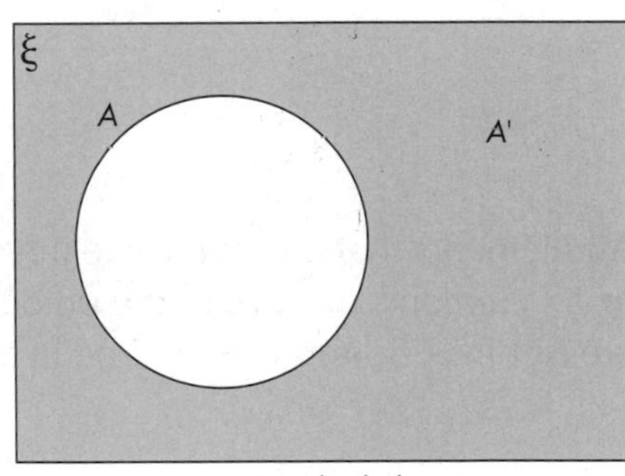

A' is shaded

All elements that are in the intersection of A and B, so they are in both A and B, will be in the shaded area.

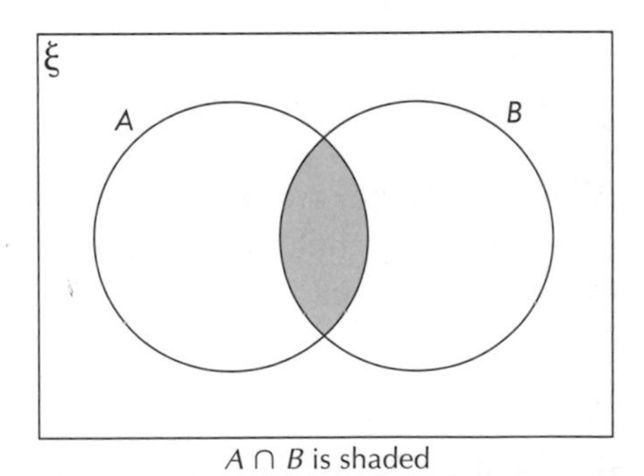

$A \cap B$ is shaded

All elements that are in the union of A and B, so they are in A or B or both, will be in the shaded area.

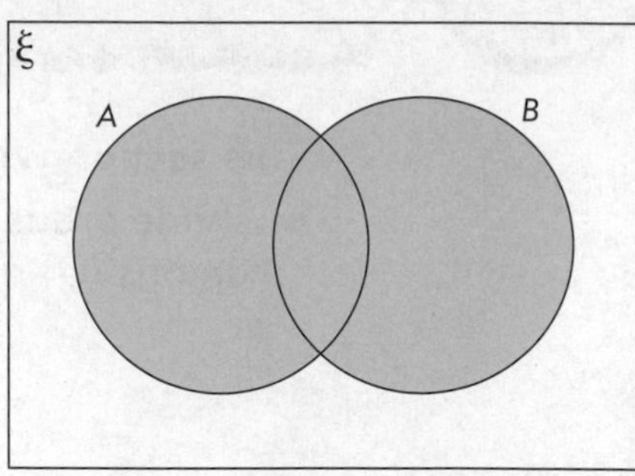

$A \cup B$ is shaded

All elements that are in the intersection of A' and B, so they are in B but **not** in A, will be in the shaded area.

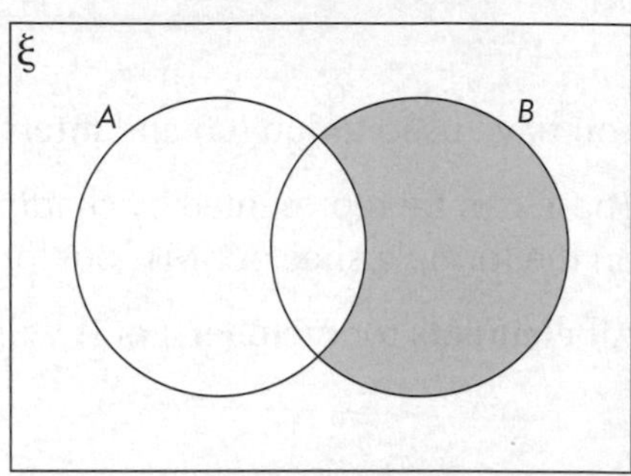

$A' \cap B$ is shaded

All elements that are in the union of A' and B, so they are **not** in A but in B, will be in the shaded area.

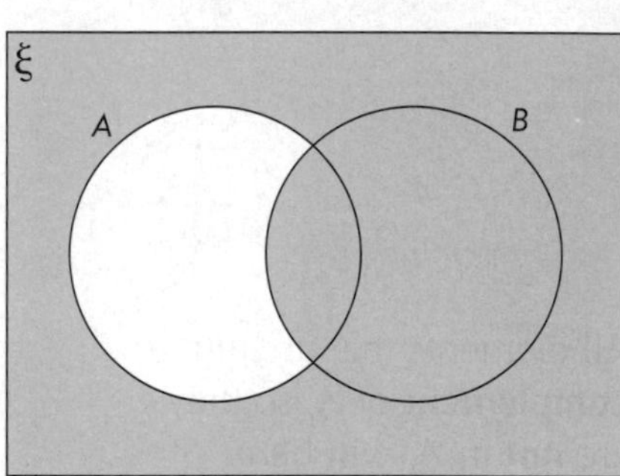

$A' \cup B$ is shaded

All elements that are **not** in the intersection of A and B **or** all elements that are in the union of A' and B', so they are **not** in A or **not** in B, will be in the shaded area.

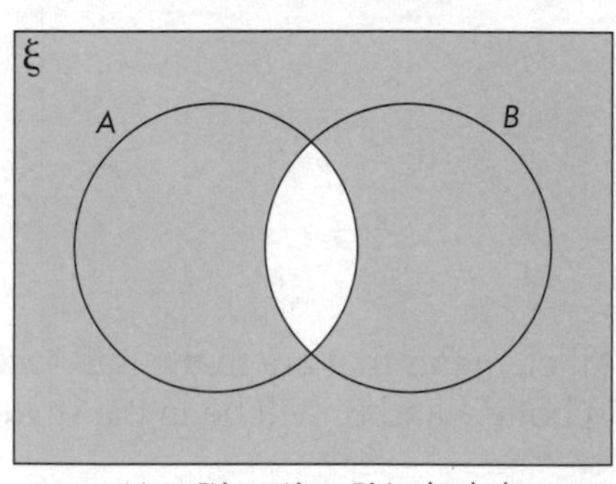

$(A \cap B)' = A' \cup B'$ is shaded

All elements that are **not** in the union of A and B **or** all elements that are in the intersection of A' and B' will be in the shaded area.

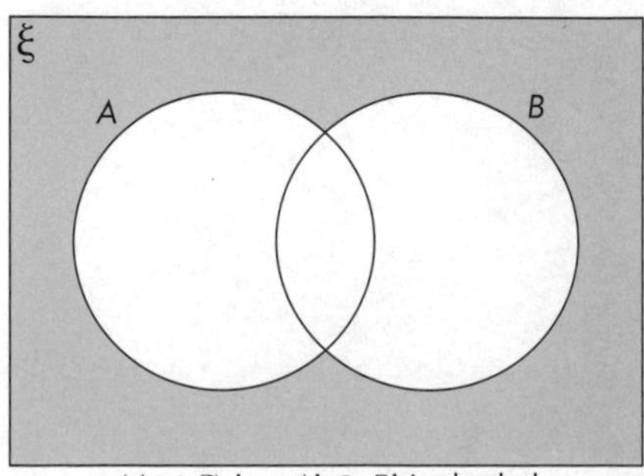

$(A \cup B)' = A' \cap B'$ is shaded

EXAMPLE 6

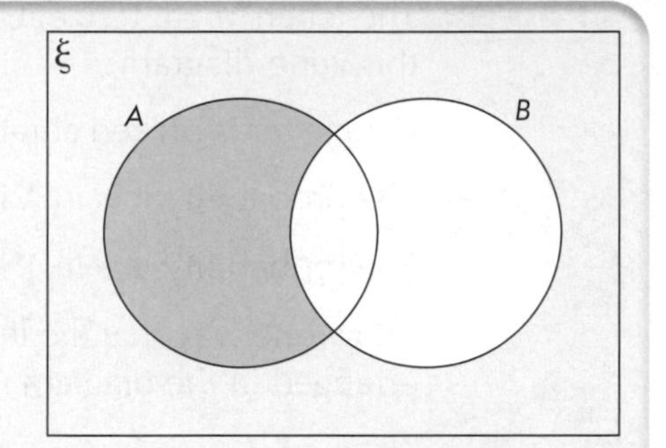

a Use set notation to describe the shaded area in the Venn diagram.

b Ten different numbers are put into the diagram so that $P(A \cap B') = 0.3$ and $P(A' \cap B') = 0.5$

How many numbers is it possible to put in $A \cap B$?

SOLUTION

a The shaded area covers the intersection of A and B' so it is $A \cap B'$.

b Three of the numbers will be in $A \cap B'$.

Five of the numbers will be in $A' \cap B'$.

This leaves two numbers so in $A \cap B$ there could be 0, 1 or 2 numbers.

EXERCISE 1C

1 For each part, copy the Venn diagram and shade the appropriate area.

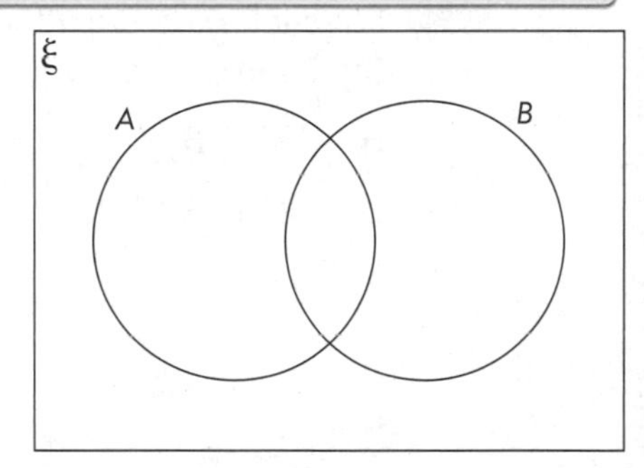

a B'

b $A \cup B'$

2 Use set notation to describe the shaded area in each of these Venn diagrams.

a

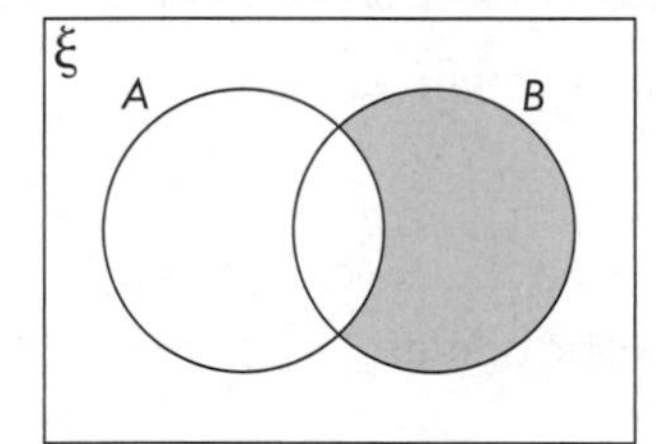

b

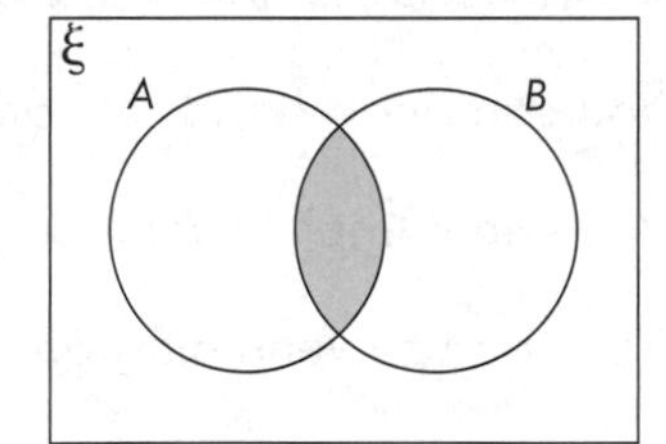

3 The numbers from 1 to 12 are put into a Venn diagram so that:

$P(A \cap B) = \frac{1}{4}$ $P(A' \cap B') = \frac{5}{12}$ $P(A) = \frac{1}{3}$

Work out $P(B)$.

4 For each part, copy the Venn diagram and shade the area described.

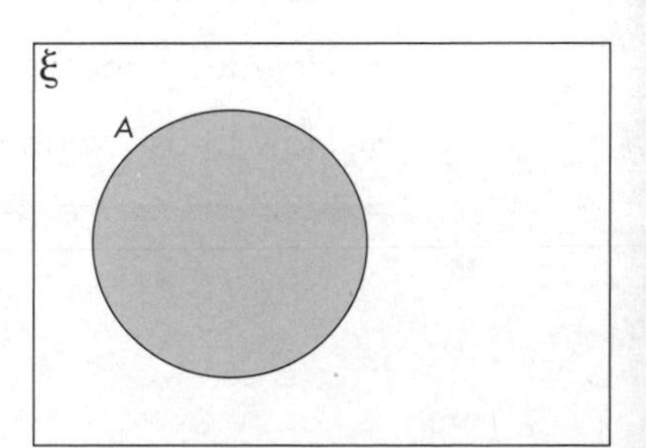

a $A \cap B$

b $A \cup B$

c $A' \cap B'$

d $A' \cup B$

B

PS **5** The letters m, a, t, h, e, t, i, c, s and v are put into this Venn diagram.

One letter is picked at random.

The probability it is in X is 0.7

The probability it is in Y is 0.5

Show one way that the letters could be arranged in the diagram.

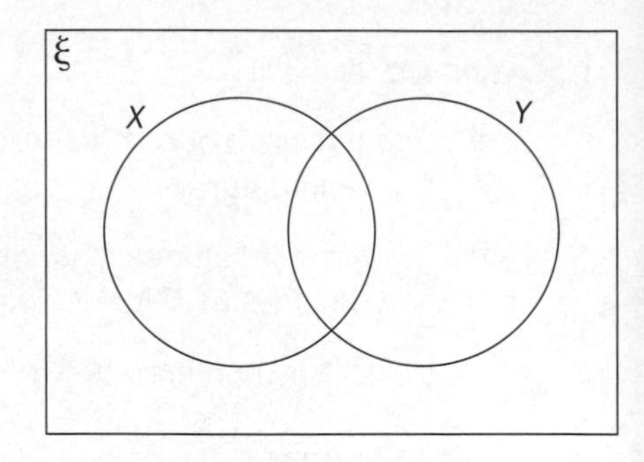

6 Use set notation to describe the shaded area in each of these Venn diagrams.

a

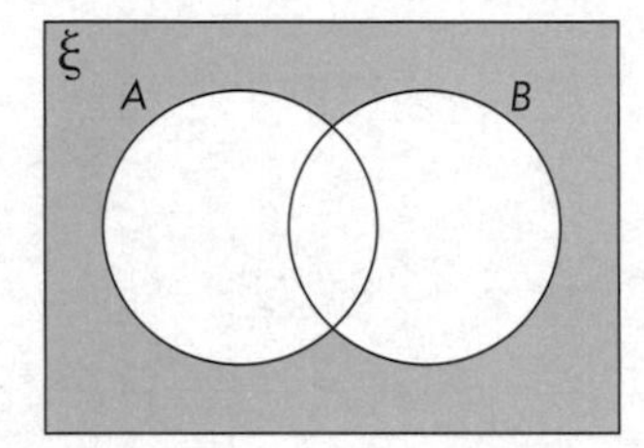

b

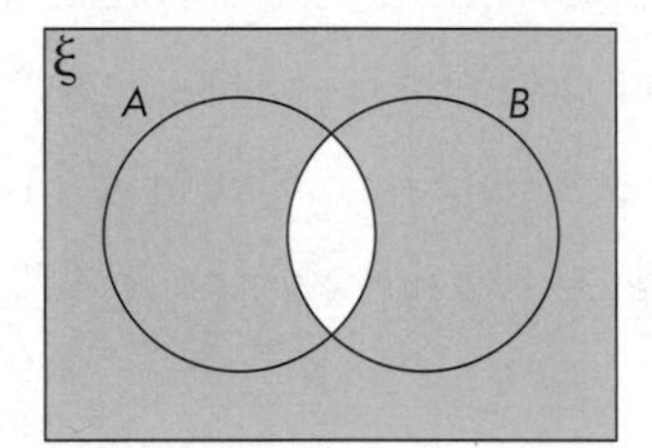

7 For each part, copy the Venn diagram and shade the appropriate area.

a $(A' \cap B')'$

b $(A' \cup B)'$

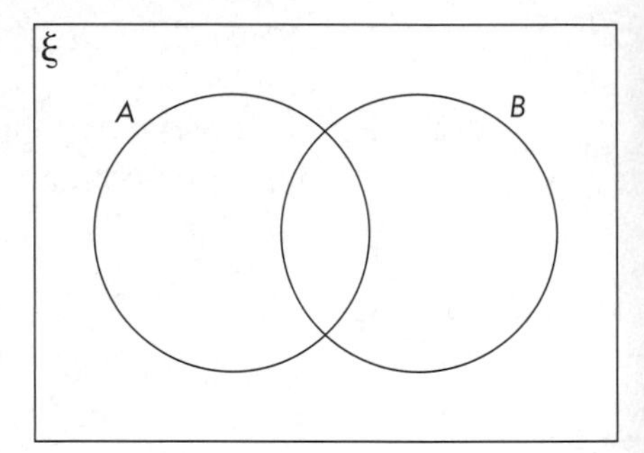

GRADE BOOSTER

- **D** You can understand set notation
- **C** You can complete Venn diagrams
- **C** You can use Venn diagrams to work out probabilities
- **B** You can shade Venn diagrams to illustrate sets

What you should know now

- How to use set notation to describe events
- How to shade Venn diagrams to illustrate sets
- How to use Venn diagrams to solve probability problems

AU 1 P(A) = 0.4   P(B) = 0.7

State whether each of the following statements could be true or could never be true.

a P(A) + P(B) = 1.1

b P(A) + P(B) = 0.7

c P(A) + P(B) = 0.9

2 ξ = {1, 2, 3, 4, 5, 6, 7, 8, 9, 10, 11, 12}

A = {multiples of 2}

B = {multiples of 3}

a List the elements of A.

b Write down P(A).

c Write down P(A′).

d List the elements of B.

e Write down P(B).

f Write down P(B′).

g What can you say about numbers that are in both A and B?

3 The manager of a shop is recording the number of people who buy shirts (S), trousers (T) or both in the shop one day.

120 people enter the shop that day.

a Copy and complete the Venn diagram to show how many people bought neither shirts nor trousers.

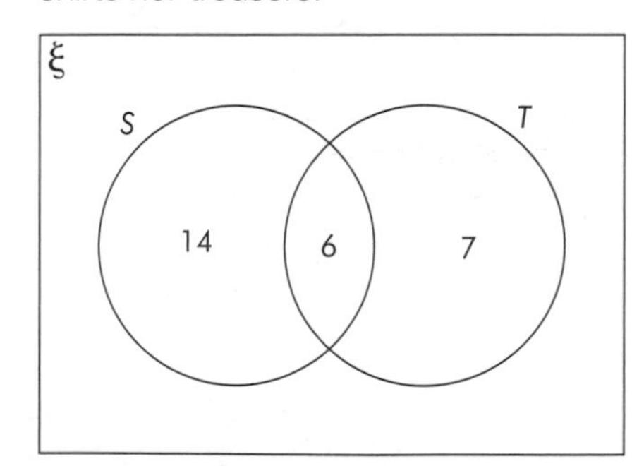

b Work out the probability that a person chosen at random bought a shirt.

Give your answer in its simplest form.

4 This Venn diagram shows some probabilities.

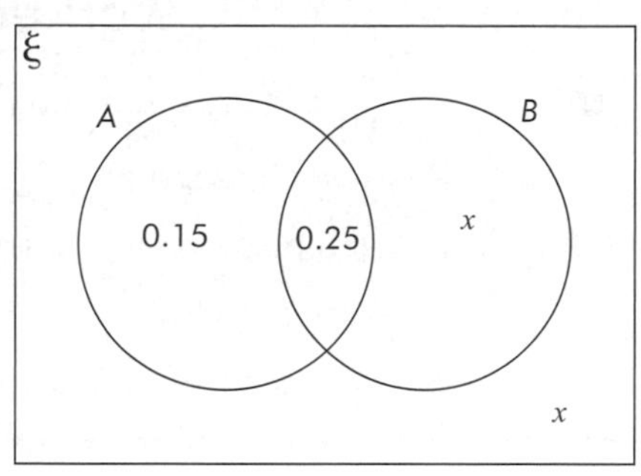

Work out the value of x.

5 This Venn diagram shows the number of students who like football (F) and the number who like hockey (H).

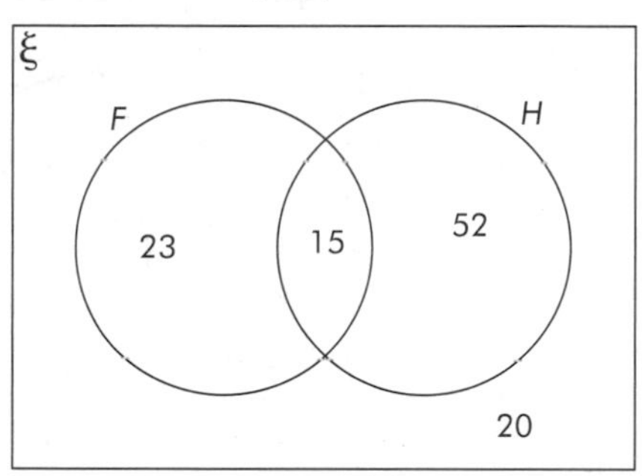

a How many students are there altogether?

b Work out P(F).

c Work out P(F ∩ H).

d A student is chosen at random.

What is the probability that the student does not like football?

PS 6 This Venn diagram shows some probabilities.

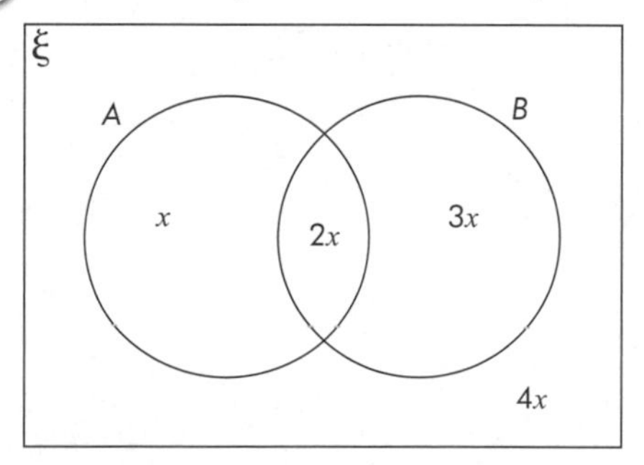

Work out:

a P(B)   **b** P(A′)

c P(A ∩ B).

Worked Examination Questions

D **1** $\xi = \{1, 2, 3, 4, 5, 6, 7, 8, 9, 10\}$

A = {even numbers}

B = {numbers greater than 6}

a Work out P(B).

b Work out P($A \cap B$).

$A = \{2, 4, 6, 8, 10\}$

$B = \{7, 8, 9, 10\}$

You will get **1 mark** for A and **1 mark** for B.

Numbers in both A and B are 8 and 10.

You will get **1 mark** for the statement about the numbers 8 and 10.

Alternatively you could get the first three marks for drawing the Venn diagram.

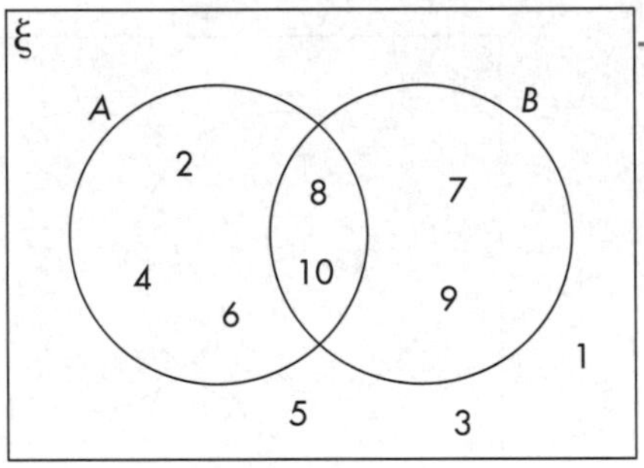

A only	2, 4, 6
$A \cap B$	8, 10
B only	7, 9
Outside A and B	1, 3, 5

a $P(B) = \frac{4}{10}$

$= \frac{2}{5}$

You will get **1 mark** for this answer.

b $P(A \cap B) = \frac{2}{10}$

$= \frac{1}{5}$

You will get **1 mark** for the final answer.

Total: 5 marks

Worked Examination Questions

C **2** The Venn diagram shows the number of students who study geography (G) and the number who study history (H).

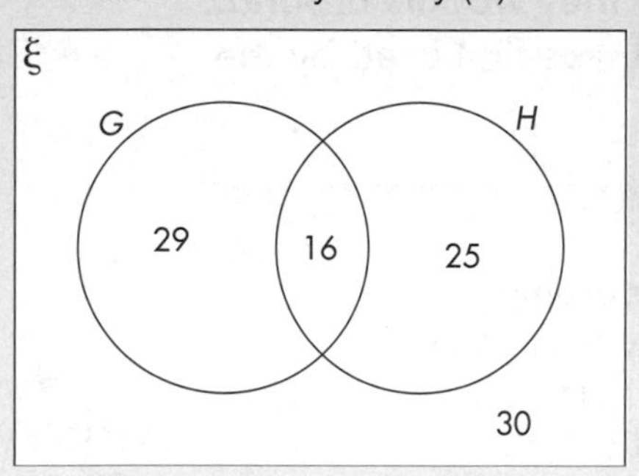

a Work out P(H).

b Describe in words what P($G \cup H$) represents.

a There are 29 + 16 + 25 + 30 = 100 students altogether.

P(H) is $\frac{16 + 25}{100} = \frac{41}{100}$

> You will get **1 mark** for setting up the calculation even if the 100 is worked out incorrectly.

> You will get **1 mark** for setting up the probability and **1 mark** for the correct answer.

b $G \cup H$ represents students who study geography, history or both.

P($G \cup H$) is the probability that a student chosen at random studies geography, history or both.

> You will get **2 marks** for a complete explanation or 1 mark for a partial explanation.

Total: 5 marks

Why this chapter matters

The triangle is a very important shape in everyday life. One reason is that a triangle keeps structures and frameworks stable. Fix three strips together and they are rigid. Fix four together and they wobble around. The study of triangles is called trigonometry and was first used by the Egyptians in the second century BC.

Trigonometry is used in many everyday applications. For example, it is used in astronomy to measure the distance to nearby stars and in geography to measure distances between landmarks. It is also used in satellite navigation systems. The sine and cosine functions are used to describe sound and light waves.

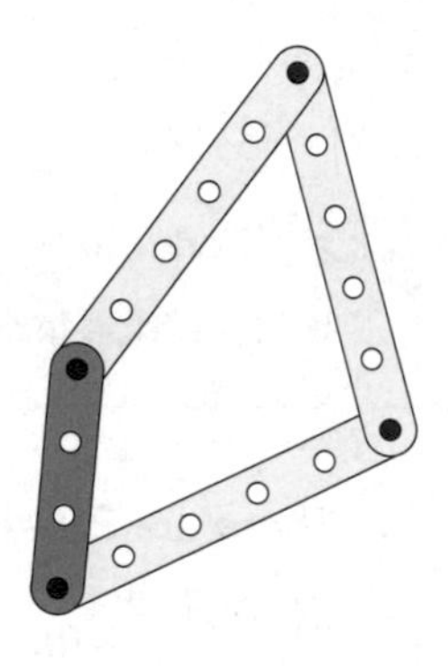

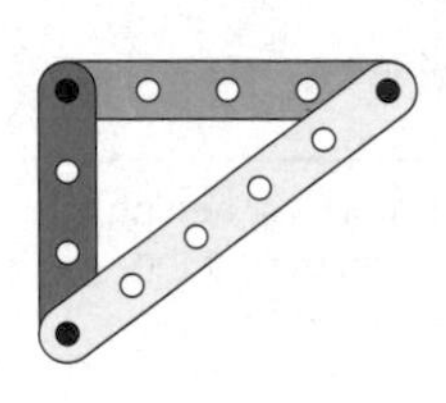

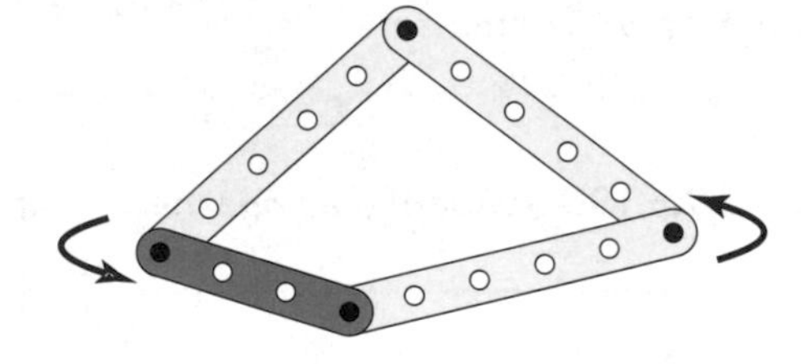

Triangles are used in structures all around us. The Egyptian pyramids are some of the most famous examples, but you can see them in buildings that are closer to home too. The building at 30 St Mary's Axe in London, which is called the *Gherkin*, is a very spectacular structure that is built as a series of triangular frameworks.

Chapter

Algebra: The circle and simultaneous equations

The grades given in this HIGHER chapter are target grades.

1 Graphs of simple loci

2 Solving simultaneous equations, one linear and one non-linear

3 The trigonometric functions

This chapter will show you ...

- **A** how to construct the graphs of simple loci such as $x^2 + y^2 = R$
- **A** how to solve simultaneous equations where one is linear and the other is of the form $x^2 + y^2 = R$
- **A** how to relate the equation of a circle centred on the origin to the trigonometric functions

Visual overview

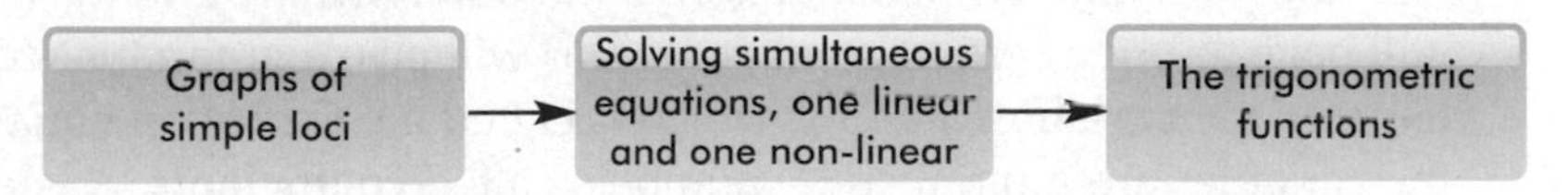

What you should already know

- How to find the equation of a straight line, using $y = mx + c$ **(KS3 level 8, GCSE grade B)**
- How to solve linear simultaneous equations **(KS3 level 7, GCSE grade A)**
- How to solve quadratic equations **(KS3 level 8, GCSE grade A)**

Quick check

1 Use an algebraic method to solve the simultaneous equations:

$$3x + 2y = 9$$
$$5x + 6y = 21$$

Do not use trial and improvement.

2 Solve the quadratic equation $12x^2 - 5x - 2 = 0$

3 Find the values of the letters in these right-angled triangles.

a

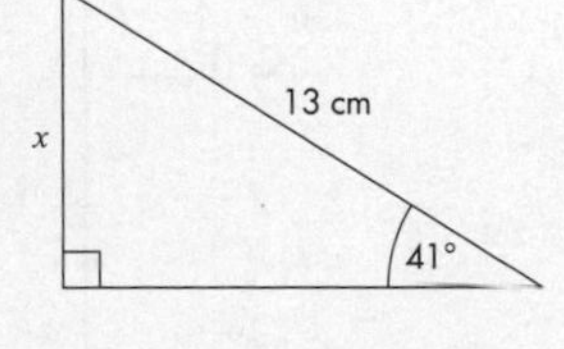

b

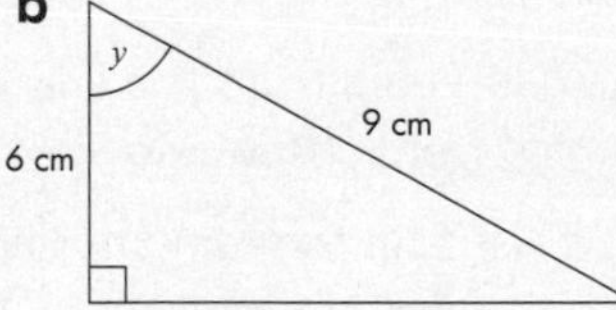

c

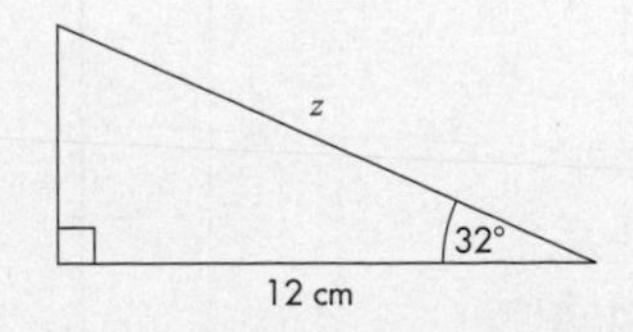

2.1 Graphs of simple loci

This section will show you how to:

- draw graphs of simple loci

Key words
Cartesian equation
Cartesian grids
coefficient
coordinate
graph
loci
locus
variable

A **locus** (plural **loci**) is the path traced by a point according to a rule, or the region described by a set of rules. In this section all the loci will be drawn on x–y grids (known as **Cartesian grids**). The rules will concern the location of the points on the grid for which the x- and y-**coordinates** will have to obey certain conditions. This will give a **graph** that can usually be expressed in terms of the **variables** x and y, which is called the **Cartesian equation**. In a Cartesian equation, the variables are multiplied by numbers called **coefficients**. A coefficient may take any value, including 0 or 1.

EXAMPLE 1

A point P moves such that its distance from the x-axis is always the same as its distance from the line $x = 1$.

a Which of the following points could be the point P? (1, 1) (2, 1) (2, 3) (4, 3)

b Plot the correct points on the graph and then draw the line that represents the locus of P.

c Give the Cartesian equation of the locus of P.

SOLUTION

a (1, 1) is on the line $x = 1$ (hence zero distance) and 1 unit from the x-axis

(2, 1) is 1 unit from the line $x = 1$ and 1 unit from the x-axis

(2, 3) is 1 unit from the line $x = 1$ and 3 units from the x-axis

(4, 3) is 3 units from the line $x = 1$ and 3 units from the x-axis

So the points that could be point P are (2, 1) and (4, 3).

b

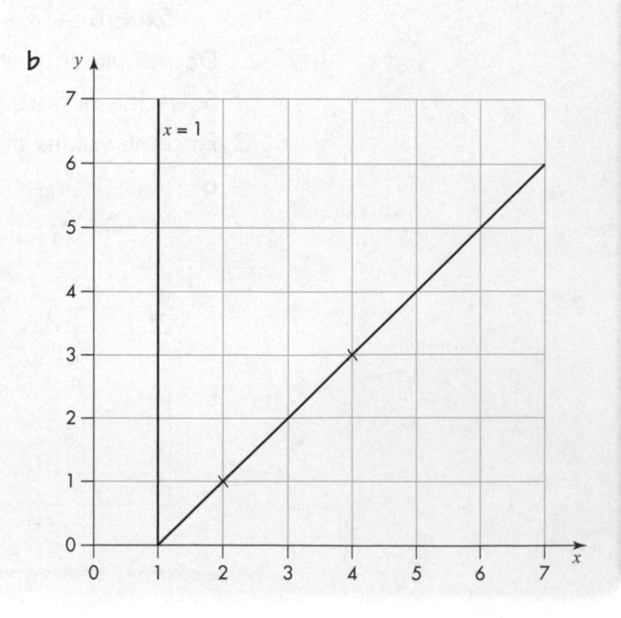

c All the points on the line have a y-coordinate that is 1 unit less than the x-coordinate, so the Cartesian equation is $y = x - 1$

EXAMPLE 2

A point R moves so that its distance from the line $x = -2$ is always twice its distance from the line $y = 3$.

a Which of the following points could be the point R? (−4, 2) (−3, 4) (0, 4) (2, 5)

b Plot the correct points on the graph and then draw the line that represents the locus of R.

c Give the Cartesian equation of the locus of R.

SOLUTION

a (−4, 2) is 2 units from the line $x = -2$ and 1 unit from the line $y = 3$

(−3, 4) is 1 unit from the line $x = -2$ and 1 unit from the line $y = 3$

(0, 4) is 2 units from the line $x = -2$ and 1 unit from the line $y = 3$

(2, 5) is 4 units from the line $x = -2$ and 2 units from the line $y = 3$

So the three points that could be point P are (−4, 2), (0, 4) and (2, 5).

b

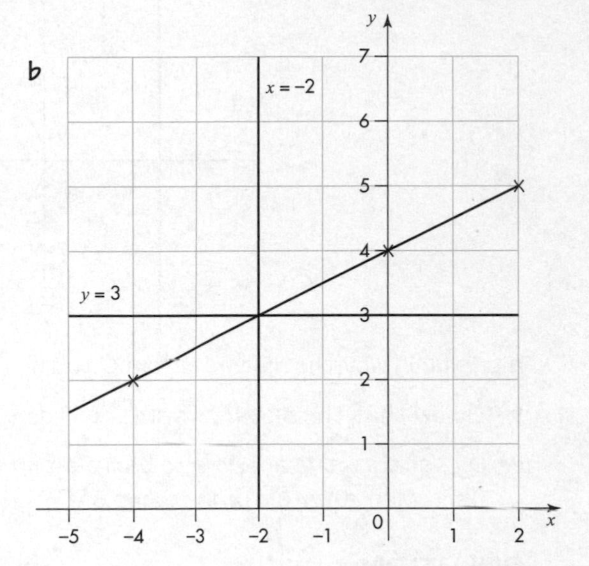

c Using the general equation $y = mx + c$, the gradient of the line is $\frac{1}{2}$ and the intercept is 4.

Hence the Cartesian equation is $y = \frac{1}{2}x + 4$

EXAMPLE 3

Find the Cartesian equation of the point S that is always 8 units away from the origin.

SOLUTION

It should be clear that this locus is a circle, radius 8 units, centred on the origin.

Taking a point A with coordinates (x, y), we can apply Pythagoras' theorem.

$$x^2 + y^2 = 8^2 \Rightarrow x^2 + y^2 = 64$$

Even taking a point for which one or both coordinates are negative, we get the same final result since squaring a negative number gives a positive value.

$$(-x)^2 + (-y)^2 = 8^2 \Rightarrow x^2 + y^2 = 64$$

Hence the Cartesian equation of the locus of a point that is always 8 units from the origin is $x^2 + y^2 = 64$

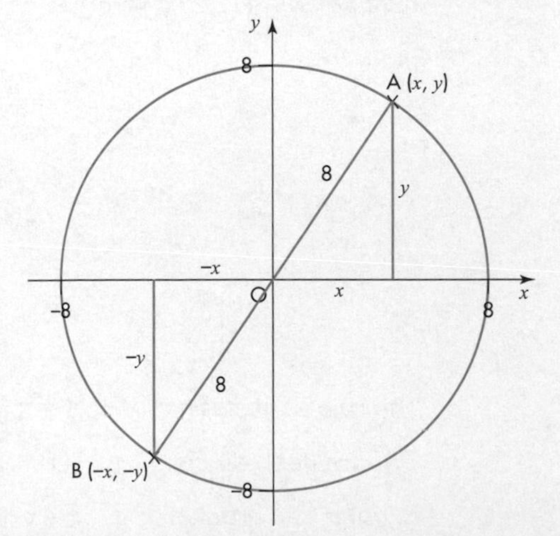

Note: This is an important result, as we can apply the same working to any circle centred on the origin. We can state the result that:

> A circle of radius r and centre the origin has the Cartesian equation $x^2 + y^2 = r^2$

EXAMPLE 4

A point Q moves so that it is always the same distance from the line $x = -2$ and the point A(2, 0).

This is a sketch of the locus of Q, with a general point Q(x, y) marked.

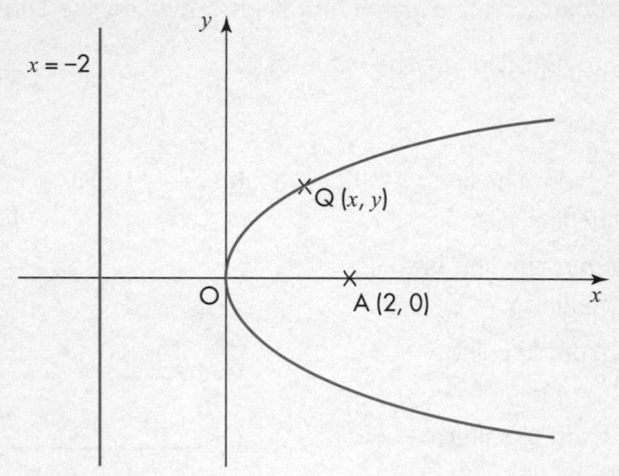

a Explain why the distance from Q to the line $x = -2$ is $x + 2$

b Show that the distance from Q to the point (2, 0) is $\sqrt{(2 - x)^2 + y^2}$

c By squaring the answers to parts **a** and **b** and putting them equal, work out the Cartesian equation of the locus of P.

SOLUTION

a The distance from $x = -2$ is 2 + the coordinate of x, which is $x + 2$.

b Using Pythagoras' theorem (see diagram), the sides of the right-angled triangle are $(2 - x)$ and y, so $AQ^2 = (2 - x)^2 + y^2$, hence AQ is $\sqrt{(2 - x)^2 + y^2}$

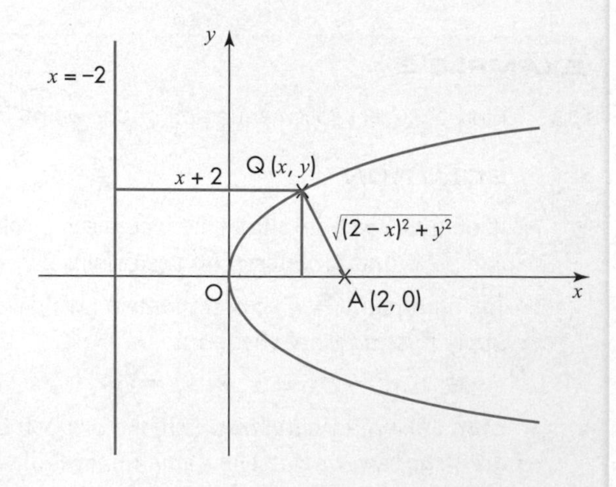

c The rule states that $x + 2 = \sqrt{(2 - x)^2 + y^2}$

Square both sides: $(x + 2)^2 = (2 - x)^2 + y^2$

Expand the brackets: $x^2 + 4x + 4 = 4 - 4x + x^2 + y^2$

Cancel terms: $4x = -4x + y^2$

Rearrange: $y^2 = 8x$

EXERCISE 2A 

1 A point P is such that its distance from the y-axis is always the same as its distance from the line $x = 4$.

a Plot the locus of P on a graph.

b Give the Cartesian equation of the locus of P.

2 A point P is such that its distance from the x-axis is always the same as its distance from the y-axis. The locus of P has a positive gradient.

a Plot the locus of P on a graph.

b Give the Cartesian equation of the locus of P.

3 A point P is such that its distance from the x-axis is always the same as its distance from the line $x = 2$.

a Which of the following points could be the point P? (2, 1) (3, 1) (4, 1) (5, 3)

b Plot the line $x = 2$ and the correct points from part **a** on a graph and then draw the line that represents the locus of P.

c Give the Cartesian equation of the locus of P.

4 A point P is such that its distance from the line $x = 3$ is always the same as its distance from the line $y = 4$.

a Which of the following points could be the point P? (0, 1) (2, 3) (3, 4) (5, 6)

b Plot the line $y = 4$ and the correct points from part **a** on a graph and then draw the line that represents the locus of P.

c Give the Cartesian equation of the locus of P.

5 A point P is such that its **vertical** distance from the line $y = x + 1$ is always the same as its **vertical** distance from the line $y = 1$.

a Plot the lines $y = x + 1$ and $y = 1$ and so identify the locus of P.

b Give the Cartesian equation of the locus of P.

6 A point P is such that its **horizontal** distance from the line $y = x$ is always the same as its **horizontal** distance from the line $x = 2$.

a Plot the lines $y = x$ and $x = 2$ and so identify the locus of P.

b Give the Cartesian equation of the locus of P.

PS **7** A point P is such that its distance from the x-axis is always three times its distance from the y-axis. The locus of P has a positive gradient.

Give the Cartesian equation of the locus of P.

PS **8** A point P is such that its distance from the line $x = 1$ is always **twice** its distance from the line $y = 2$. The locus of P has a positive gradient.

Give the Cartesian equation of the locus of P.

C

B

A

A

AU 9 Write down the Cartesian equations of:

a a circle of radius 2 units, centre the origin

b a circle of radius 6 units, centre the origin.

10 **a** Plot the locus of the following curves on the same graph.

$x^2 + y^2 = 6$

$x^2 + y^2 = 10$

b Work out the difference in area between the two curves. Give your answer in terms of π.

11 A point P moves so that it is always twice the distance from the line $x = -1$ as it is from the line $y = 2$.

a Which of the following points could be the point P?

$(-3, 1)$ $(-1, 3)$ $(1, 3)$ $(3, 4)$

b Plot the line $y = 2$ and the correct points from part **a** on a graph and then draw the line that represents the locus of P.

c Give the Cartesian equation of the locus of P.

A*

12 A point P moves so that it is always the same distance from the line $y = -1$ and the point A(0, 1).

This is a sketch of the locus of P.

A general point P(x, y) is marked.

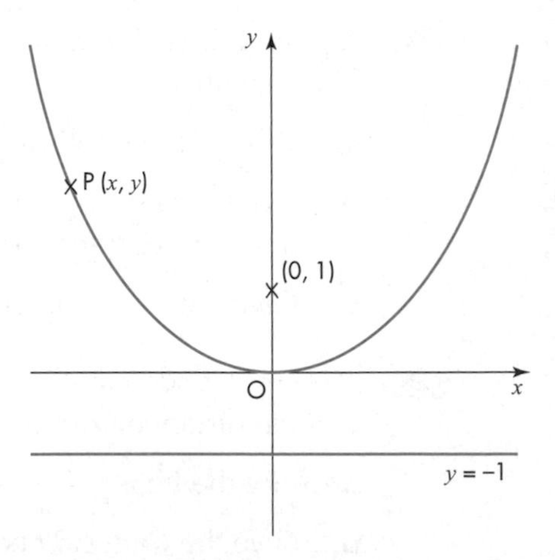

a Explain why the distance from P to the line $y = -1$ is $y + 1$

b Show that the distance from P to the point (0, 1) is $\sqrt{x^2 + (y - 1)^2}$

c By squaring the answers to parts **a** and **b** and putting them equal, work out the Cartesian equation of the locus of P.

PS 13 The Cartesian equation of a circle C_1 with a centre at (7, 7) and radius 5 units is $(x - 7)^2 + (y - 7)^2 = 25$

C_2 is the circle with a centre at (0, 0) and radius 5 units.

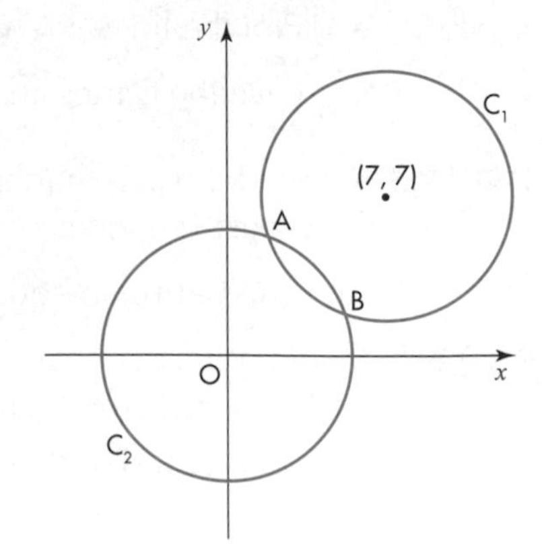

a Show that C_1 and C_2 intersect at A(3, 4) and B(4, 3)

b Write down the Cartesian equation of the line AB.

2.2 Solving simultaneous equations, one linear and one non-linear

This section will show you how to:

- solve a pair of simultaneous equations where one is linear and one is non-linear

Key words
linear
non-linear
simultaneous equations
substitution method

Previously you have solved pairs of **linear simultaneous equations**, so you should know that the solution is the unique pair of values of x and y that satisfy both equations at the same time.

In particular, you should be familiar with the **substitution method** for solving simultaneous equations.

When solving a pair of simultaneous equations, when one is linear and one is **non-linear**, the basic idea is also to find solutions that satisfy both equations simultaneously but, because the non-linear equation represents a curve, there will normally be two solutions, or two pairs of values (x, y) that are solutions for the pair of equations.

EXAMPLE 5

Solve the linear equations:

$x = 9 + 4y$ (1)

$2x + 3y + 4 = 0$ (2)

using the substitution method.

SOLUTION

Substitute equation (1) into equation (2): $2(9 + 4y) + 3y + 4 = 0$

Expand and rearrange: $18 + 8y + 3y + 4 = 0$

$\Rightarrow 22 + 11y = 0$

$\Rightarrow y = -2$

Substitute this value back into equation (1): $x = 9 - 8 = 1$

So the solution is $x = 1, y = -2$

Check in the original equations:

(1) $\Rightarrow 1 = 9 + 4 \times -2 = 9 - 8$ ✓

(2) $\Rightarrow 2 \times 1 + 3 \times -2 + 4 = 2 - 6 + 4 = 0$ ✓

EXAMPLE 6

Solve the simultaneous equations:

$y = x + 1$ (1)

$xy = 2$ (2)

SOLUTION

Substitute the linear equation (1) into the non-linear (2): $x(x + 1) = 2$

Expand and rearrange: $x^2 + x = 2$

$\Rightarrow x^2 + x - 2 = 0$

Solve the quadratic equation: $(x - 1)(x + 2) = 0$

$\Rightarrow x = 1$ or -2

Substitute back in to (1) to find the equivalent y-values.

When $x = 1, y = 1 + 1 = 2$ When $x = -2, y = -2 + 1 = -1$

So the solutions are (1, 2) and (−2, −1)

Check in the original equations:

(1) $\Rightarrow 2 = 1 + 1$ ✓ and (2) $\Rightarrow 2 \times 1 = 2$ ✓

(1) $\Rightarrow -1 = -2 + 1$ ✓ and (2) $\Rightarrow -2 \times -1 = 2$ ✓

EXAMPLE 7

Solve the simultaneous equations:

$y = x - 7$ (1)

$x^2 + y^2 = 25$ (2)

SOLUTION

Substitute the linear equation (1) into the non-linear (2): $x^2 + (x - 7)^2 = 25$

Expand and rearrange: $x^2 + x^2 - 14x + 49 = 25$

$\Rightarrow 2x^2 - 14x + 24 = 0$

$\Rightarrow x^2 - 7x + 12 = 0$

Solve the quadratic equation: $(x - 3)(x - 4) = 0$

$\Rightarrow x = 3$ or 4

Substitute back into (1) to find the equivalent y-values.

When $x = 3, y = 3 - 7 = -4$ When $x = 4, y = 4 - 7 = -3$

So the solutions are (3, −4) and (4, −3)

Check in the original equations.

(1) $\Rightarrow -4 = 3 - 7$ ✓ and (2) $\Rightarrow 3^2 + (-4)^2 = 9 + 16 = 25$ ✓

(1) $\Rightarrow -3 = 4 - 7$ ✓ and (2) $\Rightarrow 4^2 + (-3)^2 = 16 + 9 = 25$ ✓

EXERCISE 2B 

1 Use the substitution method to solve these pairs of linear simultaneous equations.

a $3x + y = 1$
$x = 5 + 2y$

b $3x - 2y = 8$
$y = 17 - 2x$

c $x = 8 + 2y$
$2x + 5y = 7$

2 Solve these pairs of simultaneous equations.

a $xy = 10$
$y = x + 3$

b $xy = 6$
$3y = x + 7$

c $xy = -6$
$x = 5 - y$

3 Solve these pairs of simultaneous equations.

a $y = x + 2$
$y = 2x^2 + 6x - 1$

b $y = 2x + 5$
$y = 6x^2 + 3x + 4$

c $y = x - 4$
$y = 4x^2 + x - 5$

4 Solve these pairs of simultaneous equations.

a $y = 2x - 1$
$x^2 + y^2 = 2$

b $2y + x = 1$
$x^2 + y^2 = 1$

c $y = 4x + 3$
$2x^2 + y^2 = 3$

d $y = x - 1$
$2x^2 + 3y^2 = 11$

e $y = 2x + 1$
$x^2 + 5y^2 = 1$

f $y + 2x = 5$
$6x^2 + y^2 = 15$

PS 5 The simultaneous equations $x^2 + y^2 = 13$ and $2y + 3x = 13$ have only one solution.

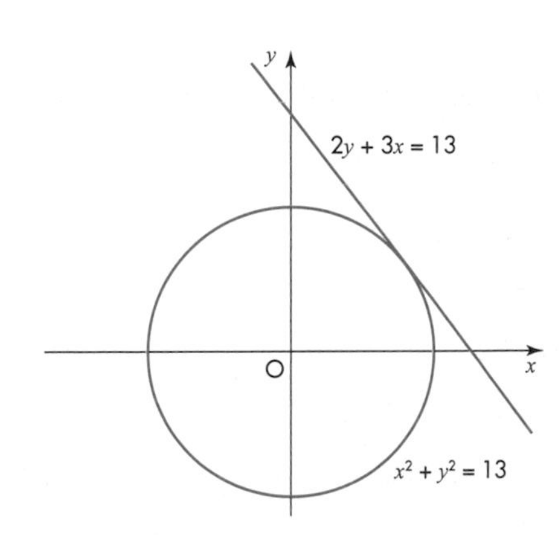

a Find the solution.

b Write down the intersections of each pair of graphs.

i $x^2 + y^2 = 13$ and $2y - 3x = 13$

ii $x^2 + y^2 = 13$ and $3x - 2y = 13$

2.3 The trigonometric functions

This section will show you how to:

- relate the trigonometric functions to circles of the form $x^2 + y^2 = R$

Key words
cosine (cos)
mathematical convention
origin
quadrant
radius
sine (sin)

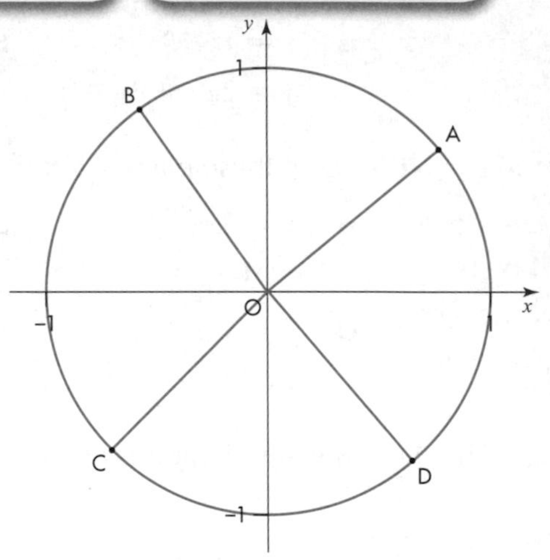

From section 2.1, you know that the locus of the Cartesian equation $x^2 + y^2 = 1$ is a circle, centre the **origin** and **radius** 1 unit.

In this section we shall be taking four points, A, B, C and D, around the circumference (one in each **quadrant**), so that the lines joining the points to the origin respectively make angles of 40°, 120°, 225° and 310° with the positive x-axis, where the angles are measured anti-clockwise. (This is known as the **mathematical convention**).

The coordinates of A

Drop the perpendicular from A to the x-axis and call the point where it meets the axis P.

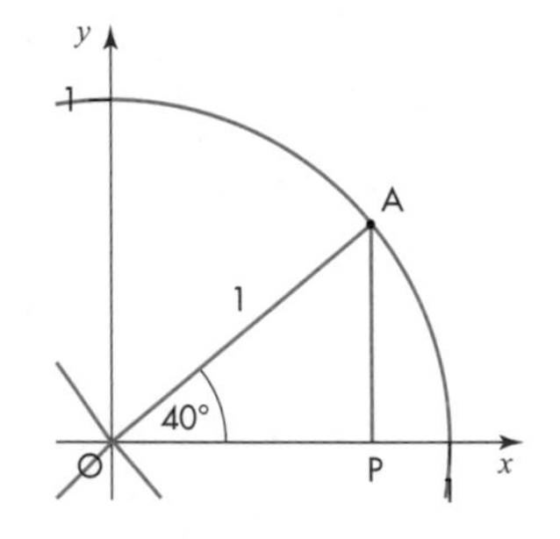

The angle between OA and the x-axis is 40°.

Using basic trigonometry:

$\text{AP} = 1 \times \textbf{sin}\ 40° = \sin 40° \approx 0.643$

$\text{OP} = 1 \times \textbf{cos}\ 40° = \cos 40° \approx 0.766$

Hence the coordinates of A are (cos 40°, sin 40°).

The coordinates of B

Drop the perpendicular from B to the x-axis and call the point where it meets the axis P.

The angle between OB and the x-axis is 60°.

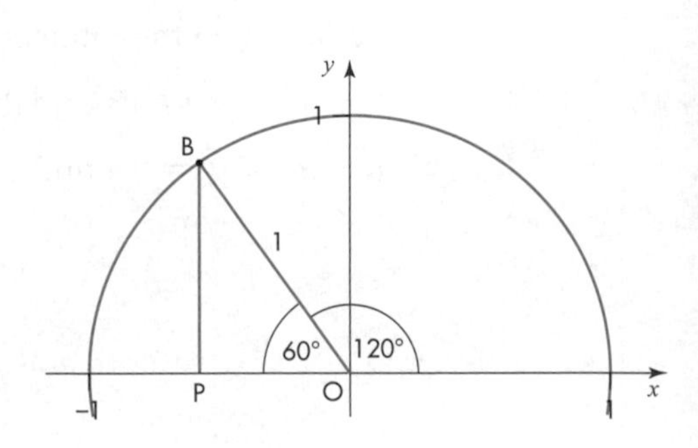

Using basic trigonometry:

$\text{BP} = 1 \times \sin 60° = \sin 60° \approx 0.866$

$\text{OP} = 1 \times \cos 60° = \cos 60° = 0.5$

The coordinates of B are approximately (−0.5, 0.866).

Using a calculator we can find that $\cos 120° = -0.5$ and $\sin 120° \approx 0.866$

Hence the coordinates of B are (cos 120°, sin 120°).

The coordinates of C

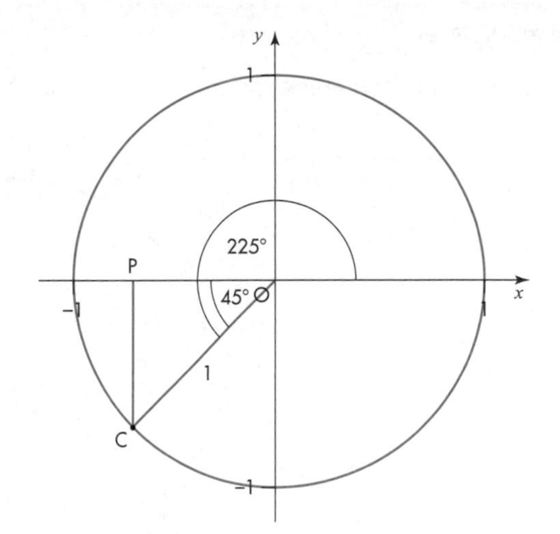

Draw the perpendicular up from C to the x-axis and call the point where it meets the axis P.

The angle between OC and the x-axis is 45°.

Using basic trigonometry:

$$CP = 1 \times \sin 45° = \sin 45° \approx 0.707$$

$$OP = 1 \times \cos 45° = \cos 45° \approx 0.707$$

The coordinates of C are approximately $(-0.707, -0.707)$.

Using a calculator, we can find that $\cos 225° \approx -0.707$ and $\sin 225° \approx -0.707$

Hence the coordinates of C are $(\cos 225°, \sin 225°)$.

The coordinates of D

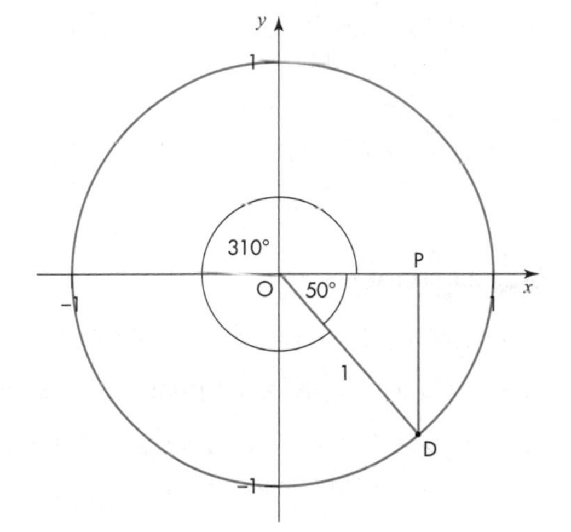

Draw the perpendicular up from D to the x-axis and call the point where it meets the axis P.

The angle between OD and the x-axis is 50°.

Using basic trigonometry:

$$DP = 1 \times \sin 50° = \sin 50° \approx 0.766$$

$$OP = 1 \times \cos 50° = \cos 50° \approx 0.643$$

The coordinates of D are approximately $(0.643, -0.766)$.

Using a calculator, we can find that $\cos 310° \approx 0.643$ and $\sin 310° \approx -0.766$

Hence the coordinates of D are $(\cos 310°, \sin 310°)$.

The coordinates of the general point

The examples above show that a point P on the circumference of a circle centred on the origin, with a radius of 1 unit, where OP makes an angle θ with the positive x-axis, has coordinates $(\cos \theta, \sin \theta)$.

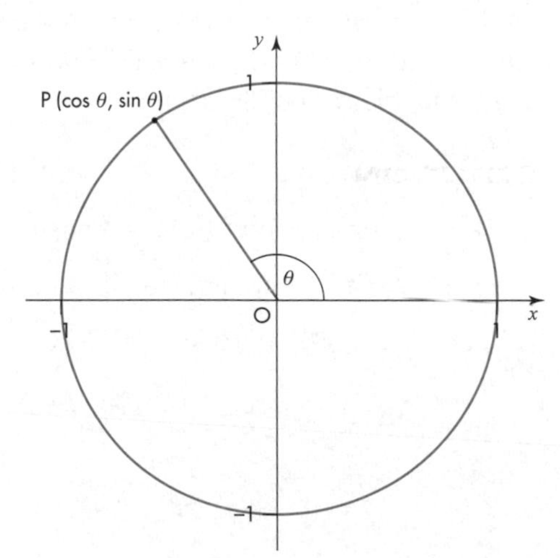

EXAMPLE 8

The point P on the circle $x^2 + y^2 = 1$ makes an angle of 220° with the positive x-axis.

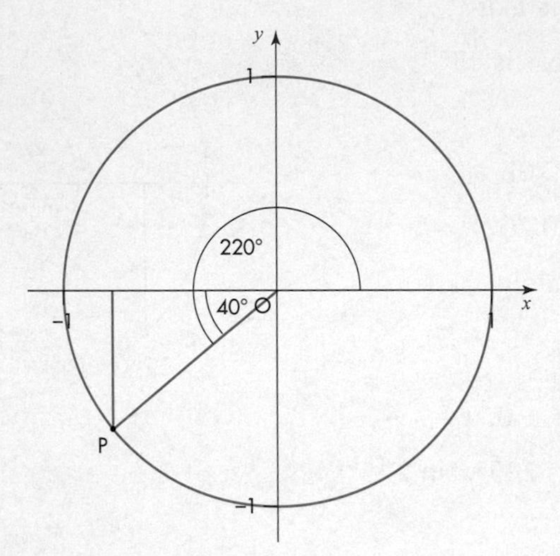

Show clearly why the coordinates of P are (–cos 40°, –sin 40°).

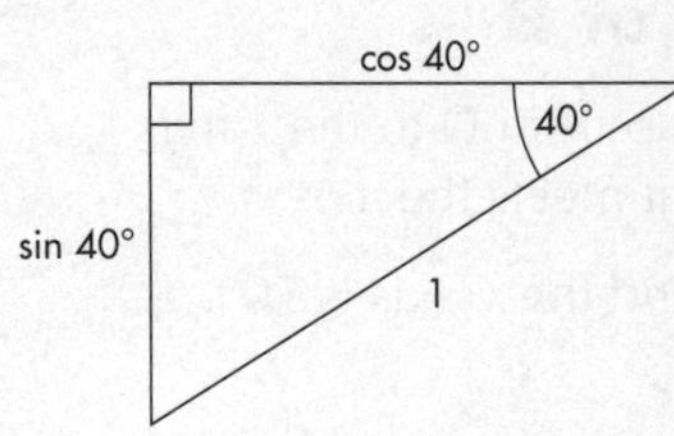

SOLUTION

The angle that OP makes with the x-axis is 40°.

This gives a right-angled triangle with a hypotenuse of 1 unit.

The lengths of the sides are cos 40° and sin 40° and, as the point P is in the third quadrant, the coordinates will be (–cos 40°, –sin 40°).

EXAMPLE 9

Find the coordinates of the point P, which is on the circumference of the circle with equation $x^2 + y^2 = 12$, where OP makes an angle of 340° with the positive x-axis measured in an anti-clockwise direction.

SOLUTION

If the circle had been $x^2 + y^2 = 1$ then the coordinates would have been (cos 340°, sin 340°).

The radius of the new circle is $\sqrt{12}$ so the coordinates of the point P are

$(\sqrt{12}\cos 340°, \sqrt{12}\sin 340°) \approx (3.26, -1.18)$.

EXERCISE 2C

A

1 The circle $x^2 + y^2 = 10$ is as shown.

The point X is such that the angle between OX and the x-axis is 50°.

Work out the coordinates of the point X.

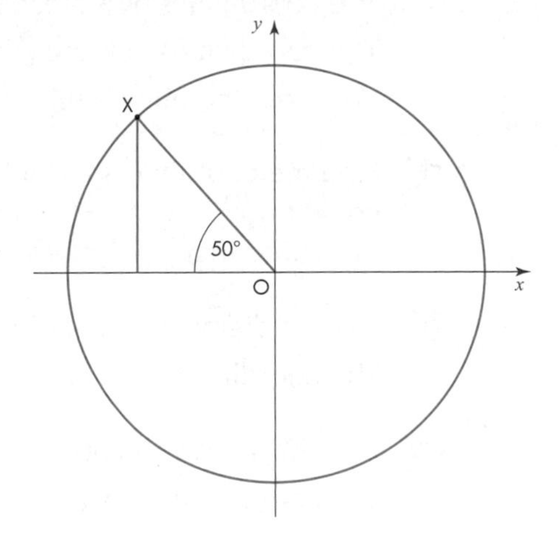

2 The circle $x^2 + y^2 = 25$ is as shown.

The point P is such that the angle between OP and the x-axis is 70°.

Work out the coordinates of the point P.

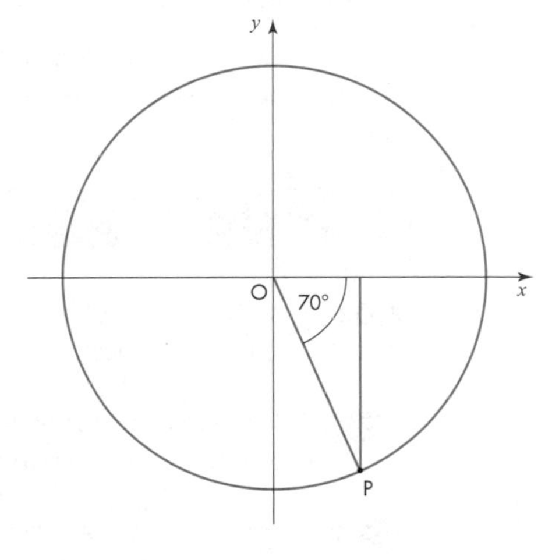

3 The point P on the circle $x^2 + y^2 = 1$ makes an angle of 300° with the positive x-axis. Explain clearly why the coordinates of P are (cos 60°, −sin 60°).

AU 4 The coordinates of the point P, which lies on the circumference of the circle $x^2 + y^2 = 1$, are (−cos 30°, sin 30°). Work out the angle between OP and the positive x-axis.

AU 5 PQ is a diameter of the circle $x^2 + y^2 = 1$. The coordinates of P are (cos 130°, sin 130°). Work out the coordinates of Q.

6 The point X lies on the circumference of a circle, centre the origin and radius 8 units.

The angle between OX and the positive x-axis is 305°.

Give the coordinates of X correct to three significant figures (3 sf).

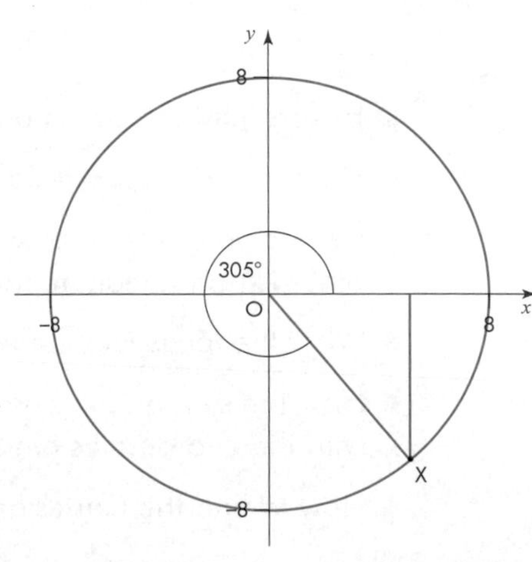

A

PS 7 Two points, A and B, lie on the circumference of the circle $x^2 + y^2 = 1$.
The coordinates of A are (cos 100°, sin 100°).
The coordinates of B are (cos 210°, sin 210°).
Work out the obtuse angle between OA and OB.

A*

8 Find the coordinates of the point P, which is on the circumference of the circle with equation $x^2 + y^2 = 20$, where OP makes an angle of 240° with the positive x-axis measured in an anti-clockwise direction.

AU 9 LM is a diameter of the circle $x^2 + y^2 = 16$.
The coordinates of L are (4 cos 75°, 4 sin 75°). Work out the coordinates of M.

PS 10 X and Y are two points on the circumference of the circle $x^2 + y^2 = 36$.
The coordinates of X are (6 cos 190°, 6 sin 190°).
The coordinates of Y are (6 cos 280°, 6 sin 280°).
The area of a circle is given by $A = \pi \times (\text{radius})^2$. Work out the area between OX and OY.
Give your answer in terms of π

GRADE BOOSTER

B You can plot the locus and find the Cartesian equation of a locus that has a linear path

A You can solve a pair of simultaneous equations where one is linear and the other is of the form $x^2 + y^2 = R$

A You can relate the trigonometric ratios sine and cosine to the coordinates of points on the circle $x^2 + y^2 = 1$

A You can plot the locus of an equation of the form $x^2 + y^2 = r^2$ and know it is a circle of radius r centred on the origin

A* You can find the Cartesian equation of a locus that has a non-linear path

What you should know now

- How to plot a locus of a point that follows a rule leading to a linear path
- How to find the Cartesian equation of the locus of a point with a linear path
- How to solve a pair of simultaneous equations where one is linear and the other is non-linear, in particular of the form $x^2 + y^2 = R$
- What the locus is of an equation of the form $x^2 + y^2 = r^2$
- How the sine and cosine functions of angles from 0° to 360° are connected with the coordinates of the circle $x^2 + y^2 = 1$
- How to find the Cartesian equation of a locus that has a non-linear path

1 The point P is such that it is always the same distance from the line $x = 3$ and the line $x = 7$.

a Copy the grid and sketch the locus of P.

b Write down the equation of the locus of P.

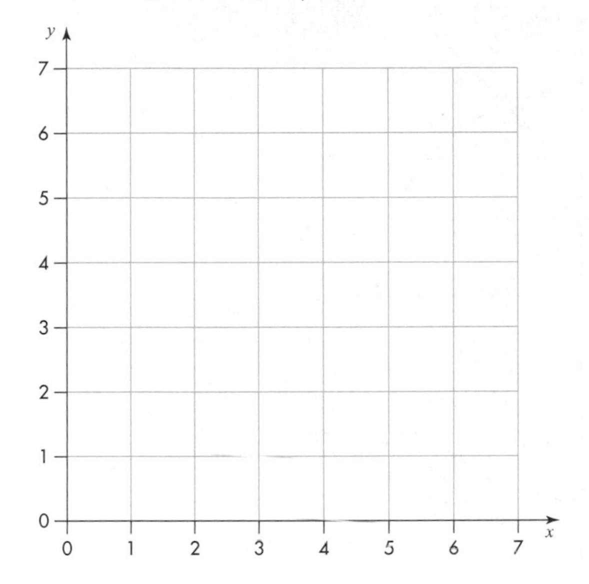

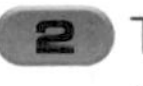
2 The point X lies on the circumference of the circle $x^2 + y^2 = 1$.

Write down the coordinates of the point X.

Give your answer correct to 3 significant figures.

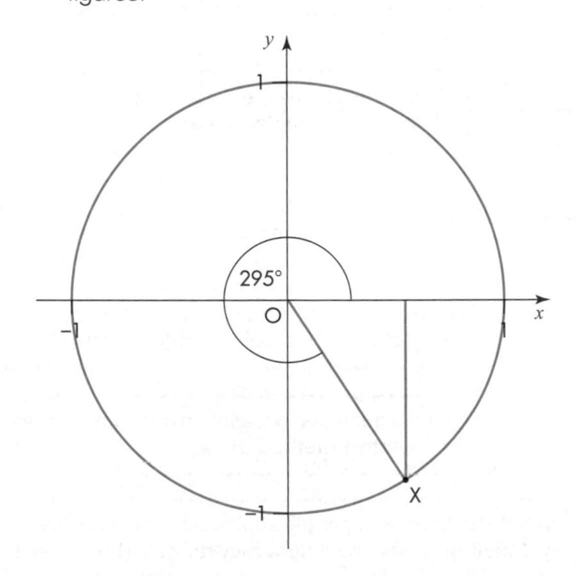

3 The points X and Y lie on the circumference of the circle $x^2 + y^2 = 49$.

The angle XOY is 90°.

X is the point $(-7 \cos 50°, 7 \sin 50°)$. Write down two possible sets of coordinates of Y.

4 LM is a diagonal of a circle, centre the origin. The coordinates of L are $(3 \cos 205°, 3 \sin 205°)$.

a What is the equation of the circle?

b What are the coordinates of the point M?

PS 5 A point P has a locus such that it bisects the acute angle between the lines $y = 2x$ and $y = \frac{1}{2}x + 3$.

Find the equation of the locus of P.

You may find it helpful to use a grid like this.

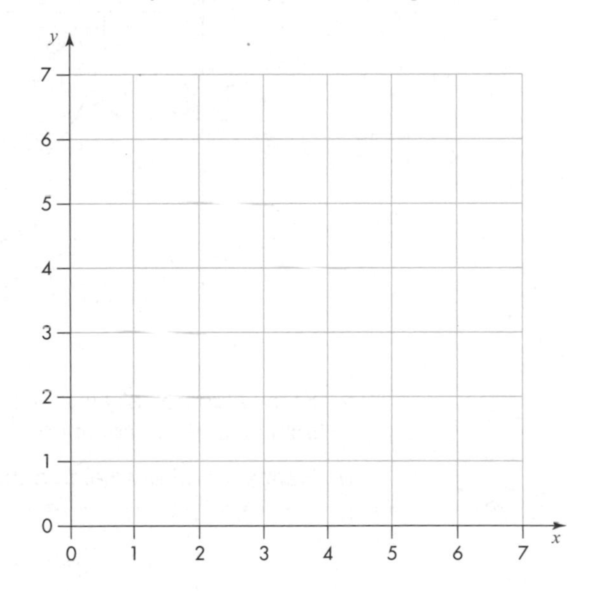

6 Solve the simultaneous equations:

$$y = 2x + 3$$
$$x^2 + y^2 = 2$$

November 2009, Specification B, Module 5, Paper 2, Question 14

7 Solve the simultaneous equations:

$$x = 3 + 2y$$
$$x^2 + 2y^2 = 27$$

June 2009, Specification A, Higher Paper 2, Question 29

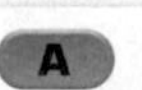

C A A*

Worked Examination Questions

A **1** The circle $x^2 + y^2 = 29$ and the line $y = x + 3$ intersect at the points A and B.

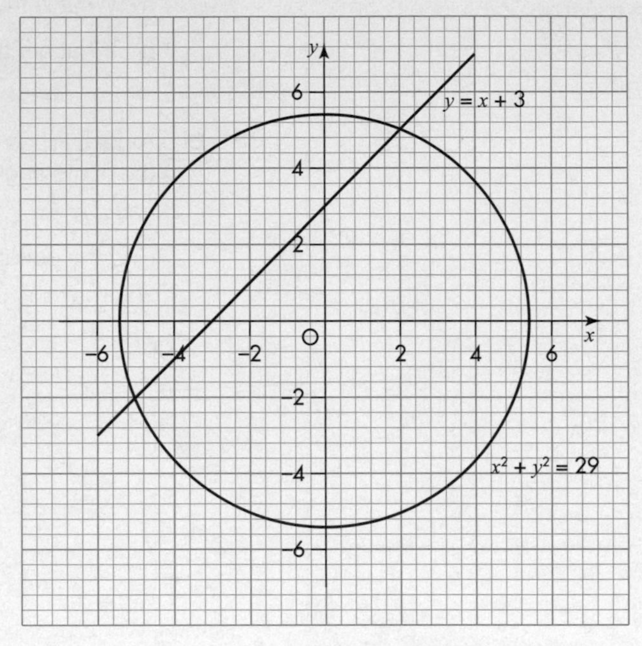

June 2007, Specification A, Higher Paper 2, Question 20

a Show algebraically that the x-coordinates of the points A and B are the solutions of the equation $x^2 + 3x - 10 = 0$

b Hence, or otherwise, find the coordinates of A and B.

a $x^2 + (x + 3)^2 = 29$

Substitute the linear equation into the non-linear. This is worth **1 method mark**.

$x^2 + x^2 + 6x + 9 = 29$

$2x^2 + 6x - 20 = 0$

Expand the bracket accurately. This is worth **1 accuracy mark**.

Divide by 2: $\Rightarrow x^2 + 3x - 10 = 0$

Rearrange into a quadratic equation and show that a factor of 2 cancels. This is worth **1 method mark**.

b $(x + 5)(x - 2) = 0 \Rightarrow x = -5$ or 2

Solve the quadratic equation to get both x-values. This is worth **1 method mark**.

The coordinates are $(-5, -2)$ and $(2, 5)$.

Substitute back to get the equivalent y-value. Gains **1 method mark** and **1 accuracy mark** for both answers.

Total: 6 marks

Worked Examination Questions

A* **2** **a** Write down the equation of the curve shown on the grid.

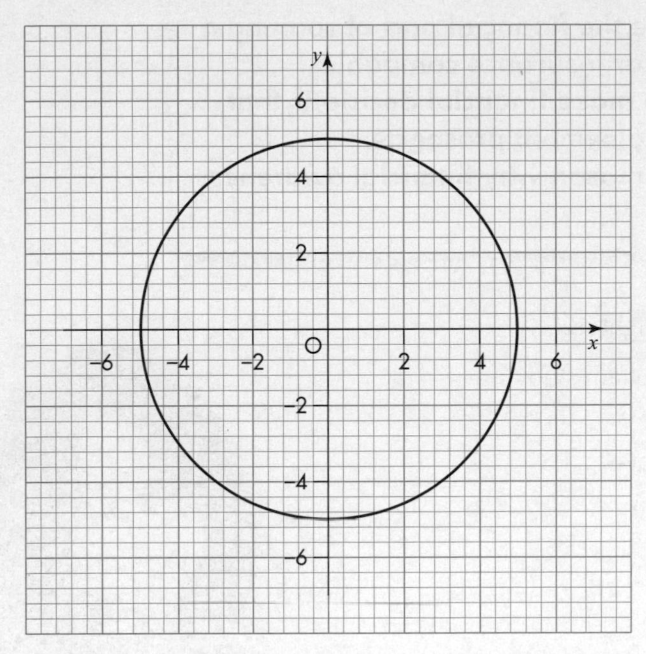

b A point P is such that it has a positive gradient and is the same distance from the x-axis as it is from the line $x = 1$.

Sketch the locus of P on the grid.

c Hence, or otherwise, find the points where the locus of P intersects the circle.

a $x^2 + y^2 = 25$

The curve is a circle, centre the origin, radius r, so has an equation of $x^2 + y^2 = r^2$ The correct equation gets **1 mark**.

b

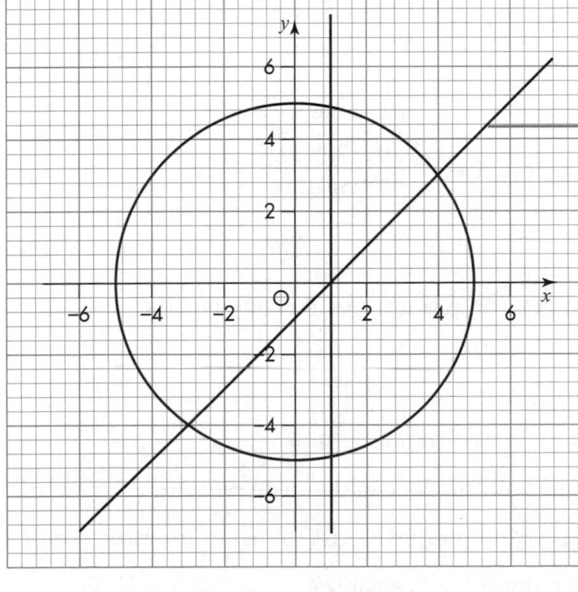

Draw the line $x = 1$ (this gets **1 mark**) and then draw the locus of P. This gets another **1 mark**.

If the drawing is accurate you can read these values from the graph but it is worth checking in the equation of the circle and the Cartesian equation of P which is $y = x - 1$. The 'otherwise' would be to do an algebraic solutions, which is lengthy and likely to lead to a mistake. You get **1 mark** for each solution.

c The points are (4, 3) and (−3, −4).

Total: 5 marks

Why this chapter matters

Financial mathematics is the mathematics of everyday life, usually where money is involved.
Business applications could be the financial part of running a business, dealing with banks or insurance companies.
Increasingly, students have to make financial decisions that affect their education and employment prospects.
Good decision-making is the reason why financial awareness is important.

Here are some questions students might ask.

- Should I get a part-time job while I am studying for my examinations?
- How much money do I need to save?
- Where will the money come from?
- How much tax will I pay?
- Should I take out a loan?
- Do I need insurance and how much will this cost?
- How much does this job pay?
- Which is the best deal?
- Which bills are due for payment?
- Can I afford it?

Here are some questions the owner of a small business might ask.

- Am I making a profit?
- Is my bank manager happy?
- Can I afford to employ more people?
- How much does it cost to insure my premises?
- Are my prices sensible?
- Are my customers happy?

Finance: Financial and business applications

The grades given in this FOUNDATION & HIGHER chapter are target grades.

1. Profit and loss
2. Value added tax (VAT)
3. Interest rates, simple interest, savings and loans
4. Income tax
5. Appreciation and depreciation
6. Savings: annual equivalent rate (AER)

This chapter will show you ...

- G–F how to work out profit or loss in real-life situations
- F–E how to calculate value added tax
- E how to calculate interest on savings and loans
- E how to compare savings accounts and the costs of borrowing money
- E how to work out income tax, using one tax band
- E how to work out the effect of the value of goods appreciating or depreciating in value
- C how to work out percentage appreciation or percentage depreciation
- B how to work out income tax, using more than one tax band
- B how to use the formula to calculate the annual equivalent rate (AER)

Visual overview

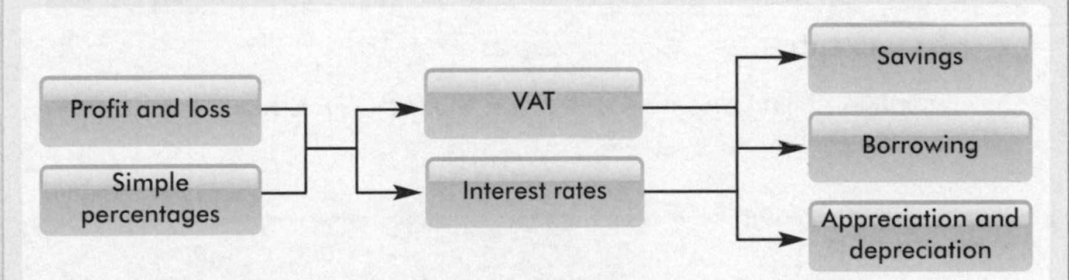

What you should already know

- Multiplication tables up to 10 × 10 **(KS3 level 4, GCSE grade G)**
- How to do simple calculations with percentages **(KS3 level 5, GCSE grade F)**
- How to increase or decrease by a percentage **(KS3 level 6, GCSE grade D)**
- How to work out a percentage increase or decrease **(KS3 level 7, GCSE grade C)**

Quick check

1 Find 10% of:

a 50 **b** 300 **c** 25

2 Work out:

a 20% of 90 **b** 25% of 80 **c** 5% of 60

3 a Increase £2 by 10% **b** Decrease £8 by 25%

3.1 Profit and loss

This section will show you how to:
- work out profit or loss from the cost price and selling price
- work out profit or loss when items are bought in bulk and sold individually
- work out profit or loss from the cost price and the percentage profit

Key words
cost price
loss
percentage
profit
selling price

The price a shop pays when it buys an item from a wholesaler (or warehouse) is called the **cost price**. The price for which a shop sells the item to a customer is called the **selling price**.

Profit is the amount of money that is made when an item is sold for more than it cost. This is sometimes written as a positive amount, for example + £3.50 means £3.50 profit. **Loss** is the amount of money that is lost when an item is sold for less than it cost. This is sometimes written as a negative amount, for example –£2.90 means £2.90 loss.

Profit (or loss) = selling price − cost price

EXAMPLE 1

An item is bought for £5.60 and sold for £7.80. Work out the profit.

SOLUTION

Profit is selling price − cost price = £7.80 − £5.60 = £2.20

EXAMPLE 2

A shop buys a box of 10 rulers for £1.50 and sells them for 20 pence each. How many rulers do they need to sell in order to make a profit?

SOLUTION

150p ÷ 20p = 7.5. So the shop needs to sell 8 rulers to make a profit.

Profit and loss are often expressed in terms of **percentages**.

EXAMPLE 3

A woman buys a dress from a shop for £20. She later sells it, making a loss of 25%. How much did she sell it for?

SOLUTION

25% of £20 = £5. So she sold it for £20 − £5 = £15

EXERCISE 3A 

G

1 Copy and complete the table. The first two parts have been done for you.

	Cost price	Selling price	Profit (+) or loss (−)
a	£3.60	£2.15	−£1.45
b	£2.70	£4.50	+£1.80
c	£3.20	£6.80	
d	£9.30	£7.10	
e	£8.50	£10.90	
f	£6.32	£8.99	
g	£5.11	£4.25	
h	£14.75	£20.00	
i	£18.75	£17.50	

2 In 2009 a company made a loss of £252 000

In 2010 the company made a profit of £164 000

How much profit does the company need to make in 2011 to break even over the three years?

F

3 A shop buys pens in packs of 10 for 60p per pack.

They are sold for 10p each.

a In one week the shop sells 80 pens.
How much profit do they make on the pens sold?

PS **b** In the following week they make £8 profit on the pens.
How many pens did they sell in that week?

4 Copy and complete the table.

	Cost price	Selling price	Profit (+) or loss (−)
a	£5.50		+£1.20
b	£6.30		+£1.69
c	£2.75		−95p
d	£4.62		−63p
e	£18.50		−51p
f	£15.13		−£1.36
g		£14.50	+£2.50
h		£18.99	+£3.49
i		£12.40	−£3.40
j		£16.35	−£6.25

F

5 A shop buys magazines for £1.40 each and sells them for £3.50 each.

- **a** How much profit is made on each magazine?
- **b** How much profit would be made if 30 magazines were sold?
- **c** How many magazines need to be sold to make at least £100 profit?

AU **6** The cost price of a bar of chocolate is 60p.

The bars are sold to make a profit of 25%.

How many bars of chocolate need to be sold to make a profit of £30?

E

7 A dressmaker buys a 20-metre roll of ribbon for £3.50.

She sells half the roll of ribbon for 30p per metre.

She then reduces the price to 15p per metre and sells the rest.

How much profit does she make?

FM **8** A salesman buys boxes of 25 T-shirts for £60. His target is to make a profit of more than 20%. He sells them for £2.99 each. Does he reach his target?

D

PS **9** The manager of a motorists' store buys a 50-metre roll of cable for £7.50.

He sells 18 metres of the cable for 25p per metre.

What is the lowest amount to which he could reduce the price on the remaining cable so that he does not make a loss overall?

3.2 Value added tax (VAT)

This section will show you how to:
- work out value added tax (VAT)

Key words
rate
standard rate
value added tax

Value added tax or VAT is a tax added to the price of some goods and services. It was introduced in Britain in 1973 at a **standard rate** of 10%.

Over the years different **rates** of VAT have been used and there are some complex rules. Here are a few of them.

- There is no VAT on food, books, newspapers, magazines, children's clothing and footwear and some equipment for disabled people.
- Exceptions to this are, for example, snacks and chocolate biscuits as these are considered non-essential luxuries so VAT is charged.
- The standard rate of VAT from January 2011 will be 20%.
- VAT on fuel bills and children's car seats is 5%.

EXAMPLE 4

An electricity company charges 7p for every unit of electricity used.

The meter read 18 047 for the previous bill and now reads 19 667.

VAT is charged at 5%.

How much is the final bill?

SOLUTION

Number of units used = 19 667 − 18 047 = 1620

Cost of 1620 units at 7p each = 1620 × £0.07 = £113.40

5% of £113.40 = $\frac{5}{100}$ × £113.40 = £5.67

[Or 10% of £113.40 = £11.34, so 5% of £113.40 is £11.34 ÷ 2 = £5.67]

Final bill = £113.40 + £5.67 = £119.07

EXAMPLE 5

The cost of an item is £18 excluding VAT.

Work out the cost including VAT at 20%.

SOLUTION

10% of £18 = £1.80

So 20% of £18 is £1.80 × 2 = £3.60

Cost including VAT is £18 + £3.60 = £21.60

EXERCISE 3B

1 Copy and complete the table for calculating VAT rates.

The first one has been done for you.

	Amount	10%	5%	20%
a	£1	10p	5p	20p
b	£2	20p	10p	
c	£3	30p		
d	£4			
e	£5			
f	£10			
g	£20			
h	£50			
i	£100			

F

F

2 VAT on fuel bills is 5%.

Use the table in question 1, or another method, to work out the VAT on each of the following bills.

a Gas £32 **b** Electricity £45

c Gas £75 **d** Electricity £99

e Gas £120 **f** Electricity £164

g Gas £186 **h** Electricity £215

E

AU 3 Liza was told the repairs to her car would be £320 plus VAT.

The VAT rate is 20%.

She has £400.

Can she afford the repair?

4 **a** Coldstore advertise a freezer at £350 excluding VAT.
An advertisement says Coldstore has frozen the cost of VAT at the old rate of $17\frac{1}{2}$%.
How much is the freezer being sold for?

PS **b** It is illegal to charge a different rate of VAT than that stated by the government.
Is the actual price at Coldstore excluding VAT more than £350 or less than £350?
Show how you decide.

FM **c** Icebox advertises the same freezer for £340 excluding VAT.
The VAT rate is 20%.
Which shop is cheaper, Coldstore or Icebox?

5 Copy and complete the following table. The first one has been done for you.

	Amount excluding VAT	VAT rate	Amount of VAT	Amount including VAT
a	£15.50	20%	£3.10	£18.60
b	£12.50	20%		
c	£19.60	20%		
d	£32.40	20%		
e	£35.90	20%		
f	£16.80	20%		
g	£12.60	5%		
h	£15.40	5%		
i	£19.80	5%		
j	£17.60	5%		
k	£67.40	5%		
l	£121.20	5%		

6 Copy and complete the table. Give answers to the nearest penny. The first one has been done for you.

	Amount excluding VAT	VAT rate	Amount of VAT	Amount including VAT
a	£13.42	20%	£2.68	£16.10
b	£24.16	20%		
c	£13.49	20%		
d	£16.78	20%		
e	£19.43	20%		
f	£28.82	20%		
g	£124.63	20%		
h	£43.22	5%		
i	£71.54	5%		
j	£19.34	5%		
k	£22.87	5%		
l	£73.12	5%		
m	£384.32	5%		

3.3 Interest rates, simple interest, savings and loans

This section will show you how to:

- work out simple percentages
- work out simple interest

Key words
interest
investment
loans
per annum
principal savings
simple interest

Interest is money paid on **savings** or **investments**, or charged on borrowings or **loans**. **Per annum** means per year, for example, 5% per annum means 5% interest is paid on an investment or loan in one year. Note that these amounts are again based on percentages.

EXAMPLE 6

Asif saves £2000 in a bond paying 5% per annum.

a How much interest does he earn in one year?

b How much is his investment now worth?

SOLUTION

a He earns 5% of £2000

which is $\frac{5}{100} \times £2000 = £100$

b His investment is now worth £2100

Simple interest, although not often used, is a fixed amount of interest paid each year. It is found by working out the interest paid for 1 year and then multiplying by the number of years invested.

The formula to calculate simple interest is:

$$I = \frac{PRT}{100}$$

where I is the interest, P is the **principal** or amount that is invested, R is the rate of interest and T is the number of years of the investment.

EXAMPLE 7

£4000 is invested at simple interest of 2% per annum for 3 years. Work out the interest earned.

SOLUTION

Method 1

Amount earned in 1 year is

$2\% \text{ of } £4000 = \frac{2}{100} \times £4000$

$= £80$

Simple interest for 3 years is

$3 \times £80 = £240$

Method 2: using the formula

$P = £4000 \quad R = 2\% \quad T = 3$ years

Simple interest $I = \frac{PRT}{100}$

$= £\frac{4000 \times 2 \times 3}{100}$

$= £240$

EXERCISE 3C

E

1 Copy and complete this table for simple interest investments.

The first one has been done for you.

	Amount invested (P)	Rate of interest per annum (R)	Simple interest per annum	Number of years invested (T)	Total amount of simple interest (I)	Total value of investment
a	£500	2%	£10	3	£30	£530
b	£500	3%		2		
c	£600	3%		2		
d	£800	2%		4		
e	£1000	4%		2		
f	£1500	3%		2		
g	£5000	1%		4		
h	£8000	2%		2		
i	£10 000	3%		3		

E

2 Copy and complete this table for simple interest investments.

Hint: Use the formula for simple interest: $I = \frac{PRT}{100}$

	Amount invested (*P*)	Rate of interest per annum (*R*)	Number of years invested (*T*)	Total amount of simple interest (*I*)
a	£500	3%	2	
b	£500		3	£15
c	£700	2%	4	
d		2%	4	£80
e	£1000	5%		£200
f	£1500		2	£300
g		1%	4	£200
h	£8000	3%		£720
i	£10 000		3	£600

PS 3 Here are two offers.

Give **two** reasons why Offer 2, from Your Bank is better than Offer 1, from The Bank.

Offer 1

Invest £1000 for 2 years and earn 4% simple interest each year.

The Bank

Offer 2

▶YOUR BANK

Invest £1000 for 1 year and earn 4% interest.

AU 4 James invests £3500 in a bond that pays 4% simple interest. He says that in three years it will be worth over £4000. Is he correct? You **must** show your working.

5 Copy and complete this table for simple interest loans. The first one has been done for you.

	Amount borrowed (*P*)	Rate of interest per annum (*R*)	Simple interest per annum	Number of years borrowed (*T*)	Total amount of simple interest (*I*)	Total cost of loan
a	£500	2.3%	£11.50	2	£23	£523
b	£500	3.3%		2		
c	£600	4.33%		2		
d	£800	2.75%		4		
e	£1000	4.8%		2		
f	£1500	1.3%		2		
g	£5000	0.2%		4		
h	£8000	1.9%		2		
i	£10 000	1.8%		3		

E

FM 6 Chloe takes out a loan of £500 for 2 years at 3% simple interest.

At the same time, she invests the £500 in a bond for 2 years at 4% simple interest.

To set up the loan she pays a fee of £50.

Does she make a profit on her dealings?

3.4 Income tax

This section will show you how to:

- work out income tax for a single rate of income tax
- work out income tax for more than one band of income tax

Key words
additional rate
basic rate
higher rate
income tax
personal allowance
personal tax allowance
salary
tax band
taxable income

When people earn above a certain amount each year, they pay **income tax**. The amount a person is allowed to earn without paying tax is called the **personal allowance** or **personal tax allowance**.

The amount a person earns in one year is called the **salary**. The personal allowance is deducted from the salary to give the **taxable income**. After that, tax is calculated as a percentage of the taxable income.

EXAMPLE 8

A man earns £27 000 a year. His personal allowance is £6200

His taxable income is taxed at 20%. How much income tax does he pay?

SOLUTION

He pays tax on £27 000 − £6200 = £20 800

$$20\% \text{ of } £20\,800 = \frac{20}{100} \times £20\,800 = £4160$$

He pays £4160 income tax.

Tax is calculated according to how much the person earns.

The percentage rate goes up, according to **tax bands**. The first band is the **basic rate**. After that, tax is levied at **higher rates**.

EXAMPLE 9 (HIGHER TIER ONLY)

A woman earns £60 000 a year. Her personal allowance is £5400

The first £37 400 of taxable pay is taxed at the basic rate of 20%.

The rest of her taxable pay is taxed at the higher rate of 40%.

How much income tax does she pay?

SOLUTION

She pays tax on £60 000 − £5400 = £54 600

She pays the basic rate on £37 400

20% of £37 400 = $\frac{20}{100}$ × £37 400 = £7480

She pays the higher rate on £54 600 − £37 400 = £17 200

40% of £17 200 = $\frac{40}{100}$ × £17 200 = £6880

She pays income tax of £7480 + £6880 = £14 360

For people on very high salaries, the top part of their income is taxed at an **additional rate**.

EXAMPLE 10 (HIGHER TIER ONLY)

A director of a company earns £250 000 a year.

His personal allowance is £7500

The table shows the tax bands.

How much income tax does he pay?

Income tax rates and tax bands	
Basic rate 20%	£0–37 400
Higher rate 40%	£37 401–150 000
Additional rate 50%	Over £150 000

SOLUTION

He pays tax on £250 000 − £7500 = £242 500

He pays the basic rate on £37 400

This leaves pay taxable at a higher rate as £242 500 − £37 400 = £205 100

He pays the higher rate on £150 000 − £37 400 = £112 600

This leaves pay taxable at the additional rate as £205 100 − £112 600 = £92 500

He pays the additional rate on £92 500

Personal allowance	£7 500
Basic rate 20%	£37 400
Higher rate 40%	£112 600
Additional rate 50%	£92 500
Total pay	£250 000

20% of £37 400 = $\frac{20}{100}$ × £37 400 = £7480

40% of £112 600 = $\frac{40}{100}$ × £112 600 = £45 040

50% of £92 500 = $\frac{50}{100}$ × £92 500 = £46 250

He pays income tax of £7480 + £45 040 + £46 250 = £98 770

EXERCISE 3D

E

1 Copy and complete the table to work out the amount of income tax paid. In all parts the rate of income tax is 20%. The first two parts have been done for you.

	Salary	Personal allowance	Amount to be taxed	Amount of income tax
a	£6000	£7000	Nil	£0
b	£8000	£7000	£1000	£200
c	£5000	£7000		
d	£9000	£7000		
e	£7500	£6200		
f	£9600	£6800		
g	£15 000	£5400		
h	£16 350	£5970		
i	£22 500	£4910		
j	£26 000	£5300		

2 Copy and complete the table to work out the amount of income tax paid for different rates. The first two parts have been done for you.

	Salary	Personal allowance	Amount to be taxed	Rate of income tax	Amount of income tax
a	£6000	£7000	Nil	22%	£0
b	£8000	£7000	£1000	25%	£200
c	£5000	£7000		26%	
d	£9000	£7000		23%	
e	£7500	£6200		22%	
f	£9600	£6800		22%	
g	£15 000	£5400		25%	
h	£16 350	£5970		24%	
i	£22 500	£4910		28%	
j	£26 000	£5300		27%	

3 Kerry pays income tax at 20% on a taxable income of £8000.

How much income tax does she pay?

4 Jack earns £11 000 a year.

He pays income tax at 22% on half of his salary.

How much income tax does he pay?

E

AU **5** Aby and Ben both work for the same company and pay income tax at the same rate.

Aby has a salary of £35 000 and a personal tax allowance of £7500.

Ben has a salary of £38 000 and a personal tax allowance of £4600.

Who pays more income tax?

You **must** show your working.

FM **6** The government decides to increase the rate of income tax from 20% to 23%.

How much more would a person with a taxable income of £18 000 pay?

PS **7** Two people both have the same taxable income of £17 000.

They have different salaries.

Their personal allowances have a difference of £1000.

Write down **two** possible salaries.

AU **8** Two people both have the same salary but pay different amounts of income tax.

Explain how this can happen.

B

9 The table below shows the tax bands.

Income tax rates and taxable bands	
Basic rate 20%	£0–37 400
Higher rate 40%	Over £37 400

Copy and complete the table to work out the amount of income to be taxed at each rate.

The first two have been done for you.

	Salary	Personal allowance	Total amount to be taxed	Amount to be taxed at basic rate (20%)	Amount to be taxed at higher rate (40%)
a	£55 000	£7000	£48 000	£37 400	£10 600
b	£41 000	£6000	£35 000	£35 000	Nil
c	£29 000	£7000			
d	£56 000	£8500			
e	£75 000	£7200			
f	£32 400	£9300			
g	£8000	£8000			
h	£36 500	£7500			

10 Calculate the amount of tax to be paid for each part of question 9.

B

11 The table opposite shows the tax bands.

Copy and complete the table to work out the amount of income to be taxed at each rate.

The first two have been done for you.

Income tax rates and taxable bands	
Basic rate 20%	£0–37 400
Higher rate 40%	£37 401–150 000
Additional rate 50%	Over £150 000

	Salary	Personal tax allowance	Total amount to be taxed	Amount to be taxed at basic rate (20%)	Amount to be taxed at higher rate (40%)	Amount to be taxed at additional rate (50%)
a	£105 000	£7000	£98 000	£37 400	£60 600	Nil
b	£180 000	£6000	£174 000	£37 400	£112 600	£24 000
c	£190 000	£7000				
d	£120 000	£8500				
e	£300 000	£7200				
f	£98 000	£9300				
g	£158 000	£8000				
h	£160 000	£7500				

12 Calculate the amount of tax to be paid for each part of question **11**.

AU **13** The table below shows the tax bands for 2009 and 2010.

In 2009 Mr Paye had a taxable income of £160 000. In 2010 Mr Paye had a taxable income of £161 000.

Income tax rates and taxable bands		
	2009	2010
Basic rate 20%	£0–37 400	£0–37 400
Higher rate 40%	Over £37 400	£37 401–150 000
Additional rate 50%	Not applicable	Over £150 000

a Work out the income tax that he paid in:

i 2009 **ii** 2010.

b In which year was his pay after deduction of income tax greater?

You **must** show your working.

PS **14** Use the table in question **13** to answer this question.

How much more income tax will a person with a taxable income of £200 000 pay in 2010 than in 2009?

AU **15** Explain why a person with an income of £155 000 might not pay income tax at the additional rate.

3.5 Appreciation and depreciation

This section will show you how to:

- solve problems involving appreciation and depreciation
- work out percentage appreciation or percentage depreciation

Key words
appreciation
depreciation

Some items, such as antiques, increase in value as they get older. Other items, such as cars, usually decrease in value as they get older.

An increase in value over a period of time is called **appreciation**.

A decrease in value over a period of time is called **depreciation**.

EXAMPLE 11

A new car costs £15 000.

In the first year, the value depreciates by 30%.

In the second year, the value depreciates by 10%.

How much is the car worth after two years?

SOLUTION

30% of £15 000 is $\frac{30}{100} \times$ £15 000 = £4500.

So its value at the end of the first year is £15 000 − £4500 = £10 500.

10% of £10 500 = £1050.

So its value after two years is £10 500 − £1050 = £9450.

EXAMPLE 12

The value of a painting appreciates by 10% each year for five years.

Explain why this is not an increase of 50% of the original value.

SOLUTION

It is 10% of the value at the end of each year so, for example, take the original value as £1000.

At the end of year 1 the value would increase by £100, so it now has a value of £1100.

At the end of year 2 the value would increase by £110, so it is not 10% of the original value and so on.

At the end of year 5 the value would be £1610.51 not £1500.

EXERCISE 3E

EXAMPLE 13

The value of a business falls from £2.5 million to £1.5 million in one year. Calculate the percentage depreciation.

SOLUTION

The fall in value is £1 million, so the percentage depreciation is

$$\frac{\text{fall in value}}{\text{original value}} \times 100\,\% = \frac{\text{£1 million}}{\text{£2.5 million}} \times 100\% = 40\%$$

E

1 An antique is valued at £5000. One year later the value has appreciated by 25%.

How much is it then worth?

2 One year ago a car was valued at £5000. Now, the value has depreciated by 15%.

What is the value now?

3 My shares were valued at £3500. One year later my shares increased in value by 15%. The following year they decreased in value by 20%.

How much are the shares worth after the two years?

D

AU 4 Use an example to show that if an item appreciates by 10% and then depreciates by 10% it will not go back to its original value.

C

5 A car falls in value from £10 000 to £7000.

Work out the percentage depreciation.

FM 6 One year, a painting increases in value from £1200 to £1600.

- **a** Work out the percentage appreciation.
- **b** The percentage appreciation is the same for the following year.

  Work out the new value for the painting.

7 The price of a share in a company rises from 108p to 144p. Work out the percentage increase.

8 The value of a luxury coach falls from £250 000 to £200 000 in one year.

Calculate the percentage depreciation.

PS 9 A piece of jewellery is valued at £500. One year later, it is valued at £650. After a further year it appreciates by 20%.

Work out the percentage increase (percentage appreciation) in value over the 2 years.

3.6 Savings: annual equivalent rate (AER)

This section will show you how to:

- understand the term 'annual equivalent rate' (AER)
- work out the annual equivalent rate, using the formula AER $= \left((1 + \frac{r}{100n})^n - 1\right) \times 100\%$ where r is the gross interest rate and n is the number of times that interest is paid each year

Key words
savings
annual equivalent rate (AER)
gross interest rate

When **savings** are invested, interest is usually paid. The **gross interest rate** is the amount you would be paid if the interest was paid annually before any deductions, for example, tax, have been made. Banks and building societies must state the interest rate they pay as an **annual equivalent rate (AER)**. This is what the interest would be if it was paid annually. If interest is only paid once a year the AER and the gross interest rate will be the same. The AER is used so that customers can compare different accounts. Generally, the higher the AER, the better the deal for the customer.

If you have to work out the AER as a percentage in the examination, this formula will be given.

$$\text{AER} = \left((1 + \frac{r}{100n})^n - 1\right) \times 100\%$$

where r is the gross interest rate and n is the number of times that interest is paid.

EXAMPLE 14

Rob invests £1000 in a savings account. The annual rate of interest is 4.5%. Interest is calculated and paid monthly.

Use the formula AER $= \left((1 + \frac{r}{100n})^n - 1\right) \times 100\%$ to work out the annual equivalent rate.

SOLUTION

In this case, $r = 4.5$ and $n = 12$ so AER $= \left((1 + \frac{4.5}{1200})^{12} - 1\right) \times 100\% = 4.6\%$

EXAMPLE 15

BIG SAVINGS ACCOUNT: 5.3% with interest paid annually
BIG BANK

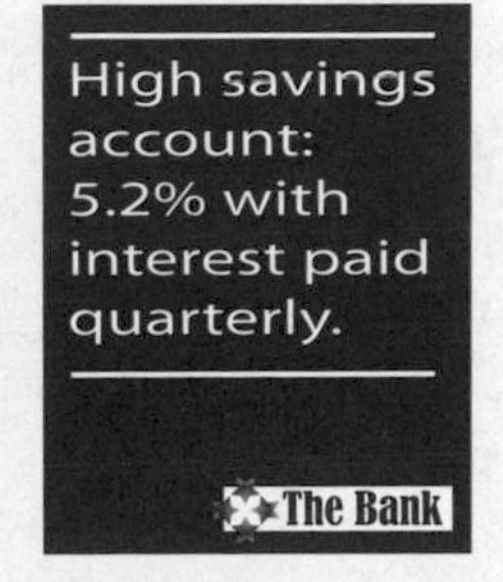

Which is the better rate of interest?

SOLUTION

The Big saving account pays interest annually so the AER will be 5.3%.

The High savings account pays interest quarterly so there will be four payments per year.

Using the formula, in this case, $r = 5.2$ and $n = 4$

AER is $\left((1 + \frac{5.2}{100 \times 4})^n - 1\right) \times 100\% = 5.302\%$

So the High savings account pays slightly more.

EXAMPLE 16

The annual equivalent rate for a saving account is 6.5%.

Interest is paid monthly.

Use the formula AER $= \left((1 + \frac{r}{100n})^n - 1\right) \times 100\%$ and trial and improvement to find the gross rate of interest.

Give your answer to two decimal places.

SOLUTION

Trial, r	AER	Comment
6	6.168	Too low
6.4	6.591	Too high
6.3	6.485	Too low
6.33	6.517	Too high
6.32	6.506	Too high
6.31	6.496	Too low
6.315	6.501	Too high

The gross rate of interest is 6.31% correct to two decimal places.

EXERCISE 3F

B

1 Use the formula to work out the AER for each of the following accounts. Give your answers correct to two decimal places.

	Gross rate of interest	Interest paid		Gross rate of interest	Interest paid
a	5.1%	Monthly	**g**	1.2%	Quarterly
b	3.5%	Monthly	**h**	0.2%	Quarterly
c	2.8%	Monthly	**i**	4.0%	Every 6 months
d	5.8%	Monthly	**j**	3.5%	Every 6 months
e	3.9%	Quarterly	**k**	7.1%	Every 6 months
f	2.4%	Quarterly	**l**	0.8%	Every 6 months

AU **2** Compare the following savings accounts.

▶YOUR BANK

Terrier account

- Minimum deposit £500
- Minimum term 2 years
- A withdrawal penalty equivalent to 180 days loss of interest applies
- Gross interest 4% per annum

Blue and white account

* No minimum deposit
* No minimum term
* Gross interest 3.5% per annum
* Interest paid monthly

B

3 Which gives the best rate of return over a year?

You **must** show all working.

a An annual rate of interest of 3.5% with interest paid monthly.

b An annual rate of interest of 3.5% with interest paid every three months.

c An annual rate of interest of 3.5% with interest paid twice yearly.

d An annual rate of interest of 3.5% with the interest paid annually.

4 **a** Use the formula to work out the AER for each of the following accounts.

Give your answers correct to two decimal places.

b Use your answers to part **a** to work out the value of the investment after 1 year.

Hint: Do not use the rounded figures for your calculations in part **b**.

	Gross rate of interest	Interest paid	Amount invested
i	2.7%	Monthly	£1000
ii	6.1%	Monthly	£1500
iii	3.2%	Monthly	£2000
iv	2.8%	Quarterly	£1200
v	1.5%	Quarterly	£800
vi	0.6%	Quarterly	£3000
vii	4.5%	Every 6 months	£2750
viii	3.3%	Every 6 months	£7125
ix	6.0%	Every 6 months	£8025

PS **5** Amy has £1500 to invest for 1 year.

She knows that she will need to withdraw £500 after 6 months.

How should she invest the money if she chooses from these accounts?

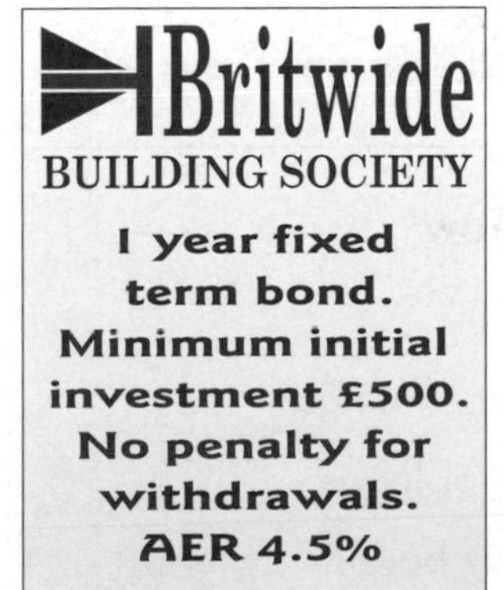

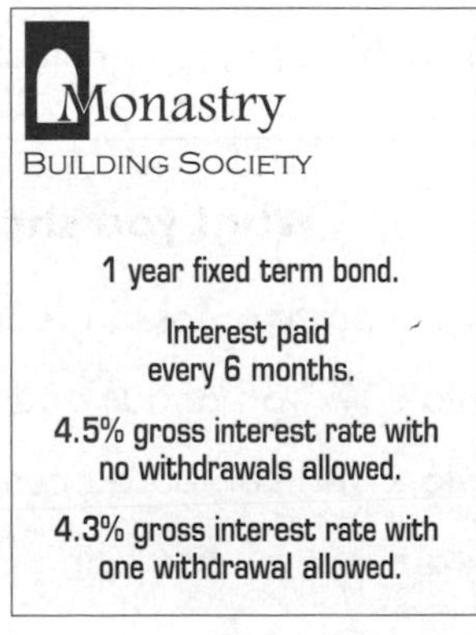

B

6 Use the AER formula and trial and improvement to work out the gross rate of interest for each of the following.

Give your answers correct to two decimal places.

	AER	Interest paid
a	1.6%	Monthly
b	5.0%	Monthly
c	2.1%	Monthly
d	1.7%	Quarterly
e	0.6%	Quarterly
f	5.5%	Quarterly
g	3.4%	Every 6 months
h	7.2%	Every 6 months
i	9.0%	Every 6 months

GRADE BOOSTER

G – **F** You can work out profit or loss in real-life situations

F – **E** You can calculate value added tax (VAT)

E You can calculate interest on savings and borrowings

E You can compare savings accounts and the costs of borrowing money

E You can work out income tax, using one tax band

E You can work out the effect of the value of goods appreciating or depreciating in value

C You can work percentage appreciation or percentage depreciation

B You can work out income tax, using more than one tax band

B You can use the formula to calculate the annual equivalent rate (AER)

What you should know now

- How to work out profit or loss in real-life situations
- How to calculate VAT on goods and income tax on earnings
- How to calculate and compare interest on savings or loans
- How to work out income tax, using one or more tax bands
- How to work out percentage appreciation or percentage depreciation
- How to use the formula to calculate the annual equivalent rate (AER)

1 The cost price of a shirt is £7.50.

The selling price of the shirt is £20.00.

Work out the profit if I buy 30 shirts and sell half of them at the full price and the rest at half price.

2 A computer game cost £25, a hairdryer cost £15 and a book cost £3. I sold them for £12, £7 and £4.

How much was my loss?

3 A shop buys chocolate bars in boxes of 48.

Each box costs £10.56.

The chocolate bars are sold for 45p each.

a How much profit is made on one box?

PS **b** How many bars do I need to sell to get back at least the cost of the box?

4 A lawnmower is priced at £85 excluding VAT.

VAT is 20%.

a Work out the price including VAT.

AU **b** In a sale the price is reduced by 10%.

Show how you can tell the reduction is more than £8.50.

5 **a** A bank has a savings account that pays 3% simple interest per annum.

How much would you need to invest to make £60 in 2 years?

PS **b** A £2000 bond pays 5% interest for 1 year.

If the interest rate in the second year drops to 1%, how much interest would be lost?

6 A picture was valued at £5000. Now, the value has appreciated by 8%.

What is the value now?

7 Molly has a personal tax allowance of £7200.

Her salary is £15 500.

The tax rate is 20%.

a How much tax does she pay?

PS **b** In the following year her personal tax allowance increases to £7500.

Her salary increases to £15 800.

Does her tax bill increase, decrease or stay the same?

You **must** show your working.

8 Tyrone has a personal tax allowance of £5600.

His salary is £14 000. The tax rate is 20%.

a How much tax does he pay?

PS **b** In the following year his personal tax allowance increases by £500. His salary increases by £500.

How does his tax bill change?

You **must** show your working.

9 Last year Jac earned £38 000. She had a pesonal tax allowance of £6900. The rate of income tax was 22%.

This year she earns £42 000. She has a personal tax allowance of £7300. The rate of income tax has decreased to 20%.

Does she pay more income tax last year or this year?

You **must** show your working.

10 A bracelet with a value of £1200 appreciates by 15% each year.

a How much is it worth after one year?

AU **b** A necklace with a value of £1500 appreciates by 10% each year.

After how many years is the bracelet worth more than the necklace?

Show your working.

PS **11** An artist has two paintings both valued at the same amount.

One appreciates by 10% each year for two years.

The other depreciates by 10% each year for two years.

Has the total value of his paintings increased or decreased over the two years?

Show how you know.

12 The table below shows the tax bands for 2009 and 2010.

Income tax rates and taxable bands		
	2009	**2010**
Basic rate 20%	£0–37 400	£0–37 400
Higher rate 40%	Over £37 400	£37 401–150 000
Additional rate 50%	Not applicable	Over £150 000

a Emma had a taxable income of £148 000 in 2009.

Work out the amount of tax she paid in 2009.

b In 2010 her taxable income increased by 10%.

Work out the amount of tax she paid in 2010.

AU **13** Wayne earns £260 000.

His personal allowance is £8500.

Here are the tax bands for this year.

Basic rate 20%	£0–37 400
Higher rate 40%	£37 401–150 000
Additional rate 50%	Over £150 000

Wayne says that he pays more than a third of his pay in income tax.

Is he correct?

You **must** show your working.

AU **14** Which gives the best rate of return over a year?

You **must** show all working.

a An annual rate of interest of 4.85% with interest paid monthly.

b An annual rate of interest of 4.9% with interest paid every three months.

c An annual rate of interest of 4.95% with interest paid twice yearly.

d An annual rate of interest of 5% with the interest paid annually.

FM **15** You are given that

$$\text{AER} = \left(\left(1 + \frac{r}{100n}\right)^n - 1\right) \times 100\%$$

Ruby invests £1000 in an account at 4% AER.

Emmy invests £1000 in an account paying 3.9% gross interest with interest paid monthly.

Who will have more money after 1 year?

You **must** show your working.

Worked Examination Questions

AU E

1 A shop sells rolls of carpet by the metre.

A roll is 25 metres long and costs the shop £1200.

How much would the shop need to sell the carpet for to make a 25% profit?

25% of £1200 is $\frac{25}{100}$ × £1200 = £300

You will get **1 mark** for setting up the calculation and **1 mark** for the answer.

Selling price per roll = £1200 + £300

= £1500

Selling price per per metre = £1500 ÷ 25

= £60

You will get **1 mark** for setting up the calculation and **1 mark** for the answer.

The shop will need to sell the carpet at £60 per metre.

Total: 4 marks

FM D

2 Sarah has a personal tax allowance of £6300.

Her salary is £18 200.

The income tax rate is 20%.

She estimates that she will pay about £200 each month in tax.

Is she correct?

You **must** show your working.

Her taxable income is

£18 200 – £6300 = £11 900

You will get **1 mark** for setting up the calculation and **1 mark** for the answer.

Tax to pay is 20% of £11 900

= $\frac{20}{100}$ × £11 900

You will get **1 mark** for method. You could do this as 0.2 × £11 900 or, as 10% is £1190, then £1190 × 2

= £2380

You will get **1 mark** for this answer.

Monthly payment

= £2380 ÷ 12

You will get **1 mark** for method. You could use £200 × 12.

= £198.33

So she is correct as this is about £200.

You will get **1 mark** for an answer with the correct conclusion. You could do this using £2400 for the year.

Total: 6 marks

Why this chapter matters

Knowing about personal and domestic finance will make your everyday life run more smoothly. It affects everything, from the money you earn at work, to running your home and planning your summer holidays.

Teacher	Shop work
Annual salary starting from £23 500	Pay: £8.70 per hour. Saturday paid at time and a half, Sunday is double time

Car sales person
Salary £1800 per month + annual bonus

As students finish their time in education, they may start to ask some of the following questions.

- Which jobs should I go for?
- What qualifications do I need?
- How much will I earn?

Someone planning a holiday for their family might ask questions like these.

- Where can I go on holiday?
- Where can I get the best deal?
- How much spending money should I take?

Chapter

Finance: Personal and domestic finance

The grades given in this FOUNDATION chapter are target grades.

This chapter will show you ...

- **F** how to work out wages and salaries from rates of pay
- **F** how to interpret questions, given net or gross amounts
- **E** how to use exchange rates and calculate commission
- **E** how to do calculations about loan repayments
- **D** how to work through a flow chart
- **D–C** how to understand and use the Retail price index (RPI) and the Consumer price index (CPI)
- **D–C** how to understand and use spreadsheets to model financial, statistical and other numerical situations
- **C** how to construct simple flow charts

Visual overview

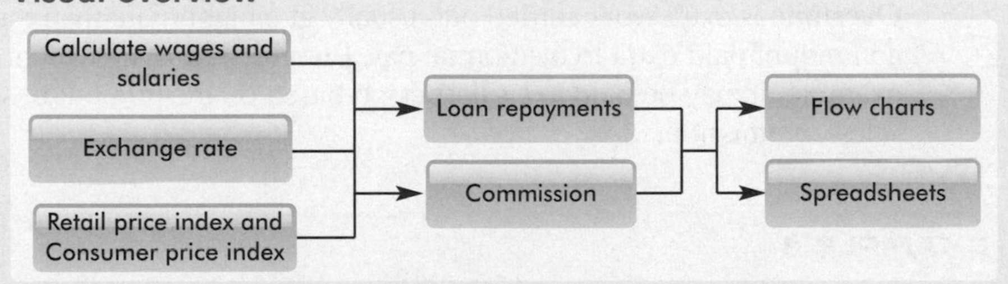

What you should already know

- How to interpret a formula in words **(KS3 level 4, GCSE grade F)**
- How to do calculations with percentages **(KS3 level 4, GCSE grade F)**
- How to write simple formulae **(KS3 level 5, GCSE grade E)**
- How to increase or decrease by a percentage **(KS3 level 7, GCSE grade D)**
- How to substitute into a formula **(KS3 level 5, GCSE grade E)**
- How to generate a simple sequence **(KS3 level 4, GCSE grade F)**

Quick check

1 Work out each of these amounts.

a $20 \times £6.50$ **b** $8 \times £7.50$ **c** $6 \times £10.75$

2 **a** Add 10% on to £56 **b** Subtract 20% from £90

c Add 25% on to £84

3 Work out the value of $3x + 2$ when:

a $x = 1$ **b** $x = 2$ **c** $x = 3$.

4.1 Wages and salaries

This section will show you how to:

- work out wages or salaries, given hourly, weekly or monthly rates of pay
- work out net pay from gross pay
- work out bonus rates or commission
- calculate overtime rates

Key words

bonus commission
deductions
gross net
overtime
rate of pay
salary wage

Some people are paid weekly and others are paid monthly. Weekly pay is called a **wage** and depends on an hourly **rate of pay**. A wage is usually calculated using this formula:

Wage (£) = number of hours worked × hourly rate of pay (£)

Annual or monthly pay is called a **salary**. It is usually a fixed amount paid each month.

Gross pay is the amount earned before any **deductions** have been made. **Net** pay is also known as 'take-home pay'. It is the amount left after deductions such as income tax, national insurance and pension payments.

Net pay = gross pay − deductions

Overtime is extra work carried out, usually at a higher rate than the basic rate of pay. A **bonus** is an amount paid extra to the normal pay. **Commission** is a payment based on performance, for example, it may be paid to a salesperson based on the number of sales. This is sometimes called sales commission.

EXAMPLE 1

Aisha earns a gross wage of £277 each week.

Deductions are £43.20.

Work out her net wage.

SOLUTION

Net wage is £277 − £43.20 = £233.80.

EXAMPLE 2

Eddie works 37 hours each week. His basic rate of pay is £12.50 per hour. He also works 2 hours overtime. Overtime is paid at 'time and a half'.

How much is his total pay?

SOLUTION

His basic pay is 37 × £12.50 = £462.50. Overtime paid at time and a half means for every 1 hour worked he is paid the equivalent of 1.5 hours at the basic rate. This is the same as being paid 1.5 times the basic rate for each hour of overtime. (£12.50 × 1.5 = £18.75)

So the overtime is 3 hours at £12.50 per hour or 2 hours at £18.75 per hour.
Overtime pay = £37.50. Total pay is £462.50 + £37.50 = £500.

EXAMPLE 3

A salesperson's annual salary is £21 000.

Each month, commission is added based on 10% of the value of sales.

In January the value of sales was £3600.

How much was the total pay for January?

SOLUTION

Basic salary for January is £21 000 ÷ 12 = £1750.

Commission is 10% of £3600 = £360.

Total pay is £1750 + £360 = £2110.

EXERCISE 4A

1 The gross pay of a worker is £325.60.

Deductions of £23.70 and £18.50 are made.

What is the net pay?

AU **2** Which is better, an annual salary of £16 000 or a weekly wage of £300?

Show how you decide.

3 Copy and complete the table for basic pay.

The first one has been done for you.

	Number of hours worked	Rate of pay per hour	Basic pay
a	10	£6.40	£64.00
b	20	£7.50	
c	30	£6.90	
d	40	£10.20	
e	35	£11.30	
f	37	£9.90	
g	15	£12.80	

AU **4** Jack is paid £11.20 per hour for a 37-hour working week.

He sees a similar job advertised at £20 000 per year plus 5% sales commission.

He estimates he will sell £3000 worth of goods each month.

Based on this information, should he change jobs?

Show how you decide.

G

F

E

D

5 Emma works a basic 35 hours each week.

Her basic rate of pay is £9.40 per hour.

She also works 3 hours of overtime on Saturday at double time.

Work out her total pay.

AU 6 Rebecca is comparing two similar jobs.

The first job has a salary of £20 000 plus 4% sales commission.

The second job has a salary of £18 000 plus 10% sales commission.

a Show that, if she sells £50 000 of goods, she will be better off with the second job.

b Give a reason why the second job has more risk involved.

PS FM 7 Ahmed works from 8 am to 5 pm each day from Monday to Friday, with one unpaid hour for lunch.

His rate of pay is £8.70 per hour.

Overtime is paid at time and a half.

How many hours of overtime will he need to work so that his total pay for a week is greater than £400?

8 Copy and complete the table for jobs with sales commission.

The first one has been done for you.

	Basic salary	Amount of sales	Sales commission rate	Commission paid	Total salary
a	£9200	£20 000	10%	£2000	£11 200
b	£8500	£30 000	10%		
c	£8000	£20 000	10%		
d	£11 500	£30 000	2%		
e	£12 500	£50 000	3%		
f	£10 600	£75 000	1%		
g	£14 800	£60 000	2%		

FM 9 Vic and Bob work for the same company.

Vic has a basic salary of £12 000 plus sales commission of 5%.

Bob is paid £1100 per month.

a Show that if Vic sells £2000 of goods in one month, they both get the same pay.

b How much more than Bob will Vic be paid in a month if he sells £8000 of goods?

4.2 Exchange rates and commission

This section will show you how to:
- change amounts of money from one currency to another
- calculate commission rates on currency transactions

Key words
commission
currency
exchange rate
variable exchange rate

There are over 150 different currencies in the world. Some countries use the same **currency**, for example France, Spain, Ireland and other European countries use the euro (€).

How many currencies can you think of? Here are just a few: British pound (£), US dollar ($), Japanese yen (¥), Indian rupee (₹).

When money is changed from one currency to another, an **exchange rate** is used. For example, the exchange rate between the pound and euro could be £1 = €1.20.

Sometimes the exchange rate will change as one currency becomes stronger or weaker. This is called a **variable exchange rate**. Sometimes, when currency is changed, a **commission** is charged. This can be a fixed amount or a percentage of the value of the transaction.

EXAMPLE 4

Matt went to Ireland. He changed £1000 into euros at an exchange rate of £1 = €1.14.

a How many euros did Matt receive?

b When he returned he had €200 left which he changed back into pounds at the exchange rate of £1 = €1.32. How many pounds did he get back?

SOLUTION

a Using £1 = €1.14 £1000 is equivalent to €1.14 × 1000 = €1140

b Using £1 = €1.32 €200 is equivalent to £(200 ÷ 1.32) = £151.52 (to the nearest penny)

EXAMPLE 5

Vrai visits the USA and uses his credit card to buy things.

Charges for using this card 2.5% on every transaction. Minimum charge of $2.50.

He pays for a $120 meal. He also spends $50 on the credit card.

The exchange rate is £1 = $1.54.

How much is charged to his card, **in pounds**?

SOLUTION

For the meal: 2.5% of $120 is $$\frac{2.5}{100} \times 120 = \$3$

So $123 is charged to his card.

$123 is £(123 ÷ 1.54) = £79.87 (to the nearest penny)

For the credit card, 2.5% of $50 is £1.25 so the minimum charge of $2.50 is made. So $52.50 is charged to his card.

$52.50 is £(52.50 ÷ 1.54) = £34.09 (to the nearest penny)

Total charged to his card is £79.87 + £34.09 = £113.96

EXAMPLE 6

A bank charges £2.50 or 0.5% commission, whichever is greater, for each transaction.

The post office does not charge commission.

Natasha wants to change £1200 into Japanese yen (¥).

The bank exchange rate is £1 = ¥134.

The post office exchange rate is £1 = ¥127.

Where should she change her money?

You **must** show your working.

SOLUTION

At the bank she would pay commission of 0.5% of £1200 = £6 (since 1% is £12).

£1200 would give ¥134 × 1200 = ¥160 800.

So she would be charged £1206 for ¥160 800.

At the post office:

£1200 would give ¥127 × 1200 = ¥152 400.

£6 would buy her ¥6 × 127 = ¥762 at the post office so for £1206 she would get ¥153 162.

So she should buy from the bank.

EXERCISE 4B

E

1 Copy and complete the table for converting from pounds (£) to different currencies.

The first one has been done for you.

	Currency wanted	Exchange rate	Amount to be converted	Amount received
a	Euros €	£1 = €1.15	£500	€575
b	Euros €	£1 = €1.15	£700	
c	US dollars $	£1 = $1.50	£700	
d	US dollars $	£1 = $1.50	£750	
e	Japanese yen ¥	£1 = ¥130	£900	
f	Japanese yen ¥	£1 = ¥130	£1200	
g	Australian dollar $	£1 = $1.68	£1250	
h	Australian dollar $	£1 = $1.68	£1500	
i	Swedish krona	£1 = 10.78 kr	£1750	
j	Swedish krona	£1 = 10.78 kr	£2000	

E

2 Copy and complete the table for converting from different currencies back to pounds.

The first one has been done for you.

	Currency wanted	Exchange rate	Amount to be converted	Amount received
a	Euros €	£1 = €1.15	€200	£173.91
b	Euros €	£1 = €1.15	€500	
c	US dollars $	£1 = $1.50	$800	
d	US dollars $	£1 = $1.50	$950	
e	Japanese yen ¥	£1 = ¥130	¥20 000	
f	Japanese yen ¥	£1 = ¥130	¥35 000	
g	Australian dollar $	£1 = $1.68	$1250	
h	Australian dollar $	£1 = $1.68	$1600	
i	Swedish krona	£1 = 10.78 kr	10 000 kr	
j	Swedish krona	£1 = 10.78 kr	12 000 kr	

AU **3** Which is better?

Changing £500 to US dollars at an exchange rate of £1 = $1.50 or changing £480 at an exchange rate of £1 = $1.53 with a £20 commission.

AU **4** Ingrid visits Sweden.

a She exchanges £600 for Swedish krona (kr).

The exchange rate is £1 = 10.5 kr. How much does she receive?

b Later, she exchanges £500 for Swedish krona.

The exchange rate is now £1 = 10.8 kr. How much more will she receive at the new rate than she would at the old rate?

c On her return to England she has 250 kr left.

The exchange rate in England is £1 = 11.25 kr. How much will she get back?

D

Olivia and Josh visited the USA.

Josh exchanged cash at a bank for $300. Josh pays no commission and his exchange rate is £1 = $1.42.

Olivia used her credit card.

The credit card company charges 2% for using the card with a minimum charge of £2.

Olivia used the credit card to obtain $300 cash. The exchange rate was £1 = $1.45.

Which is the better deal?

You **must** show your working.

D

6 The graph shows the exchange rate for the dollar against the pound for each month in one year.

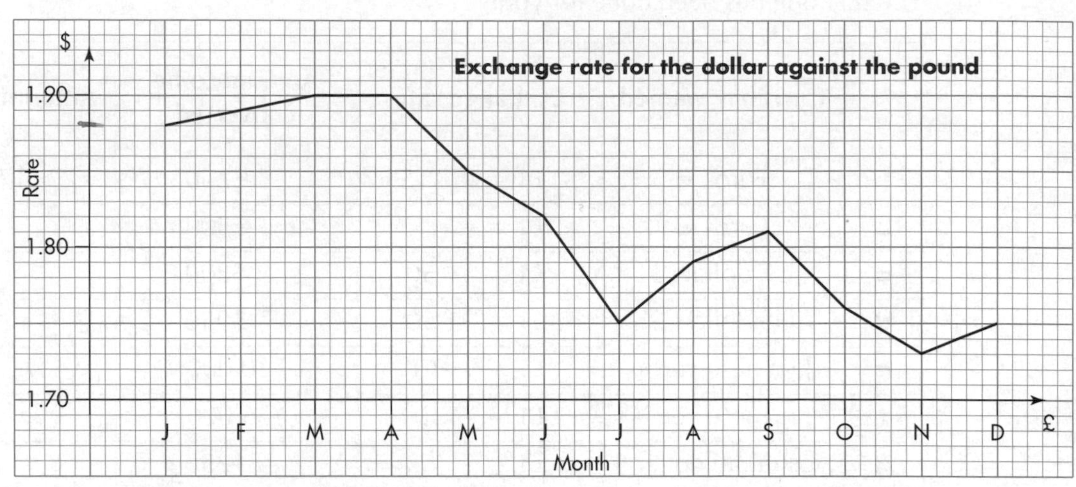

a What was the exchange rate in January?

b Between which two consecutive months did the exchange rate fall most?

c Explain why you could not use the graph to predict the exchange rate in January in the following year.

4.3 Loan repayments

This section will show you how to:

- understand the term 'annual percentage rate' (APR)
- understand the terms 'fixed rate' and 'variable rate'
- calculate the amount to be repaid when taking out a loan

Key words

annual percentage rate (APR)
fixed rate
loan
variable rate

When a **loan** is taken out, interest is usually charged. A company offering a loan must state the interest rate they charge as an **annual percentage rate (APR)**. This is what the interest would be if it was paid annually. The APR includes the cost of any set-up fees or other charges.

The APR is used so that customers can compare different loans. Generally, the lower the APR, the better the deal for the customer. You will **not** be required to work out the APR for the examination.

Most loans use a **fixed rate** of interest, although some longer-term loans, such as mortgages for buying houses, use a **variable rate** of interest. A fixed rate means that the interest rate stays the same throughout the loan. A variable rate means the interest rate can change, so the repayments can increase or decrease.

EXAMPLE 7

A company has these £1000 loans.

Monthly repayment	Term	Loan amount	Typical APR
£91.33	1 year	£1000	18.7%
£49.57	2 years	£1000	18.7%
£35.79	3 years	£1000	18.7%

a Give an advantage of paying off the loan in 1 year.

b Give a disadvantage of paying off the loan in 1 year.

c How much is the total repayment for a loan of £1000 paid back over 2 years?

d How much is the interest for a loan of £1000 paid back over 3 years?

SOLUTION

a The advantage is that the debt is repaid sooner or the total repaid is less.

b A disadvantage is that payments each month are higher than for a longer-term loan.

c £49.57 × 24 = £1189.68

d £35.79 × 36 − £1000 = £288.44

EXERCISE 4C

1 Patrick takes out a loan of £5000 to be repaid over 2 years.

The repayments are £240.20 per month. Work out the total repayment.

2 Emma takes out a loan for £3000 to be repaid over 5 years.

The repayments are £69.89 per month. Work out the amount of interest to be repaid.

3 Copy and complete the table for different loans.

	Amount of loan	Term in years	Term in months	Monthly repayment	Total amount of repayment	Total interest paid
a	£1000	2	24	£48.46	£1163.04	£163.04
b	£1500	2		£72.68		
c	£2000	3		£69.27		
d	£2500	3		£86.58		
e	£5000	2		£242.28		
f	£7500	5		£178.17		
g	£8000	7		£154.08		
h	£10 000	10		£160.93		

E

E

FM **4** Laura is buying a new car for £9500.

She pays a deposit of £1500 and then pays the balance over 3 years using this loan.

Pay over 3 years
Only 9.6% APR
Only £31.89 per month
for each £1000 borrowed

How much will she have to pay in total for the car?

PS **5** Sita sees this advertisement for a sofa.

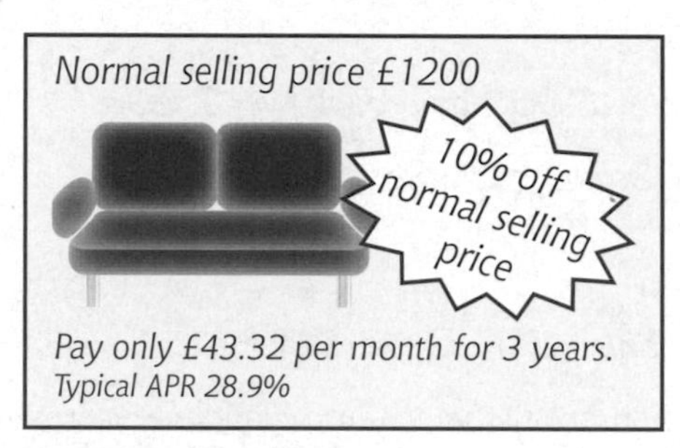

How much more does it cost to pay using the loan than to pay cash?

D

AU **6** A £1000 loan over 12 months with an APR of 21% will cost £92.26 per month.

A £1000 loan over 2 years with an APR of 21% will cost £50.51 per month.

Explain why the loan repaid over 12 months is less than double the amount of a loan of 2 years.

7 Joe sees these advertisements.

Quick Loan
£2500 with repayment over 3 years at £86.58 per month.
APR 16.0%

Fast Loan
£2500 over 4 years.
Repayment £78.37 per month.
APR 24.0%

Which loan costs more and by how much?

8 Joe sees the same furniture advertised in two different shops.

Seats For You
£1200
2 years interest free
Pay only £50 per month

Comfort furniture
Single payment of £1100
or 36 monthly instalments
of £36.85 *(APR 12.6%)*

At which shop is it more expensive to pay in instalments?

You **must** show your working.

4.4 Retail price index (RPI) and consumer price index (CPI)

This section will show you how to:

- understand the term 'retail price index' (RPI) and be able to do related calculations
- understand the term 'consumer price index' (CPI) and be able to do related calculations

Key words

Consumer price index (CPI)
Retail price index (RPI)

The **Retail price index (RPI)** is a measure of how much the cost of living increases or decreases over time. It measures the average change each month in the price of goods and services.

One year is chosen as the base year and given an index of 100. The last base of 100 was set in January 1987. Values of the RPI after this give a comparison with the base year. So, for example, in September 2006 the index was at 200, the prices had doubled since January 1987.

The **Consumer price index (CPI)** is a different measure of prices of goods and services purchased by households and is the main measure of inflation. The base year for this index is 2005.

EXAMPLE 8

The base year for the CPI is 2005 when the index was set at 100.

In 2010 the overall CPI was 114.2. The CPI for food was 126.4. The CPI for new cars was 108.8. The CPI for second-hand cars was 93.0. Compare the CPI for these items.

SOLUTION

The overall index has risen by 14.2% (100 to 114.2) since 2005.

Since the index for food is above this, food prices have risen more than average prices.

Prices of new cars have risen but by less than the average.

Prices of second-hand cars have fallen by 7%, compared to 2005, and in relative terms this means the difference in prices between new and second-hand cars has increased.

EXAMPLE 9

The base year for the CPI is 2005 when the index was set at 100. In 2010 the CPI for household fuel bills was 130. In 2005 the average bill was £676 per annum. How much is the average bill in 2010?

SOLUTION

A CPI of 130 means that average fuel bills have risen 30%.

30% of £676 is $\frac{30}{100} \times £676 = £202.80$

So the average bill is $£676 + £202.80 = £878.80$

EXERCISE 4D 

D

1 The RPI increased by 50% from its base figure.

What is the new figure?

2 In 2004 the cost of a litre of petrol was 78p. Using 2004 as a base year, the price of petrol for each of the next five years is shown in this table.

Year	2004	2005	2006	2007	2008	2009
Index	100	103	108	109	112	120
Price	78p					

Work out the price of petrol in each subsequent year.

Give your answers correct to 1 decimal place.

3 The general index of retail prices started in January 1978, when it was given a base number of 100. In January 2006 the index number was 194.1.

If the 'standard weekly shopping basket' cost £38.50 in January 1978, how much would it have cost in January 2006?

AU 4 In 2010 the CPI for costs of communications was as shown.

Jan	Feb	Mar	Apr	May	Jun
97.6	97.9	98.0	100.8	100.6	100.5

a In which months did prices fall?

b Compare this data with the base year.

5 In 2005 a first-class stamp cost 30p.

A newspaper headline at the time stated that by 2010 the price would rise by 20%.

A first-class stamp cost 41p in 2010.

Was the newspaper correct?

You **must** show your working.

6 In 2005 the average price of a CD album was £9.

The price index in 2010 was 105.

What is the average price of a CD in 2010?

7 In 2000 a piece of land was valued at £25 000.

In 2010 the same piece of land was valued at £87 500.

Using 2000 as the base year, work out the index number for 2010.

8 The graph shows CPI inflation from June 2008 to June 2010.

a In which month was CPI inflation the greatest?

b Does the graph show that prices fall or that inflation falls from September 2008 to September 2009?

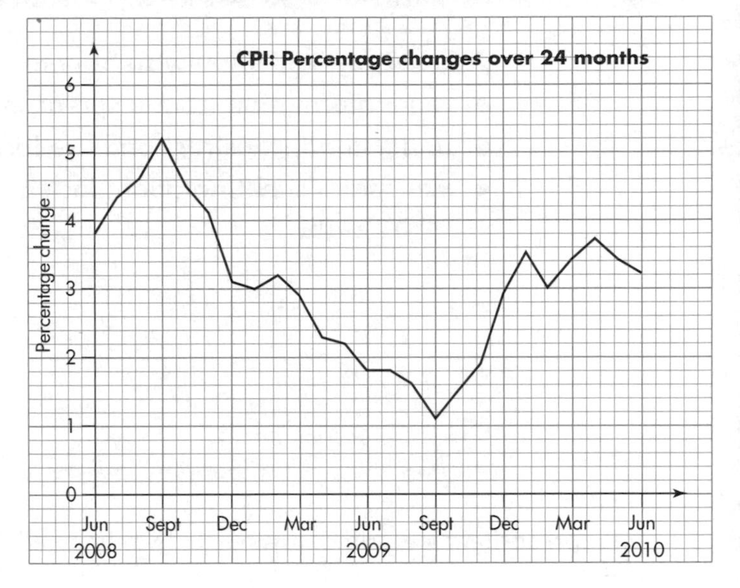

AU 9 The retail price index measures how much the cost of living increases or decreases. If 2008 is given a base index number of 100, then 2009 is given 98. What does this mean?

PS 10 On one day in 2009, Alex spent £51 in a supermarket. She knew that the price index over the last three years was:

2007	2008	2009
100	103	102

How much would she have paid in the supermarket for the same goods in 2008?

FM 11 The time series shows car production in Britain from November 2008 to November 2009.

a Why was there a sharp drop in production in June 2009?

b The average production over the first three months shown was 170 000 cars.

i Work out an approximate value for the average production over the last three months shown.

ii The base month for the index is January 2005, when the index was 100. What was the approximate production in January 2005?

4.5 Spreadsheets

This section will show you how to:
- read information from a spreadsheet
- understand, construct and use formulae
- use spreadsheet notation, including cell referencing

Key words
cell
column
fixed or absolute cell references
formula
relative cell references
row
spreadsheet

A **spreadsheet** is a grid of **rows** (usually numbered 1, 2, 3, …) and **columns** (usually lettered A, B, C, …).

Spreadsheets are used to display information clearly in a table. They allow you to organise what can often be large amounts of data and to make calculations. They are very useful for financial planning.

Spreadsheets used to be written on paper but, these days, they are normally produced by computer programs.

Each small box in the table is called a **cell**.

Each cell may contain text, numbers or a **formula** to carry out calculations.

Cells are referenced by the column letter followed by the row number, for example, A1, B27, D5, so A1 means column A and row 1.

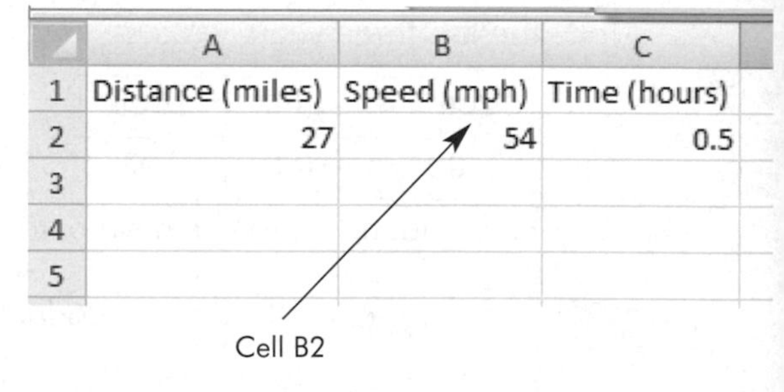

Cells can be referred to in groups, using a colon, for example E6:E10 means cells E6 to E10 inclusive.

Cells can be set up to work in different formats, for example, currency, time or general numbers.

Spreadsheets and formulae

Different spreadsheets use different conventions but the most common is to use an equals sign (=) to start a formula.

In formulae the following mathematical signs are used.

+ Addition

− Subtraction

* Multiplication (Notice that a standard multiplication sign is not used.)

/ Division (Notice that a standard division sign is not used.)

∧ Power or index

( ) Brackets

The only function used in the examination will be SUM, for example:

=SUM(A1:A5)

All the cell references used so far are called **relative cell references** as they change when they are copied and pasted to other cells. For example, if cell B5 said =A5 and this was then copied into cell B6 it would paste as =A6 to reflect the new position.

Fixed or **absolute cell references** can be used to keep the formula unchanged. The dollar sign is used to indicate no change. For example, if cell B5 said =A5 and you want the formula to copy without changing you would change cell B5 to =$A$5

When this is then copied into cell B6 it would paste as =$A$5 and would give the value as in cell A5.

EXAMPLE 10

Key the word 'Hello' in cell A3.

Then key =A3 in cell A5.

This will result in the contents of cell A3 appearing in cell A5.

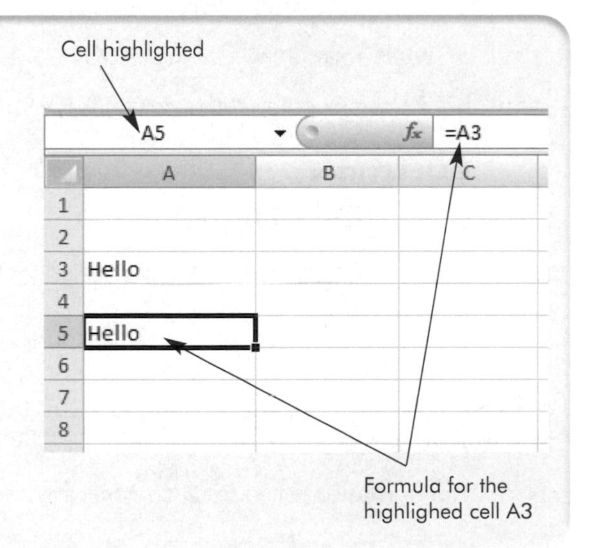

EXAMPLE 11

Key £1.50 into cell A1.

Key £2.40 into cell A2.

Key £5.32 into cell A3.

Key =SUM(A1:A3) into cell A5.

This will result in the contents of cells A1, A2 and A3 being added together and the total appearing in cell A5.

A5 =SUM(A1:A3)

	A	B	C	D	E
1	£1.50				
2	£2.40				
3	£5.32				
4					
5	£9.22				
6					

EXAMPLE 12

A manager of a music shop uses a spreadsheet to check his stock levels and prices.

D2 | fx =B2*C2

	A	B	C	D
1	Item	Number of items	Price per unit	Value of items
2	CD	8	£8.99	£71.92
3	DVD	6	£11.99	£71.94
4	Blu-ray	14	£14.99	£209.86
5				
6				

The formula used in cell D2 is =B2*C2

a What formula will be in cell D3?

b If the spreadsheet is continued, what formula will be in cell D10?

SOLUTION

a =B3*C3

The number of items is multiplied by the price per unit.

b =B10*C10

EXAMPLE 13

The spreadsheet is used to generate terms of the sequence with nth term $5n - 2$

B6 | fx =5*A6-2

	A	B	C	D
1	*n*	5*n* - 2		
2	1	3		
3	2	8		
4	3	13		
5	4	18		
6	5	23		
7	6	28		
8	7	33		
9	8	38		
10	9	43		

The formula used in cell A3 is =A2+1

The formula used in cell B6 is =5*A6–2

a What formula is used in cell A4?

b What formula is used in cell B2?

c What formula is used in cell B10?

d The value in cell B20 is 98.

What value is in cell B24?

SOLUTION

a =A3+1

b =5*A2−2

c =5*A10−2

d $98 + 4 \times 5 = 118$

EXAMPLE 14

The spreadsheet shows how a group of students travel to school.

The spreadsheet is being used to calculate the angles to make a pie chart.

The formula =B3/$B$8*360 is used in cell C3 to calculate the angle for those who walk.

C3 | *fx* =B3/$B$8*360

	A	B	C
1	Method of transport	Number of students	Angle
2			
3	Walk	24	120
4	Bus	36	180
5	Car	10	50
6	Other	2	10
7			
8	Total	72	360
9			

a Write down a formula for the total number of students.

b What does $B$8 mean?

c What is the formula used in cell C4?

SOLUTION

a =SUM(B3:B6)

b The contents of cell B8 are always used, even if this is copied and pasted to a different cell.

c =B4/$B$8*360

EXERCISE 4E

In this exercise the spreadsheets could be completed using an actual spreadsheet with formulae or by using calculations. Alternatively, spreadsheets could be used to check answers to calculations.

1 Copy and complete the table, using spreadsheet formulae to show how to carry out each operation. The first two have been done for you.

	Operation	Spreadsheet formula
a	Add 3 to the value in cell A1	=A1+3
b	Multiply the value in cell A2 by the value in cell C5	=A2*C5
c	Subtract 5 from the value in cell E2	
d	Divide the value of cell B9 by the value of cell A9	
e	Add the value in cell A6 to the value in cell A7	
f	Sum all the values in column B from cell B2 to cell B12	
g	Cube the value in cell C2	
h	Add the value in cell A3 to the value in cell A4 and then divide the answer by 12	
i	Divide the value in cell B3 by 100, add 1 to the answer and then raise this answer to the power 4	
j	Divide the value in cell F7 by the value in cell E7 and then subtract 20 from the answer	

D

2 The spreadsheet is used to convert distances covered (in miles) and times taken (in minutes) into speeds (in miles per hour, mph).

	A	B	C
1	Distance (miles)	Time (minutes)	Speed (mph)
2	20	30	40
3	40	20	
4	60	60	

The formula used for cell C2 is =A2*60/B2

a What formula is used for cell C3?

b Complete the values in the spreadsheet.

AU 3 The spreadsheet shows the costs when a group visit the cinema. There is a discount of £1 per person for this visit.

	A	B	C	D
1	Item	Number required	Cost per item	Total cost per item
2	Adult bus fare	4	£5.50	
3	Child bus fare	8	£1.80	
4	Adult cinema ticket	4	£5.60	
5	Child cinema ticket	8	£3.50	
6	Early bird discount	12	£1	

a Write down a formula for working out the total cost of the adult bus fares.

b Complete the values in the spreadsheet.

c Here is a formula: =Sum(D2:D5)–D6

Explain what this formula is working out.

4 The diagram shows a spreadsheet. The formula in cell B3 is =$A$12

	A	B	C
1	8		
2	12		
3	9	12	
4	5		
5	7		
6	11		

a The formula in cell B3 is copied and pasted into cell B4.

What value will appear in cell B4?

b The formula in cell A1 is =$B$7.

What will appear in cell B1?

PS **5** This spreadsheet is used to generate terms of sequences with nth term $4n + 2$ and with nth term $5n - 13$. The formula in cell B2 is =A2*4+2

	A	B	C
1	Term number n	nth term $4n + 2$	nth term $5n - 13$
2	1		
3	2		
4	3		
5	4		
6	5		

a What formula should be put in cell C2?

b The formula in cell A3 is =A2+1. What formula is in cell A6?

c Copy and complete the values in the table.

d Continuing the sequences, what value of n gives the same value for the nth term in both?

AU **6** The spreadsheet is used to work out the value added tax (VAT) for some clothes. The rate of VAT is 20%. The formula for cell C2 is =B2*20/100

	A	B	C	D
1	Item	Price excluding VAT	VAT	Price including VAT
2	Shirt	£20	£4	£24
3	Dress	£35		
4	Coat	£84		
5	Shoes	£45		
6	Tie	£11		

a Write down a formula for cell C3.

b One formula for cell D3 is =B3+C3. An alternative [formula] is B3*1.2. Why?

c Copy and complete the values in the table.

7 Khalid uses a spreadsheet to work out the amount of income tax that employees of the company have to pay. The rate of income tax is 20%.

	A	B	C	D	E
1	Name	Salary	Taxable allowance	Taxable pay	Amount of income tax to pay
2	A Smithies	£15 000	£8000	£7000	£1400
3	L Peltier	£11 500	£6000		
4	S Black	£24 000	£7500		
5	U Gul	£18 500	£3450		
6	A Chopra	£16 000	£8750		

a The formula for cell D2 is =B2–C2. What is the formula for cell D3?

b Work out a formula for cell E2.

c Write down a formula for cell E3.

d Copy the table and complete the values.

C

8 This spreadsheet is used to convert between pounds and other currencies.

The exchange rates are £1 = $1.49 and £1 = €1.15.

	A	B	C
1	Amount in pounds	Amount in US dollars	Amount in euros
2	£10		
3	£25		
4	£50		
5	£100		

a What formula should be put in cell B2?

b What formula should be put in cell C2?

c Copy and complete the values for columns B and C.

d The formula =B2/C2 is written in cell D2.

Explain what this formula is working out.

9 The spreadsheet is used to work out simple interest, using the formula:

simple interest $I = \frac{PRT}{100}$

	A	B	C	D
1	Amount invested (£P)	Rate of interest per annum (R%)	Time invested (T years)	Amount of interest (I)
2	100	2.1	2	
3	200	3.6	3	
4	500	1.3	2	
5	1000	2.5	4	

a What formula should be put in cell D2?

b Copy and complete the values in the spreadsheet.

10 The spreadsheet shows the calculation of monthly repayments for some loans.

	A	B	C	D	E	F
1	Monthly repayment	Term	Loan amount	Typical APR	Total amount repayable	Interest payable
2	£91.33	1 year	£1000	18.7%		
3	£49.57	2 years	£1000	18.7%		
4	£35.79	3 years	£1000	18.7%		

The formula for cell E2 is =A2*12 The formula for cell F2 is =E2−C2

a **i** What is the formula for cell E3? **ii** What is the formula for cell F3?

b **i** What is the formula for cell E4? **ii** What is the formula for cell F4?

c Copy and complete the values in the spreadsheet.

4.6 Flow charts

This section will show you how to:
- understand and use shapes as symbols in a flow chart
- work through a flow chart
- construct flow charts for simple calculations, which may be financial and may involve a repeated action

Key words
action
data input
data output
decision
end
flow chart
process
start

A **flow chart** is a diagram that uses a step-by-step series of operations or instructions to work through a problem.

Flow charts use different shapes for different types of operation.

A rounded rectangle is used at the **start** or **end** of a flow chart.

A parallelogram is used when **data** is **input** or **output**. A rhombus (in the shape of a diamond) is used for a **decision**, for example. 'Is x greater than 6, yes or no?'

You would write '$x > 6$?' inside the box, which has two exit lines for 'yes' and 'no'. This is the only sort of box that has more than one entry or exit line. A rectangle is used when a **process** or **action** is required.

The boxes are joined by straight lines, from one box to the next. The exit line from a decision box may loop back or on to join another line.

EXAMPLE 15

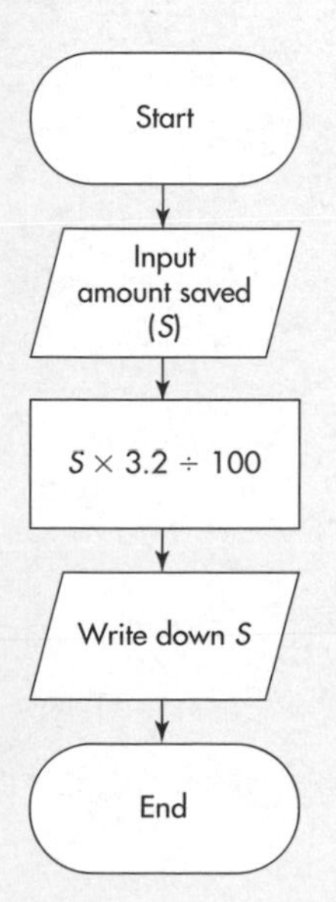

This flow chart is used to work out the interest paid on a savings account.

a What is the interest rate?

b How much interest is paid when £800 is invested?

SOLUTION

a The calculation to work out the interest is to multiply by 3.2 and divide by 100, so the interest rate is 3.2%.

b Interest paid on savings of £800 is
£800 × 3.2 ÷ 100 = £25.60.

EXAMPLE 16

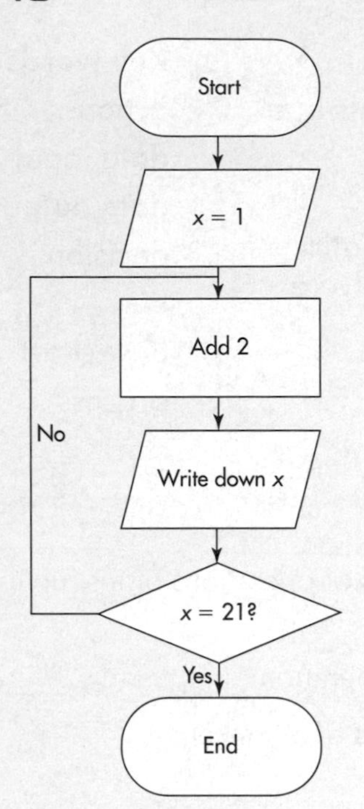

Which set of numbers is output from this flow chart?

SOLUTION

The numbers to write down are 3, 5, 7, 9, 11, 13, 15, 17, 19, 21.

So this flow chart generates the set of odd numbers from 3 to 21 inclusive.

EXAMPLE 17

Here is a flow chart.

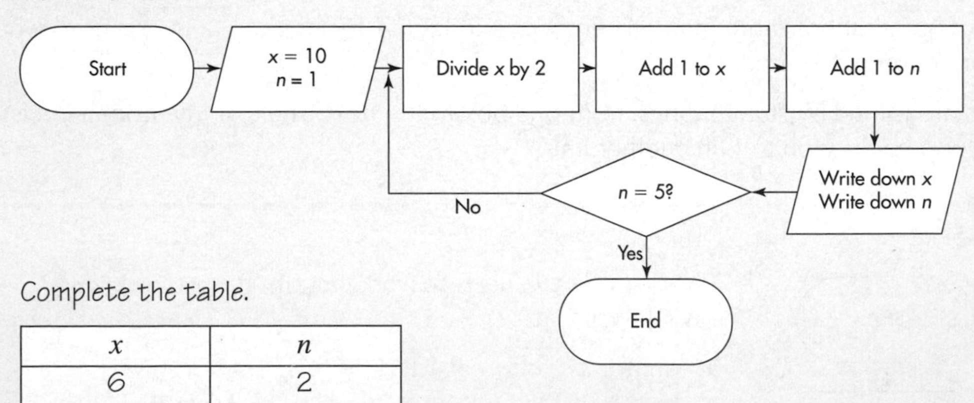

Complete the table.

x	n
6	2

SOLUTION

x	n
6	2
4	3
3	4
2.5	5

EXERCISE 4F

1 Here is a flow chart.

Write down the terms given by this flow chart.

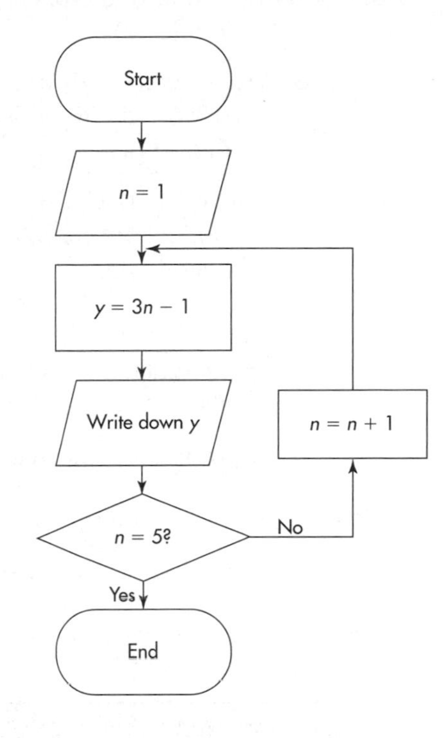

AU 2 Here is a flow chart for working out income tax.

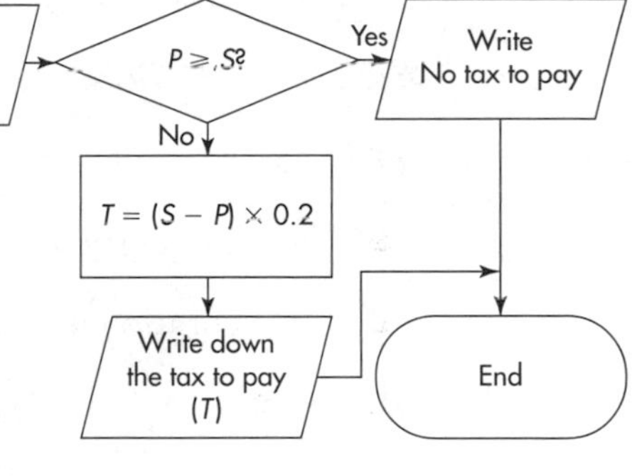

a Work out the tax payable for a salary of £12 000 and a personal allowance of £4500.

b Josh has a personal allowance of £6500.

He does not have to pay tax.

What do you know about his salary?

3 Construct a flow chart to multiply two numbers together.

AU 4 Here is a flow chart for converting pounds (£) to euros (€).

a Convert £900 into euros.

b Convert £400 into euros.

FM c How much is the commission?

Give a full answer.

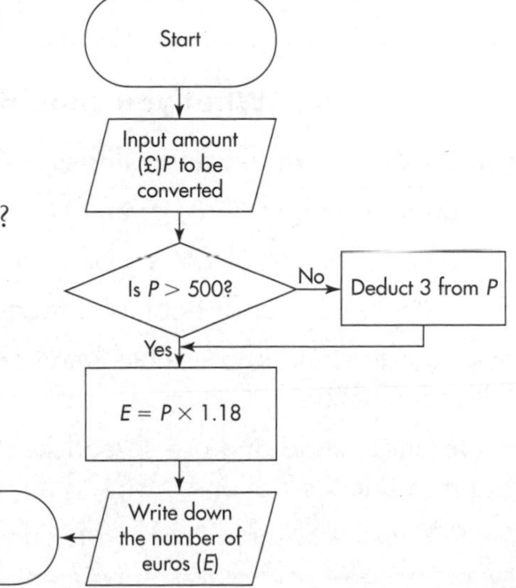

B

PS 5 This flow chart is for calculating VAT at the standard rate (20%) and at the reduced rate (5%).

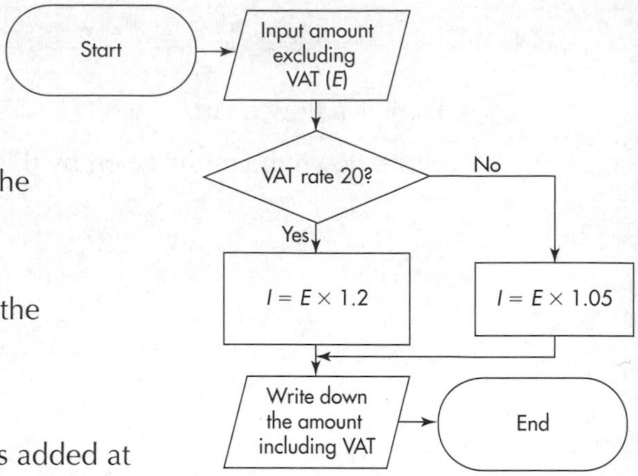

Start → Input amount excluding VAT (E) → VAT rate 20? — Yes → $I = E \times 1.2$; No → $I = E \times 1.05$ → Write down the amount including VAT → End

a A cycle costs £140 excluding VAT. Work out the price including VAT at the standard rate.

b A gas bill is £125 excluding VAT. Work out the gas bill including VAT at the reduced rate.

c An item costs £300 excluding VAT.

How much more would it cost if VAT is added at the standard rate instead of the reduced rate?

d How would you change the flow chart if you wanted to show the amount of VAT charged?

6 Draw a flow chart to work out the value of an investment if the rates of interest are:

2% for investments of less than £1000. 3% for investment of £1000 or more.

7 Draw a flow chart to generate the first 6 multiples of 8.

GRADE BOOSTER

- F You can work out wages and salaries from rates of pay
- F You can interpret questions given net or gross amounts
- E You can carry out calculations using exchange rates including commission
- E You can do calculations about loan repayments
- D–C You can understand and use the Retail price index (RPI) and the Consumer price index (CPI)
- C–B You can design and use spreadsheets and flow charts to solve problems

What you should know now

- How to work out wages and salaries from rates of pay
- How to interpret questions, given net or gross amounts
- How to carry out calculations, using exchange rates including commission
- How to do calculations about loan repayments
- How to understand and use the Retail price index (RPI) and the Consumer price index (CPI)
- How to understand and use spreadsheets to model financial, statistical and other numerical situations
- How to construct simple flow charts and work through them

AU 1 Which is better, an annual salary of £15 000 or £1300 per month?

Show how you decide.

2 Sally has an annual salary of £12 000.

Commission is added to her monthly pay, based on 10% of the value of sales for that month.

In January the value of sales was £3600.

In February the value of sales was £5000.

In March the value of sales was £2900.

a How much did she earn in the first three months of the year?

AU b In April she was given a 10% pay rise on her salary but her commission was reduced to 5% of the value of sales.

In April the value of sales was £3000.

Was she better off or worse off as a result of the new pay scheme?

You **must** show your working.

3 The exchange rate for pounds to US dollars is £1 = $1.45.

a Joe changes £500 into US dollars.

How many US dollars does he receive?

b Abby visits England from the US.

Abby changes $3000 into pounds, using the exchange rate shown above.

She then spends two-thirds of it and changes the remainder back into dollars.

The exchange rate for changing back is £1 = $1.25.

How many dollars does she get back?

4 Josh takes out a loan for £2000 to be repaid over four years.

The repayments are £54.45 per month.

a Work out the amount of interest to be repaid.

b After three years Josh pays off the loan.

He is charged £593.40

How much has he saved by paying off the loan early?

FM 5 Aaron uses his credit card to go shopping in France.

The credit card company charges 2.5% commission for using the card, with a minimum charge of €2.50.

a How much commission will he pay if he spends €60?

b How much does he need to spend for the commission to be greater than €2.50?

c Aaron spends €300 in a shop.

After the commission is added, the credit card company changes the bill back to pounds, using an exchange rate of €1 = 87p.

How much is his bill, in pounds?

AU 6 The base year for the consumer price index (CPI) was 2005.

In January 2010 the CPI was 112.4.

a Which of the following statements are definitely true for 2005 to January 2010?

The CPI has risen by 112.4%.

The CPI has risen by 12.4%.

Prices have risen every month since 2005.

b An item cost £27.50 in 2005.

Assuming the price rises in line with the CPI, what is the price of this item in January 2010?

c In December 2009 the CPI was 112.6.

What does this tell you about prices in January 2010?

7 The flow chart is about monthly savings.

It is used to work out the total amount saved, in pounds.

a How many years does it calculate?

b Work out the total amount saved.

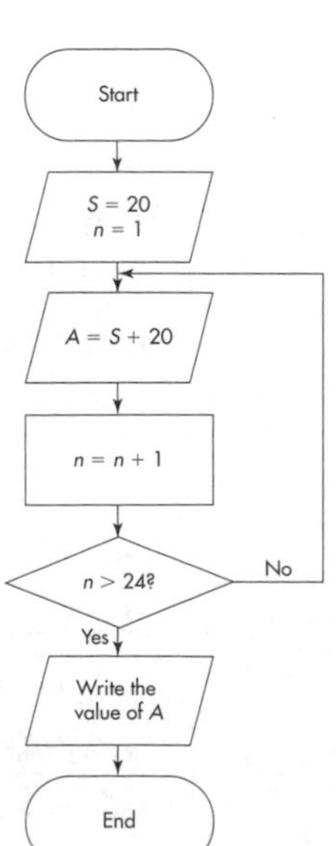

8 The base year for the Consumer price index (CPI) was 2005.

In June 2010 the Consumer price index (CPI) for food was 126.2.

Assume that these items change in line with the CPI for food in this question.

a What will the cost of a basket of food costing £25 in 2005 cost in June 2010?

b What was the cost of a loaf in 2005 if it now costs £1.08?

c If from June 2010 to June 2011 the average price of food rises by 2%, what will the CPI for food be in June 2011?

9 In January 1987 the Retail price index (RPI) was 100.

In June 2010 it was 224.1.

a Use this to work out the value of values of goods and services that cost £350 in 1987.

b Give a reason why the answer to part **a** might not be accurate.

10 The spreadsheet shows a timesheet for a worker.

She is paid time and a half for Saturday and double time for Sunday.

	A	B	C	D	E
1	Day	Rate of Pay	Number of hours worked	Amount of pay	
2	Monday	£7.30			
3	Tuesday				
4	Wednesday				
5	Thursday				
6	Friday				
7	Saturday	£10.95			
8	Sunday				
9				Total pay	

a What formula should be used in cell B3?

b **i** What formula should be used in cell B8?

ii What value will appear in cell B8?

c What formula should be used in cell D2?

d The total pay should appear in cell D9.

What formula should be put in cell D9?

11 In 2005 the Consumer price index (CPI) was 100.

In 2010 the Consumer price index (CPI) was 114.

The spreadsheet shows the costs of some goods in 2005.

	A	B	C
1	Item	2005 price	2010 price
2	Loaf	£1	£1.14
3	Skirt	£20	
4	Towel	£12	
5	Vacuum cleaner	£230	

a What is the formula for cell C3?

b Copy and complete the spreadsheet.

12 This flow chart is used to work out the new value V for items, using the consumer price index (CPI) with base year 2005.

The formula for V is **not** complete.

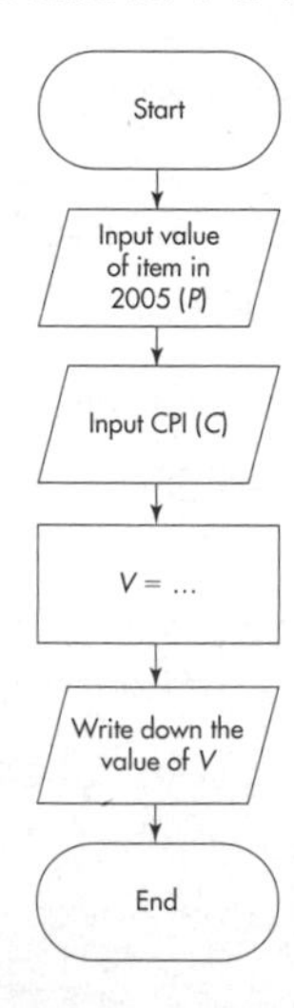

a Complete the missing formula.

b $C = 115$

Use your formula to work out the new value for an item that was valued at £90 in 2005.

Worked Examination Questions

PS F **1** Lee works a basic 40 hours each week.

His basic rate of pay is £10.70 per hour.

On one week he also works 2 hours on Saturday at time and a half.

How many hours will he need to work on Sunday that week, at double time, so his total pay is over £500?

40 hours at £10.70 per hour is 40 × £10.70 = £428

You will get **1 mark** for setting up the calculation and giving the answer.

Overtime for Saturday is 2 × 1.5 = 3 hours pay at normal rate (or 2 hours at £10.70 × 1.5 = £16.05)

You will get **1 mark** for setting up either calculation.

Saturday pay is
3 × £10.70 (or 2 × £16.05) = £32.10

You will get **1 mark** for a complete method and giving the answer.

Total pay = £428 + £32.10 = £460.10

You will get **1 mark** for quality of written communication so money notation must be correct throughout the question.

To earn £500 he needs another £39.90 so he will need to work 2 hours on Sunday to earn £42.80 and take his pay over £500.

You will get **1 mark** for the correct conclusion.

Total: 5 marks

C **2** In 2005 the consumer price index (CPI) was 100.

In June 2010 the CPI for clothing and footwear was 78.4.

a What is the percentage change?

b How much would a pair of shoes costing £35 in 2005 cost in 2010, assuming the price changes in line with this index?

a 100% – 78.4% = 21.6%

Change is a decrease of 21.6%.

You would be expected to take 78.4 away from 100 mentally or with a calculator, so you will get **1 mark** for the method and the answer.

b £35 × 0.784 = £27.44

You will get **1 mark** for the method and **1 mark** for the answer.

Total: 3 marks

Why this chapter matters

It isn't often that mathematical research gets labelled 'Top secret' but that is just what happened when the techniques of linear programming were being developed during World War 2, by British and American mathematicians, to make the production of armaments and the movement of troops more effective and efficient.

Linear programming is used to find the best solution to a problem that has various conditions, or **constraints**. The technique was kept secret until 1947 but, once it was made widely known, its development accelerated rapidly as many industries discovered valuable uses for linear programming.

The founders of the technique are generally acknowledged as being **George B. Dantzig** and **John von Neumann**. Their work allows the airline industry, for example, to schedule crews and make fleet assignments. Shipping companies can determine how many vehicles they need and where their delivery trucks should be deployed. The oil industry uses linear programming in refinery planning, as it determines how much of its raw product should become different types of fuels. Linear programming is also used in manufacturing, revenue management, telecommunications, advertising, architecture, circuit design and countless other areas.

Chapter 5

Finance: Linear programming

The grades given in this HIGHER chapter are target grades.

1 Representing inequalities graphically

2 Problem solving

3 Linear programming

This chapter will show you ...

- **C** how to set up an algebraic inequality in two variables
- **B** how to find a region on a graph that obeys a linear inequality in two variables
- **A** how to plot the region that obeys two or more different linear inequalities
- **A*** how to find the optimal solution within a region that obeys a given condition
- **A*** how to use linear programming techniques to solve a practical problem

Visual overview

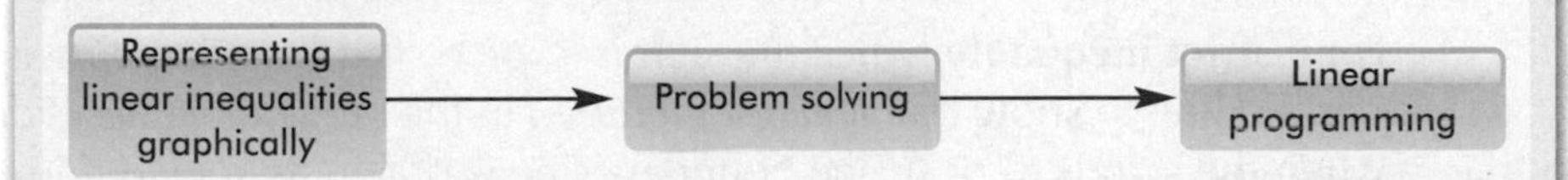

What you should already know

- How to solve linear equations **(KS3 level 6, GCSE grade D–C)**
- How to solve linear inequalities **(KS3 level 7, GCSE Grade C)**
- How to plot linear graphs **(KS3 level 7, GCSE Grade C)**

Quick check

1 Solve these equations.

a $4x + 1 = 2x + 8$ b $\frac{2x - 3}{5} = 1$

2 Solve these inequalities.

a $2x + 7 > 16$ b $\frac{2x + 3}{5} \leqslant 2$

3 Draw a grid like this and draw these lines.

a $y = 3$

b $x = 5$

c $y = 2x + 1$

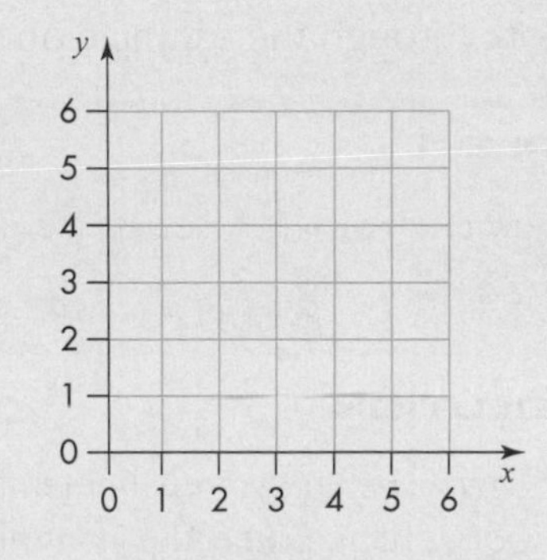

5.1 Representing inequalities graphically

This section will show you how to:

- find regions such as $y > 3$, $x \leqslant 4$ or $y \leqslant 2x + 1$
- draw straight lines with equations of the form $ax + by = c$
- find regions such as $ax + by \leqslant c$
- plot strict and inclusive inequalities in two variables
- find a region that obeys two or more inequalities in two variables

Key words
boundary line
graph
inclusive inequality
region
strict inequality
variable

A linear inequality can be plotted on a **graph**. The result is a **region** that lies on one side or the other of a straight line. You can recognise an inequality as it looks like an equation but, instead of the equals sign, it has an inequality sign: $<$, $>$, $\leqslant$ or $\geqslant$. Here are some examples of linear inequalities that can be represented on a graph.

$$y < 3 \qquad x > 7 \qquad -3 \leqslant y < 5$$

The method for graphing an inequality is to draw the **boundary line** that defines the inequality. To do this, replace the inequality sign with an equals sign.

For a **strict inequality**, when the sign is $<$ or $>$, the boundary line is drawn as a *dotted* or *dashed* line to show that it is *not included* in the range of values. For an **inclusive inequality**, when the sign is $\leqslant$ or $\geqslant$, the boundary line is drawn as a *solid line* to show that the boundary *is included.* After the boundary line has been drawn, the *required region is shaded.*

To confirm the side of the line on which the region lies, choose any point not on the boundary line and substitute in the inequality. If it satisfies the inequality, that is the side required. If it doesn't, the other side is required.

More than one inequality

To show a region that satisfies more than one inequality, it is clearer if we shade out the regions not required, leaving the required region blank.

Work through the example on inequalities, below, to see the procedure applied.

EXAMPLE 1

Show each of the following inequalities on a graph.

a $y \leqslant 3$ **b** $x > 7$ **c** $-3 \leqslant y < 5$

SOLUTION

a Draw the line $y = 3$. Remember that $y = c$ lines are horizontal. Since the inequality is stated as $\leqslant$, the line is solid. Test a point that is not on the line. The origin (0, 0) is always a good choice, if possible, as it is easy to test. Putting 0 into the inequality gives $0 \leqslant 3$. The inequality is satisfied and so the region containing the origin is the side we want. Shade it in.

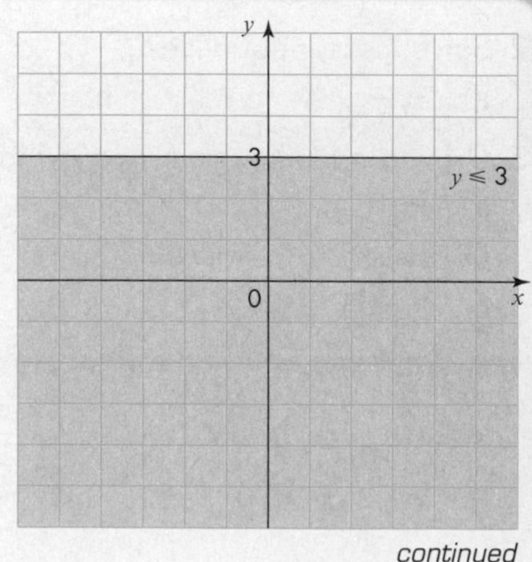

continued

b Since the inequality is stated as $>$, the line is dashed. Draw the line $x = 7$. Remember that $x = c$ lines are vertical. Test the origin (0, 0), which gives $0 > 7$. This is not true, so we want the other side of the line from the origin. Shade it in.

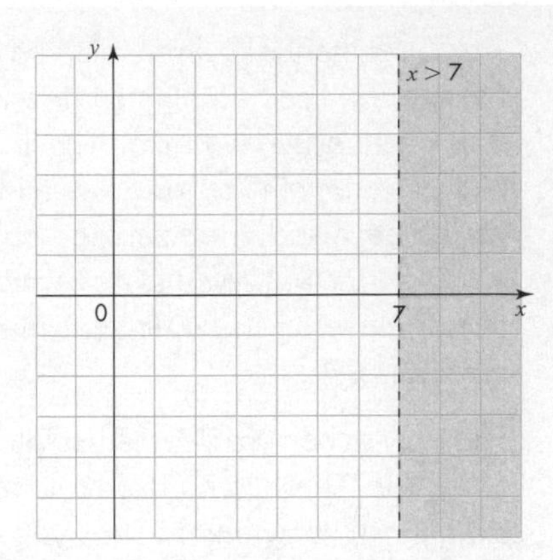

c Draw the lines $y = -3$ (solid for $\leqslant$) and $y = 5$ (dashed for $<$). Test a point that is not on either line, say (0, 0). Zero is between −3 and 5, so the required region lies between the lines. Shade it in.

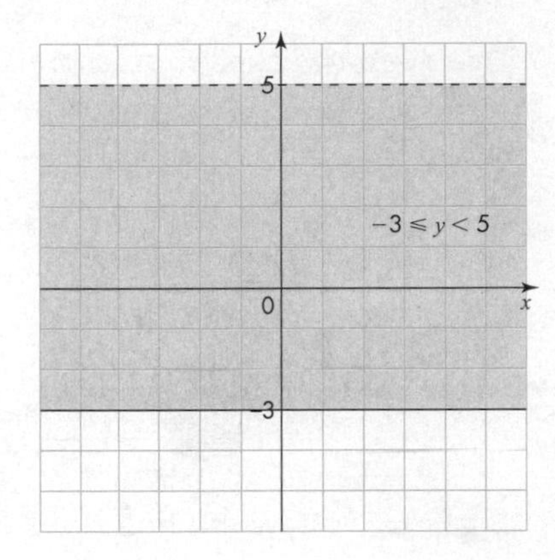

EXAMPLE 2

Show the region that obeys the inequalities $y > 2$, $x \leqslant 4$ and $y \leqslant 2x + 1$. Mark the region clearly as R.

SOLUTION

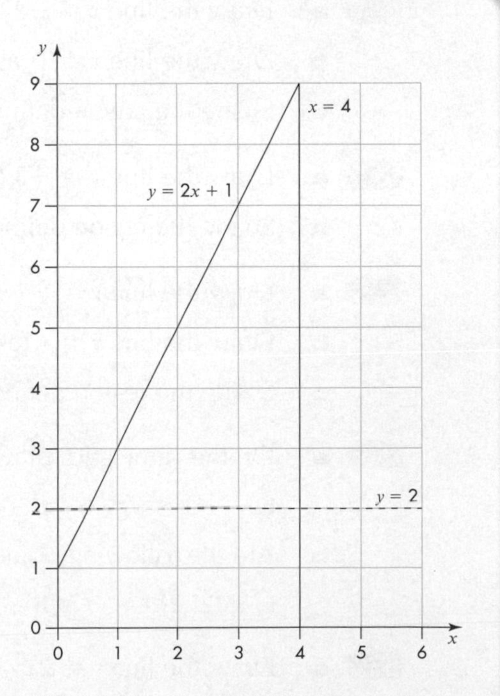

First draw the lines $y = 2$, $x = 4$ and $y = 2x + 1$ which are as shown on the grid. Note that the line $y = 2$ is dashed, as this is a strict inequality ($>$ or $<$) and the other lines are solid as they are inclusive inequalities ($\geqslant$ and $\leqslant$).

To find the region represented by $y > 2$, $x \leqslant 4$ and $y \leqslant 2x + 1$ we need to identify the appropriate sides of the boundary lines.

The first two are obvious. As $y = c$ lines are horizontal, any 'greater than' inequalities will be above the line and any 'less than' inequalities will be below the line.

As $x = c$ lines are vertical, any 'greater than' inequalities will be to the right of the line and any 'less than' inequalities will be to the left of the line.

continued

The third inequality may also be obvious but, to be sure, test a point that is not on the line. If the origin is not on the line then this is the best point to test, as all terms in the **variables** x and y are zero, so putting (0, 0) into $y \leqslant 2x + 1$ gives $0 \leqslant 1$, which is true, so the origin is on the side that obeys the inequality.

The only other problem is deciding which side to shade. There is no fixed rule for this but examiners recommend that you shade the side that is not required.

This gives the answer as shown.

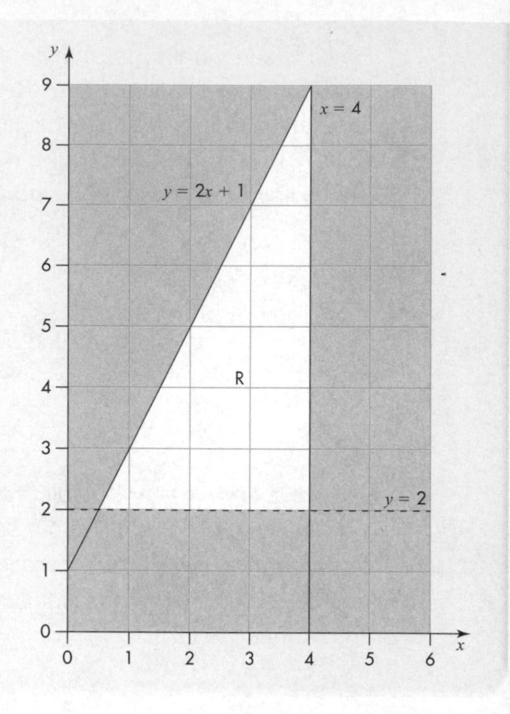

EXERCISE 5A

C

1 **a** Draw the line $x = 2$ (as a solid line).

b Shade the region defined by $x \leqslant 2$.

2 **a** Draw the line $x = -2$ (as a solid line).

b Draw the line $x = 1$ (as a dashed line) on the same grid.

c Shade the region defined by $-2 \leqslant x < 1$.

3 **a** Draw the line $y = -3$ (as a dashed line).

b Shade the region defined by $y > -3$.

4 **a** Draw the line $y = -1$ (as a dashed line).

b Draw the line $y = 4$ (as a solid line) on the same grid.

c Shade the region defined by $-1 < y \leqslant 4$.

B

5 **a** On the same grid, draw the regions defined by these inequalities.

i $-3 \leqslant x \leqslant 6$ **ii** $-4 < y \leqslant 5$

b Are the following points in the region defined by both inequalities?

i (2, 2) **ii** (1, 5) **iii** (−2, −4)

6 **a** Draw the line $y = 2x - 1$ (as a dashed line).

b Shade the region defined by $y < 2x - 1$.

7 **a** Draw the line $3x - 4y = 12$ (as a solid line).

b Shade the region defined by $3x - 4y \leqslant 12$.

8 **a** Draw the line $y = \frac{1}{2}x + 3$ (as a solid line).

b Shade the region defined by $y \geqslant \frac{1}{2}x + 3$.

9 **a** Draw the line $y = 3x - 4$ (as a solid line).

b Draw the line $x + y = 10$ (as a solid line) on the same diagram.

c Shade the diagram so that the region defined by $y \geqslant 3x - 4$ is left unshaded.

d Shade the diagram so that the region defined by $x + y \leqslant 10$ is left unshaded.

AU **e** Are the following points in the region defined by both inequalities?

i (2, 1) **ii** (2, 2) **iii** (2, 3)

10 **a** Draw the line $y = x$ (as a solid line).

b Draw the line $2x + 5y = 10$ (as a solid line) on the same diagram.

c Draw the line $2x + y = 6$ (as a dashed line) on the same diagram.

d Shade the diagram so that the region defined by $y \geqslant x$ is left unshaded.

e Shade the diagram so that the region defined by $2x + 5y \geqslant 10$ is left unshaded.

f Shade the diagram so that the region defined by $2x + y < 6$ is left unshaded.

AU **g** Are the following points in the region defined by these inequalities?

i (1, 1) **ii** (2, 2) **iii** (1, 3)

PS **11** **a** Show the region that obeys the inequalities $3x + 4y \geqslant 12$, $x + y < 6$, $x > 2$ and $y \geqslant 1$.

Label the region clearly as R.

b Which point in R, for which x and y are integers, has the maximum value for the function $2x + y$?

c Which point in R, for which x and y are integers, has the minimum value for the function $3x + 2y$?

PS **12** **a** Show the region that obeys the inequalities $2x + 3y \geqslant 6$, $5x + 4y < 20$, $y > 1$ and $x \geqslant 0$.

Label the region clearly as R.

b x and y are both integers.

Which point in R has a maximum value for the function $3x - y$?

c x and y are both integers.

For which points in R is the inequality $x^2 + y^2 < 9$ true?

5.2 Problem solving

This section will show you how to:
- set up an inequality in two variables given certain constraints
- find solutions that obey more than one inequality

Key words
condition
constraint
inequality
variable

Inequalities can arise in the solution of certain kinds of problem involving two **variables**.

EXAMPLE 3

James has £2.50 to buy drinks for himself and four friends. A can of cola costs 60 pence, a can of orangeade costs 40 pence. He buys x cans of cola and y cans of orangeade.

a Explain why:

i $x + y \geqslant 5$

ii $6x + 4y \leqslant 25$

b Write down all the possible numbers of each type of drink he can buy.

SOLUTION

a i James needs to buy at least five cans as there are five people. So the total number of cans of cola and orangeade must be at least five.

$x + y \geqslant 5$

ii x cans of cola cost $60x$ pence and y cans of orangeade cost $40y$ pence. So:

total cost $= 60x + 40y$

But he has only 250 pence to spend, so the total cost cannot exceed this.

Hence:

$60x + 40y \leqslant 250$

Dividing through by 10 gives:

$6x + 4y \leqslant 25$

b Trying different values of x and y gives four combinations that satisfy the conditions.

1 can of cola and 4 cans of orangeade.

2 cans of cola and 3 cans of orangeade.

5 cans of cola.

6 cans of orangeade.

EXERCISE 5B 

1 A company sells two types of bicycle, the Chapper and the Graffiti. A Chapper costs £148 and a Graffiti costs £125.

a How much do x Chappers cost?

b How much do y Graffitis cost?

c How much do x Chappers and y Graffiti cost altogether?

2 A school tuck shop sells Pluto bars at 19p each, Pogo mints at 15p per packet and Chews at 3p each. How much does each of the following orders cost?

a 2 Pluto bars and a packet of Pogo mints.

b x Pluto bars and y packets of Pogo mints.

c x Pluto bars, y packets of Pogo mints and z Chews.

d d Pluto bars, f packets of Pogo mints and 7 Chews.

3 A computer firm makes two types of machine, the Z210 and the Z310. The price of the Z210 is £A and that of the Z310 is £B. Find the cost of:

a x Z210s and y Z310s

b x Z210s and twice as many Z310s

c 9 Z210s and $(9 + y)$ Z310s.

AU 4 A bookshelf holds P paperback and H hardback books. The bookshelf can hold a total of 400 books. Which of the following may be true?

a $P + H < 300$

b $P \geqslant H$

c $P + H > 500$

5 A school uses two coach firms, Excel and Storm, to take students home from school. An Excel coach holds 40 students and a Storm coach holds 50 students. 1500 students need to be taken home. If E Excel coaches and S Storm coaches are used, explain why:

$4E + 5S \geqslant 150$

6 A boy goes to the fair with £6.00 in his pocket. He only likes rides on the big wheel and only eats hot dogs. A big-wheel ride costs £1.50 and a hot dog costs £2.00. He has W big-wheel rides and D hot dogs.

a Explain why:

i $W \leqslant 4$

ii $D \leqslant 3$

iii $3W + 4D \leqslant 12$

b If he cannot eat more than 2 hot dogs without being ill, write down an inequality that must be true.

c Which of these combinations of big-wheel rides and hot dogs obey all of the above conditions?

i 2 big-wheel rides and 1 hot dog.

ii 3 big wheel rides and 2 hot dogs.

iii 2 big-wheel rides and 2 hot dogs.

iv 1 big-wheel ride and 1 hot dog.

7 Pens cost 45p each and pencils cost 25p each. Jane has £2.00 with which she buys x pens and y pencils.

a Write down an inequality that must be true.

b She must have at least two more pencils than pens. Write down an inequality that must be true.

More than one constraint

Most practical and real-life problems have several **conditions** or **constraints** that must be met.

EXAMPLE 4

Aimee has £4.50 to spend on drinks for herself and five friends. A can of lemonade costs 60p and a can of ginger beer costs 75p. Two of her friends will only drink lemonade and one will only drink ginger beer. She buys x cans of lemonade and y cans of ginger beer.

a Explain why:

i $x + y \geqslant 6$ ii $4x + 5y \leqslant 30$

b Explain why:

i $x \geqslant 2$ ii $y \geqslant 1$

c Which of the following combinations of drinks (x, y) satisfy all the conditions?

i (2, 4) ii (2, 5) iii (5, 1)

iv (6, 2) v (6, 1) vi (7, 1)

d Of the combinations in **c** that obey all the conditions, which costs the least?

SOLUTION

a i Aimee needs to buy at least six cans as there are six people, so the total number of cans must be at least six.

This is expressed as $x + y \geqslant 6$

ii x cans of lemonade cost $60x$ pence and y cans of ginger beer cost $75y$ pence. The total cost of these is $60x + 75y$ pence. She only has 450 pence to spend so the total cost cannot be more than this.

This is expressed as $60x + 75y \leqslant 450$. Dividing through by 15 gives $4x + 5y \leqslant 30$

b i Two people only drink lemonade so she must buy at least 2 cans of lemonade.

This is expressed as $x \geqslant 2$

ii One person only drinks ginger beer so she must buy at least 1 can of ginger beer.

This is expressed as $y \geqslant 1$

c i 2 cans of lemonade and 4 cans of ginger beer obey all conditions.

ii 2 cans of lemonade and 5 cans of ginger beer obey three conditions but not $4x + 5y \leqslant 30$, as the total is 33, which is greater than 30.

iii 5 cans of lemonade and 1 can of ginger beer obey all conditions.

iv 6 cans of lemonade and 2 cans of ginger beer obey three conditions but not $4x + 5y \leqslant 30$, as the total is 34, which is greater than 30.

v 6 cans of lemonade and 1 can of ginger beer obey all conditions.

vi 7 cans of lemonade and 1 can of ginger beer obey three conditions but not $4x + 5y \leqslant 30$, as the total is 33, which is greater than 30.

d (2, 4) costs £4.20, (5, 1) costs £3.75, (6, 1) costs £4.35 so 5 cans of lemonade and 1 can of ginger beer is the cheapest option.

EXERCISE 5C

FM **1** Mushtaq has £3.00 to spend on apples and pears. Apples cost 30p each and pears cost 40p each. He must buy at least 2 apples and at least 3 pears, and at least 7 fruits altogether. He buys x apples and y pears.

a Explain each of these inequalities.

i $3x + 4y \leqslant 30$ **ii** $x \geqslant 2$

iii $y \geqslant 3$ **iv** $x + y \geqslant 7$

b Which of these combinations satisfies all of the above inequalities?

i 3 apples and 3 pears. **ii** 4 apples and 5 pears.

iii 0 apples and 7 pears. **iv** 3 apples and 5 pears.

FM **2** A shop sells only sofas and beds. A sofa takes up 4 m^2 of floor area and is worth £300. A bed takes up 3 m^2 of floor area and is worth £500. The shop has 48 m^2 of floor space for stock. The insurance policy will allow a total of only £6000 of stock to be in the shop at any one time. The shops stocks x sofas and y beds.

a Explain each of these inequalities.

i $4x + 3y \leqslant 48$ **ii** $3x + 5y \leqslant 60$

b Which of these combinations satisfy both of the above inequalities?

i 10 sofas and no beds. **ii** 8 sofas and 6 beds.

iii 10 sofas and 5 beds. **iv** 6 sofas and 8 beds.

FM **3** The 300 students in Year 7 are going on a trip to Adern Towers theme park. A local bus company has six 40-seater coaches and five 50-seater coaches. The school hires x 40-seater coaches and y 50-seater coaches.

a Explain each of these inequalities.

i $4x + 5y \geqslant 30$ **ii** $x \leqslant 6$ **iii** $y \leqslant 5$

b Check that each of these combinations obeys all of the inequalities above.

i 6 40-seaters and 2 50-seaters. **ii** 2 40-seaters and 5 50-seaters.

iii 4 40-seaters and 3 50-seaters. **iv** 3 40-seaters and 4 50-seaters.

c It costs £100 to hire a 40-seater coach and £120 to hire a 50-seater. Which of the combinations in part **b** would be the cheapest option?

AU **d** There is one combination that is even cheaper than the answer to part **c**. What is it?

PS FM **4** A lorry can carry two types of pre-packed pallet. Pallet A takes up 2 m^3 of space and weighs 1.25 tonnes. Pallet B takes up 3 m^3 of space and weighs 2.5 tonnes. The lorry has a capacity of 48 m^3 and a maximum load of 37.5 tonnes. The lorry is carrying x of pallet A and y of pallet B. Write down two inequalities that must be true for this situation.

PS FM **5** A farmer plans to keep ostriches and emus instead of cows. He has 300 acres of fields. Each ostrich needs 2 acres and each emu needs 1.5 acres. An ostrich gives 50 kg of meat and an emu 40 kg. The farmer can get a contract with a local meat-pie company if he can supply at least 6000 kg of meat. The ratio of the number of ostriches to the number of emus must never be more than 2 : 1, otherwise the ostriches attack the emus. The farmer buys x ostriches and y emus. Write down three inequalities that satisfy the given conditions.

5.3 Linear programming

This section will show you how to:
- solve a linear programming problem

Key words
feasible region
optimal solution
variable

Linear programming is a graphical method for solving problems in two **variables**, that have several conditions that must be satisfied simultaneously.

EXAMPLE 5

An agency wants to send at least 600 tonnes of emergency supplies via a ferry. It has eighteen 30-tonne trucks and twelve 40-tonne trucks; 24 drivers are available. The ferry company charges £200 for each 30-tonne truck and £300 for each 40-tonne truck. The agency uses x 30-tonne trucks and y 40-tonne trucks. Find:

a the cheapest option that satisfies the conditions

b the least number of trucks that could be used.

SOLUTION

First, write down the inequalities that describe the given conditions.

Trucks:	Number of 30-tonne trucks	$x \leqslant 18$	(1)
	Number of 40-tonne trucks	$y \leqslant 12$	(2)
Weight of supplies:	$30x + 40y \geqslant 600 \Rightarrow$	$3x + 4y \geqslant 60$	(3)
Number of drivers:		$x + y \leqslant 24$	(4)

Next, plot these inequalities on a graph, leaving unshaded the region that satisfies all of them. This is called the **feasible region**.

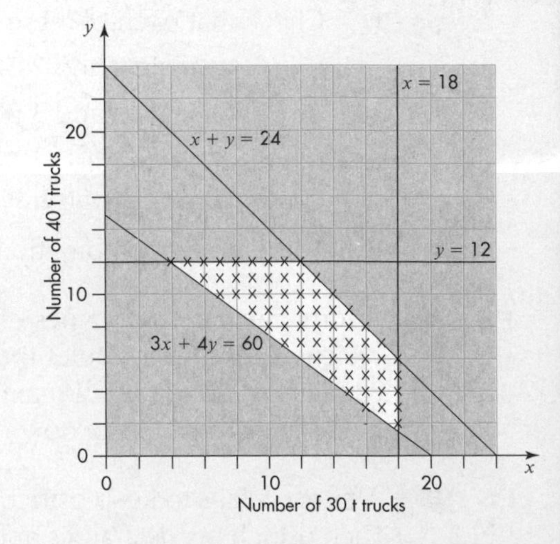

Tip: The size of the grid is usually given in an examination. In this case, we need the x-axis to go from from 0 to 24 and the y-axis from 0 to 24.

Within the feasible region, several possible pairs of values of x and y (marked with crosses) satisfy the conditions. To find the solutions to the problem we could test each of these points, but this would take too long. Instead, we test the points nearest to the corners of the region. These are (4, 12), (12, 12), (18, 6), (18, 2). The point that gives the cheapest option or the least number of trucks is called the **optimal solution**. This term is not used in examinations.

continued

a The ferry costs given by this set of test points are:

(4, 12) $4 \times 200 + 12 \times 300 = £4400$ (12, 12) $12 \times 200 + 12 \times 300 = £6000$

(18, 6) $18 \times 200 + 6 \times 300 = £5400$ (18, 2) $18 \times 200 + 2 \times 300 = £4200$

So the cheapest option is 18 30-tonne trucks and 2 40-tonne trucks.

b The least number of trucks is 16, comprising 4 30-tonne and 12 40-tonne trucks.

EXERCISE 5D

In each question, draw the graph and shade the regions not required. Label the feasible region as R.

A

FM 1 To entertain the children at a birthday party, Gianni decides to hire some DVDs. He can hire two types of DVD: cartoons, which last 30 minutes and cost 75p each to hire, and animations, which last 45 minutes and cost £1.00 each to hire. He needs to entertain the children for 3 hours and he has £6.00 to spend. So that the children do not get bored, he wants at least two of each type of DVD. He hires x cartoons and y animations.

a Explain each of these inequalities. **i** $2x + 3y \geqslant 12$ **ii** $3x + 4y \leqslant 24$

b Write down two other inequalities that must be true. Draw all four inequalities on the same grid. Take the x-axis from 0 to 8 and the y-axis from 0 to 6.

c Which combination would give the children the maximum amount of time watching DVDs?

AU d Which combination would give Gianni the cheapest possible option?

FM 2 Dave has a fish tank that holds 30 gallons of water. He wants to put two types of fish into it. Fantails need at least a gallon of water per fish and cost £2.50 each. Rainbows need at least 2 gallons of water per fish and cost £1.50 each. Dave wants more rainbows than fantails. He has £37.50 to spend, and buys x fantails and y rainbows.

a Explain each of these inequalities. **i** $x + 2y \leqslant 30$ **ii** $5x + 3y \leqslant 75$ **ii** $x < y$

b Write down two more inequalities that must be true. Draw all five inequalities on the same grid. Take the x-axis from 0 to 30 and the y-axis from 0 to 25.

c Which combination would give Dave the greatest possible number of fish?

d What combination would give Dave's wife the cheapest possible option?

PS FM 3 A school party stays for a week at a hotel that offers two types of room. The basic holds two people and the large holds four. There are 40 students in the school party. A basic room costs £100 per week and a large room costs £150 per week. The hotel has 10 basic and 7 large rooms available. The school has charged students £45 each for accommodation. The party takes x basic and y large rooms.

a Write down four inequalities that satisfy the given conditions. Draw them on the same grid. Take the x-axis from 0 to 30 and the y-axis from 0 to 20.

b What is the cheapest combination of rooms?

c The teacher in charge books the cheapest combination of rooms. When he arrives at the hotel with 23 girls and 17 boys he realises that he has made a serious mistake. What is it?

PS 4 FM An examiner has a stock of questions. Short questions carry 8 marks and take 15 minutes to do, and long questions take 3 minutes to do and carry 20 marks. There must be at least the same number of short questions as long questions. The examination lasts 3 hours, so the total time needed to do the questions must be less than this. Also, the examination must offer at least 100 marks, so the total marks available for the questions must be greater than this. The examiner set x short questions and y long questions.

a Write down three inequalities for this situation. Draw them on the same grid. Take the x-axis from 0 to 13 and the y-axis from 0 to 6.

b What is the smallest number of questions?

c How many marks would this combination give?

d How much time would this combination take?

Alf is on a diet. He eats only PowerBics, which have 4 g of fat and 300 calories, or SlimBars, which have 6 g of fat and 500 calories. He needs to keep his daily fat intake below 24 g and his daily calorie intake above 1500 calories. Because they taste so awful, Alf cannot take more than 4 PowerBics a day or more than 3 SlimBars a day. In one day he eats x PowerBics and y SlimBars.

a Write down four inequalities that satisfy these conditions. Draw them on the same grid, Take the x-axis from 0 to 6 and the y-axis from 0 to 4.

b What is the smallest total number of bars Alf could eat in a day?

c A PowerBic costs £2.50. A SlimBar costs £3.50. Which combination will cost Alf the least?

GRADE BOOSTER

- **C** You can set up an inequality such as $45x + 75y \leq 300$ based on conditions given in the question and divide through by a common factor, if possible, $(3x + 5y \leq 20)$
- **B** You can find the region on a graph that is represented by an inequality such as $3x + 5y \leq 20$
- **A** You can find a region on a graph defined by two separate inequalities
- **A*** You can find the feasible region in a linear programming problem
- **A*** You can find the optimal solution to a condition given in a linear programming problem

What you should know now

- How to set up inequalities in two variables that represent certain conditions
- How to draw regions on a graph that represent an inequality in two variables
- How to solve a linear programming problem

1 Packets of crisps cost 45p and packets of nuts cost 90p.

a How much will x packets of crisps and y packets of nuts cost?

b Pete has £3.60 to spend on crisps and nuts. Write an inequality in its simplest form, based on the above information.

2 If $x + y > 40$, which of the following may be true?

a $x > 40$ **b** $x + y \leqslant 20$

c $x - y = 10$ **d** $x \leqslant 5$

3 On three separate grids, shade the region where:

a $x \geqslant 2$ **b** $y \leqslant 4$ **c** $x + y \leqslant 4$

4 On two separate grids, shade the region representing each inequality.

a $x + y \leqslant 3$ **b** $y \leqslant 2x - 2$

5 'Creature Comforts' pet hotel is allowed to have a maximum of 10 pets at any one time. It only takes cats and dogs. A cat requires 1 unit of accommodation and a dog requires 3 units. For the hotel to make a profit, there must be at least 15 units occupied.

Let x be the number of cats and y the number of dogs in the pet hotel.

a Write down the inequalities that express the given facts.

b By shading the unwanted regions, illustrate the inequalities graphically.

6 Rob, a builder, has a plot of land with an area of 12 000 square feet. He builds two types of house, the Balmoral and the Sandringham. The Balmoral takes an area of 1500 square feet and the Sandringham 2000 square feet. Building a Balmoral requires five workers each day and building a Sandringham requires 3 workers each day. Rob has 30 workers available each day. He builds x Balmorals and y Sandringhams.

a Explain why $3x + 4y \leqslant 24$ and $5x + 3y \leqslant 30$.

b Draw these inequalities on the same diagram.

c The profit on any house is £4000. What combination of houses should Rob build, to make the maximum profit?

PS 7 **a** Show the region that obeys the inequalities $2x + 3y \geqslant 6$, $5x + 4y < 20$, $y > 1$ and $x \geqslant 0$. Label the region clearly as R.

b x and y are both integers. Which point in R has a maximum value for the function $3x - y$?

c x and y are are both integers. For which points in R is $x^2 + y^2 < 9$?

8 Julie is organising cars for her wedding. She needs to take 20 people to the ceremony. The local hire-car company has two types of limousine. The Canardley holds four people and costs £50 to hire. The Rapture holds five people and costs £60 to hire. Her budget for car hire is £300. The car-hire company has only four Canardleys and three Raptures. Julie hires x Canardleys and y Raptures.

a Write down four inequalities that are true for the above information.

b Draw these inequalities on the same diagram.

c Julie wants to spend as little as possible. What combination should she choose to do this?

9 Marc is making two types of biscuit for a fete. He has 1.5 kg of sugar and 1 kg of butter.

A batch of cookies needs 150 g of sugar and 175 g of butter.

A batch of brownies needs 200 g of sugar and 75 g of butter.

a Mark makes x batches of cookies and y batches of brownies. Use the amounts of butter described to show that $7x + 3y \leqslant 40$.

The graph shows the line $7x + 3y = 40$.

The amounts of sugar give the inequality $3x + 4y \leqslant 30$.

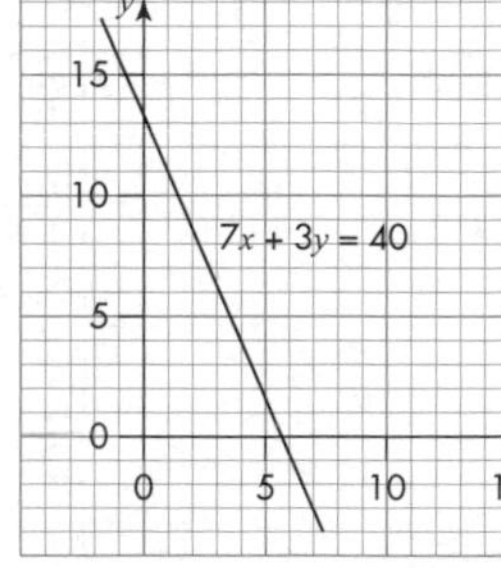

b Copy the graph and draw on it the line $3x + 4y = 30$. Label, with the letter R, the region where the following inequalities are both true.

$3x + 4y \leqslant 30$ $7x + 3y \leqslant 40$

c Marc makes three batches of cookies. Use the graph to find the largest number of batches of brownies he can make.

Worked Examination Questions

AU A* FM

1 A ferry company has two types of boat, the *Swift* and the *Swallow*.

In one day the *Swift* ferries can make a maximum of 15 crossings and the *Swallow* ferries can make a maximum of 10 crossings.

Each *Swift* can carry 200 cars and 50 lorries.

Each *Swallow* can carry 150 cars and 80 lorries.

On one day in summer they expect to carry a total of 3600 cars and 1200 lorries.

Let x be the number of *Swift* ferries crossings and y the number of *Swallow* ferries crossings.

a Explain why $4x + 3y \geqslant 72$

b Write down **three** other inequalities that must be true, from the information above.

c Using your inequalities, show on a graph the region that satisfies all the conditions.

d Each *Swift* crossing costs £16 000 and each *Swallow* crossing costs £10 000.

What is the cheapest combination of ferries?

a Using the information for cars: $200x + 150y \geqslant 3600$

Divide through by 50 to get $4x + 3y \geqslant 72$

> Work out the number of cars that each ferry can carry and state that this must be at least 3600. Show the cancelling by 50. This will gain **1 mark**.

b Using the information for lorries:
$50x + 80y \geqslant 1200$

Divide through by 10 to get $5x + 8y \geqslant 120$

Using the information about number of crossings:
$x \leqslant 15$ and $y \leqslant 10$

> Do the same thing for the number of lorries and also use the number of crossings to get two further inequalities. This will gain another **2 marks**.

c

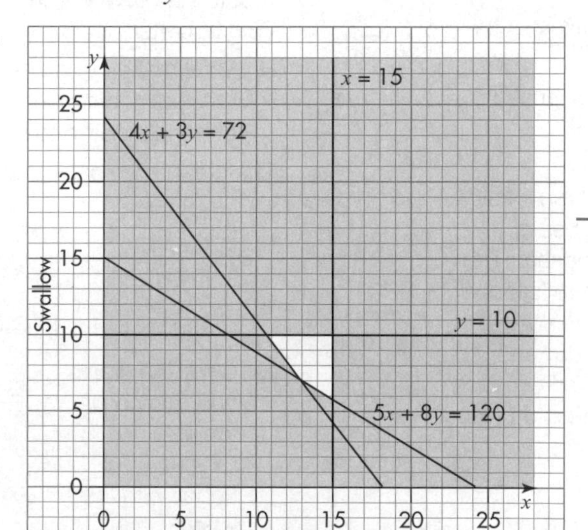

> Plot all the inequalities and shade out the areas that are not required to leave the feasible region blank. This will gain **3 marks**.

d The cheapest combination is 11 Swift and 10 Swallow which costs £276 00.

> Use the combinations nearest the corners (11, 10), (13, 7) and (15, 6) to check your answers. **1 mark** each.

Total: 8 marks

Worked Examination Questions

PS **A***
AU
FM

2 A smallholder wants to keep some chickens and some ducks.

Each chicken needs 2 m^2 of space and each duck need 3 m^2 of space.

The smallholder has an area of 24 m^2 in which to keep the birds.

Chickens cost £4.20 and ducks cost £3.50.

The smallholder has £42 to spend.

The smallholder wants at least two of each type of bird.

Let c be the number of chickens and d be the number of ducks.

a Write down **four** inequalities that describe the information above.

b Show these inequalities on a graph, leaving the region that satisfies all conditions blank.

c Chickens lay an average of 5 eggs per week and ducks lay an average of 7 eggs per week.
Which of the possible combinations of chickens and ducks will give the most eggs?

a $c \geq 2$, $d \geq 2$, $2c + 3d \leq 24$, $4.2c + 3.5d \leq 42$

$4.2c + 3.5d \leq 42 \Rightarrow 6c + 5d = 60$

There are no common factors in the first three inequalities so just write them down from the information given. This will gain **3 marks**.

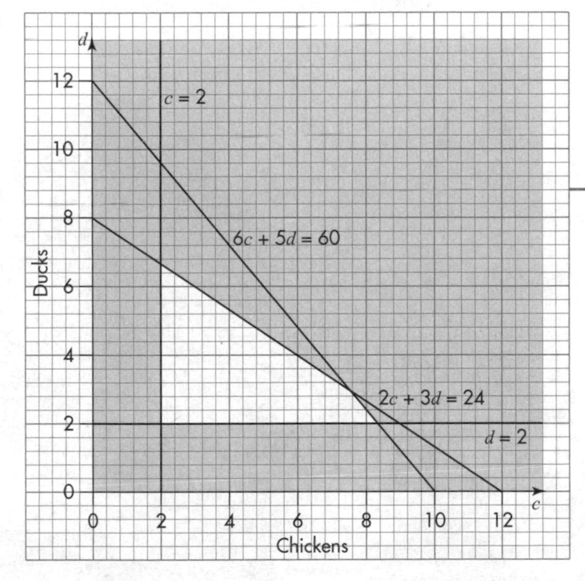

Plot all the inequalities and shade out the areas that are not required to leave the feasible region blank. This will gain **3 marks**

c 7 chickens and 3 ducks will lay 56 eggs per week.

Use the combinations nearest the corners (2, 2), (2, 6) (8, 2) and (7, 3) to check your answers. **1 mark** each.

Total: 8 marks

Why this chapter matters

Governments and companies need to plan for future conditions. To do this, they try to spot trends, using existing data to predict how things may change. Of course, this is not an exact science since nobody can predict the future accurately.

Population set to increase by 20% by 2020

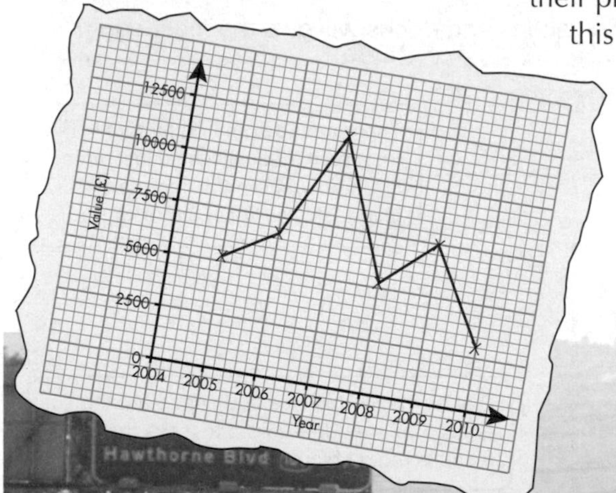

The Government needs to know in advance how many cars there will be on the roads, or how the demands on the health service will change, as it would be too late to take effective action once the roads are totally clogged or people are dying because there are not enough hospital beds for them to be treated.

Businesses need to know how to invest, so they can be sure their production facilities are ready to meet demands. Although this is a bit of a gamble, if they do not do it, and another business does, they will lose out on income.

Even examination boards can use the trend in entries for subjects to predict future entries. This allows them to plan for printing costs, for papers, and the number of examiners needed for marking.

Chapter

Statistics: Time series and moving averages

The grades given in this FOUNDATION & HIGHER chapter are target grades.

1 Time series

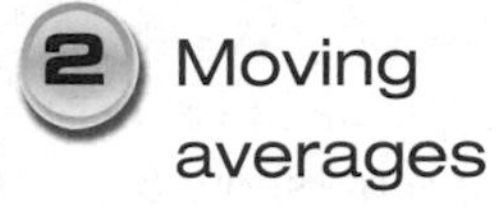

2 Moving averages

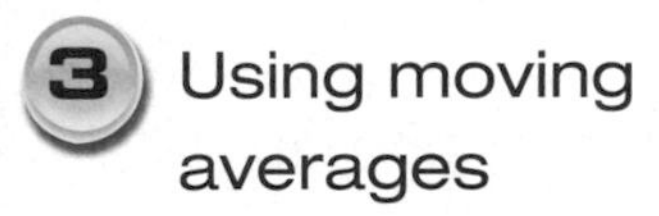

3 Using moving averages

This chapter will show you …

- **D** how to interpret and draw time-series graphs
- **B** how to calculate a moving average
- **A** how to use moving averages to predict future values

Visual overview

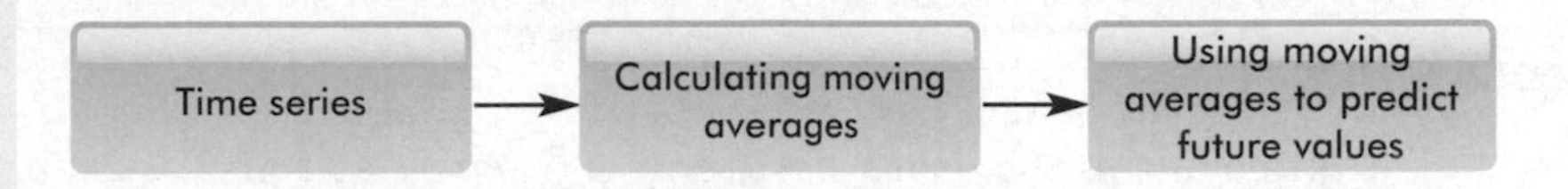

What you should already know

- How to work out the mean of a set of discrete data **(KS3 level 5, GCSE grade E)**
- How to plot points and draw a line of best fit through them **(KS3 level 6, GCSE grade D)**

Quick check

1 Calculate the mean for each set of numbers.

a 5, 8, 9, 6, 7

b 12.6, 13.8, 15.2, 22.4

c 105, 96, 121, 89, 108, 98, 123, 92

d 27.8, 19.3, 31.9, 42.6, 53.9

2 a Plot the following data, which shows the distance and cost of some flights with an airline. Use a vertical scale from 0 to 600 for cost and a horizontal scale from 0 to 6000 for distance.

Distance (miles)	600	900	1100	2600	3600	5600	6000
Cost (£)	150	170	210	300	380	500	540

b Draw a line of best fit through the data.

c Estimate the cost of a flight of distance 4500 miles.

6.1 Time series

This section will show you how to:
- plot and interpret time-series graphs

Key words
time series
trend

A **time series** is simply a set of data collected over a period of time. For example, seaside resorts publish the daily high and low temperatures as well as the average weekly or monthly temperature. These values can be plotted on a graph. Although the plotted values may be joined with a line, this line does not have the same significance as a line on a linear graph. On a linear graph such as $y = 2x + 3$, all the points on the graph can be used to find an equivalent value of x (or y) for any given value of y (or x). On a time-series graph the lines joining points show a **trend** and cannot normally be used to predict values between plotted points.

EXAMPLE 1

The graph shows the depth of water in a reservoir at the start of each month during a year.

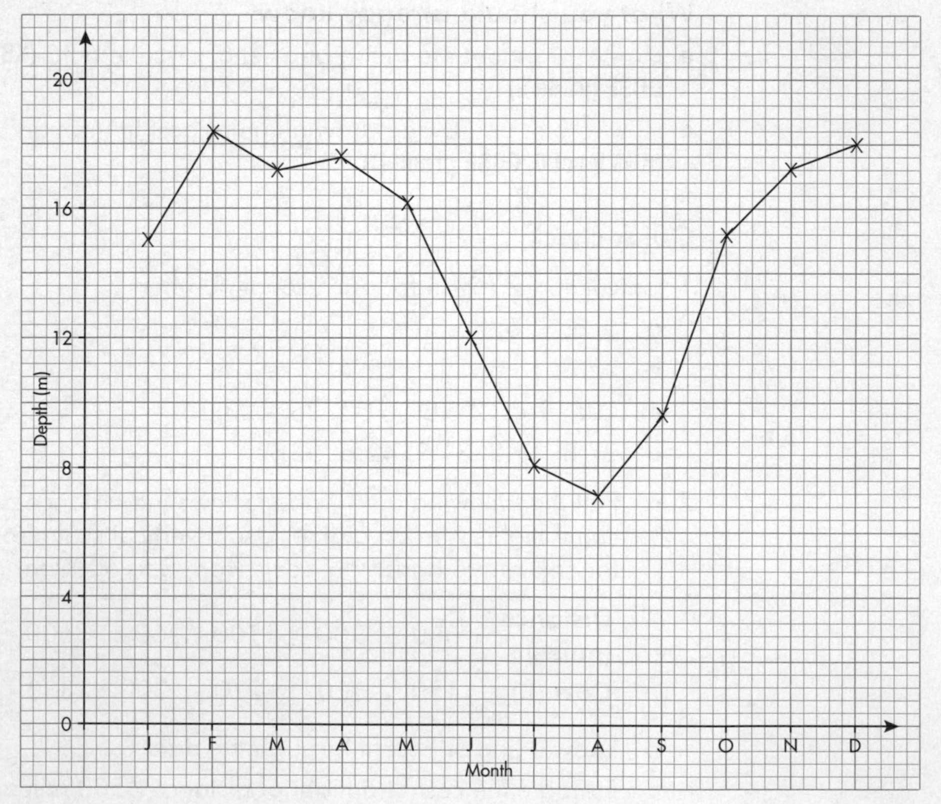

a What was the depth of water at the start of January?

b There was a severe drought for 2 months. When were these months? Explain how you can tell.

c When the depth of water falls below 10 m the water company imposes a hosepipe ban. For approximately how many weeks was the ban in force?

continued

SOLUTION

a Reading from the scale, the depth at the start of January was 15 metres.

b May and June, as these were the months when the depth fell most steeply.

c The ban started halfway through June and remained until just after the start of September. This is approximately 11 weeks.

Note: in this situation the trend line does have a meaning as the water level is likely to drop continuously.

EXAMPLE 2

The table shows the maximum and minimum temperatures in a greenhouse for a week.

Day	Sun	Mon	Tue	Wed	Thu	Fri	Sat
Max °C	22	24	28	21	19	16	14
Min °C	12	13	14	11	10	9	7

a Plot both sets of data separately on a graph.

b Describe any similarities between the trend of the maximum and minimum temperatures.

c The forecast for the following week predicts maximum temperatures over 20 °C all week. The gardener wants to plant out some orchids which will die if they get colder than 10 °C in their first week of growth. Is it safe for the gardener to plant them?

d On Sunday the maximum temperature occurred at 12 noon and the minimum at 12 midnight. Can you say what the temperature was at 6 p.m. on Sunday? Explain your answer.

SOLUTION

a

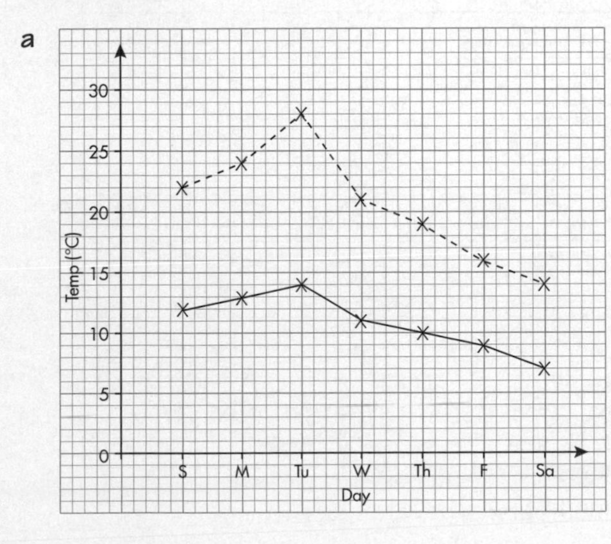

b The trend of both maximum and minimum temperatures follows a similar pattern. There is a difference of about 10 degrees.

c It should be safe for the gardener to plant her orchids as the minimum temperatures for days when the maximum is above 20 °C are well above 10 °C.

d It is not possible to say what the temperature was at 6 pm as the fall in temperature is not likely to be smooth.

Note: In this example you should see that the lines between plotted values have no meaning. They are just used to show a trend.

EXERCISE 6A 

D

1 The table shows the temperature in London, measured at 12 noon for seven days.

Day	Mon	Tue	Wed	Thu	Fri	Sat	Sun
Temp (°C)	23	18	17	21	23	25	26

a Plot the data on a graph, with days on the horizontal axis and temperature on the vertical axis.

b Can the graph be used to find the temperature at 12 midnight on Monday? Explain your answer.

2 The table shows the number of bacteria in a Petri dish, measured every hour for 8 hours.

Time (hours)	0	1	2	3	4	5	6	7
Number of bacteria	800	1600	3200	6400	8200	8100	8200	8000

a Plot the data on a graph, with time on the horizontal axis and number of bacteria on the vertical axis.

b The number of bacteria becomes critical when it reaches 8000 and then the bacteria stop multiplying. Approximately how long after the start of the experiment does this happen?

3 The graph shows how the temperature changes in a garden one day in winter.

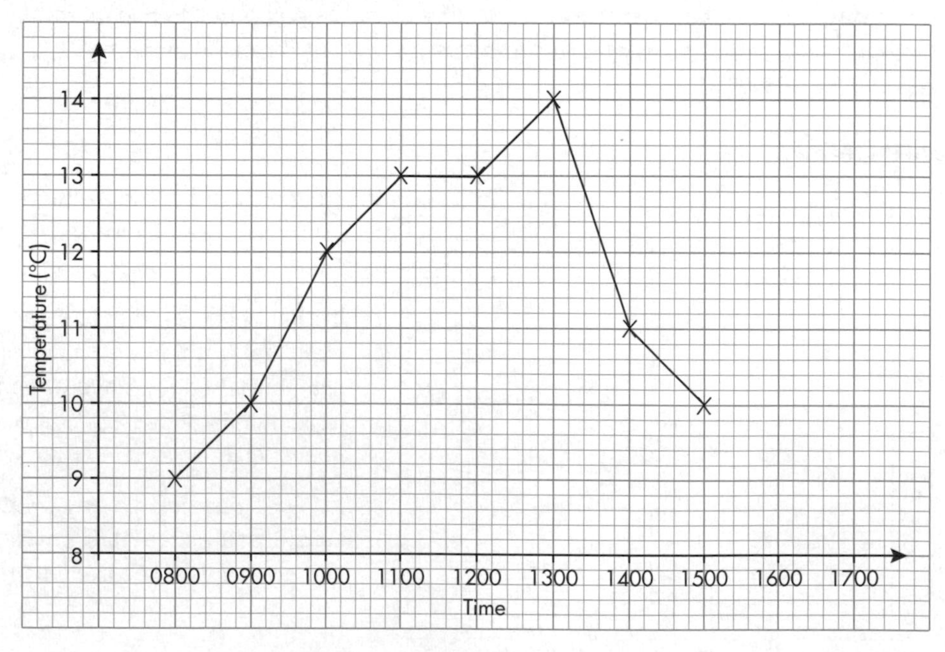

a What is the temperature at 1000 hours?

b What is the temperature at 1330 hours?

c In the afternoon there was a snow shower. What time did it start? Explain your answer.

d Use the graph to estimate the temperature at 1600 hours.

D

4 A puppy's owner recorded its mass, in kilograms, over a year.

Month	1	2	4	6	8	10	12
Mass (kg)	4.5	7.5	9	11.5	17	19	20

a Plot the data on a graph, with the months as the horizontal axis, marked from 0 to 12, and the mass on the vertical axis, marked from 0 to 25.

b What was the approximate mass of the puppy after 3 months?

c What was the approximate mass of the puppy after 9 months?

d Why is the graph not useful for estimating the mass of the puppy after 13 months?

C

PS 5 The graph shows the depth of water in a harbour, measured each hour after midnight.

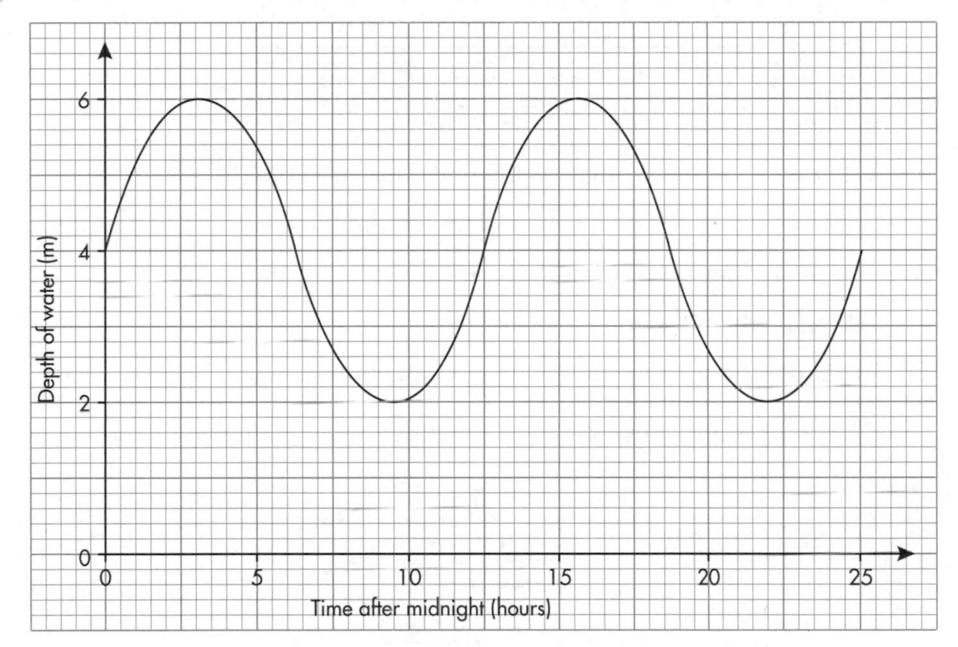

a Cargo ships can only enter the harbour when the depth is at least 5 metres. Between what times can cargo ships enter the harbour?

b A trawler sets off at midnight and fishes for 6 hours. Due to the weight of the catch the draft (depth of the hull below the water) of the trawler is 4.5 metres. What is the latest time they can come back into the harbour?

c **i** The tide has a cyclic pattern, which means it repeats itself regularly. Approximately how long is the tide cycle?

ii What is the earliest time a ship with a draft of 6 metres can enter the harbour **the next day**?

C

AU **6** The graph shows the value of a car that cost £12 000 new on 1 January 2004.

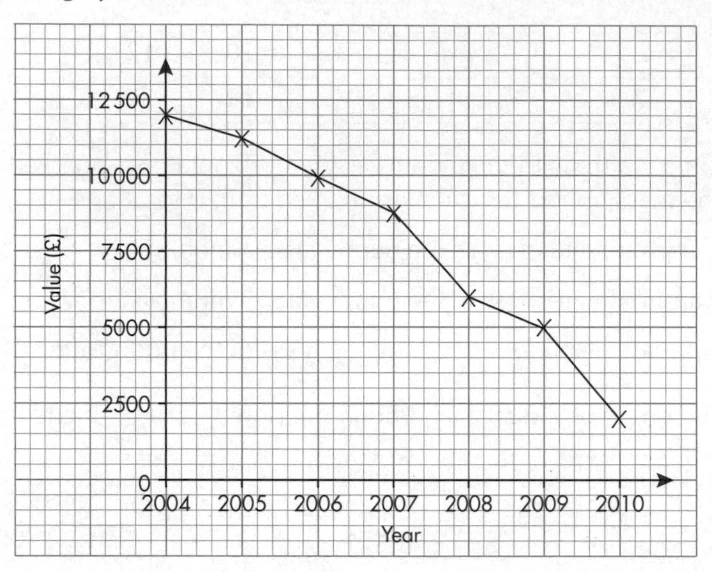

a What was the approximate value of the car in 2006?

b Approximately in which year and month did the car reach half its initial value?

c The owner decides to get rid of the car when it is worth £2000.

Estimate when this will be from the graph.

d What percentage of its original value was the car worth after 3 years?

AU **7** The spreadsheet shows the amount owing on a loan of £10 000 taken over 10 years.

	A	B
1	Start of year	Capital owed
2	0	10 000
3	1	9500
4	2	8700
5	3	7900
6	4	7000
7	5	6300
8	6	5600
9	7	4700
10	8	3800
11	9	2100
12	10	0

a Plot the data on a graph, with time on the horizontal axis and capital owed on the vertical axis.

b For x years the interest rate was the same, then it went down considerably.

How many years did the interest rate stay the same?

Give a reason for your answer.

C

8 The graph shows the price of petrol, adjusted to be relative to the 1999 price per gallon of £3.46.

a In approximately which year was petrol relatively the most expensive?

b In approximately which year was petrol relatively the least expensive?

c During which years was the price of petrol the most stable, with the least variation in price?

d The price of petrol in 2010 is approximately £5.40 a gallon. Does the graph support this?

6.2 Moving averages

This section will show you how to:

- calculate an n-point moving average

Key words
mean
moving average
n-point moving average
seasonal trend

Some variables, such as quarterly gas bills, will change considerably over the course of a year. The bill will be higher in winter, when the heating is on more, and lower in the summer when the weather is warmer and less gas is used.

These are **seasonal trends** in the data. As gas prices tend to increase it becomes difficult to predict the next bill. This can be overcome by using the overall trend of the data.

Quarter-by-quarter variations can be evened out by using a **moving average**. You may see these referred to as ***n*-point moving averages**. Over a year, when bills are received quarterly, it would be sensible to use a four-point moving average, calculated as the **mean** of every four consecutive bills.

There is no rule about how many pieces of consecutive data need to be used to calculate a moving average. It depends on the type of data. For example, GCSE mathematics examinations take place in November, March and June so a three-point moving average would be sensible. Absence rates in a school might best be analysed with a five-point average, as there are five school days in a week.

EXAMPLE 3

The table shows the quarterly gas bills for a household for three years.

Year	Winter	Spring	Summer	Autumn
2009	£210	£170	£124	£192
2010	£222	£186	£136	£208
2011	£238	£194	£144	£220

The first four-point moving average is in the table below.

Time period	1	2	3	4	5	6	7	8	9
Moving average	174								

a Calculate the other four-point moving averages.

b Plot the original values and the moving averages on a graph.

SOLUTION

a The first moving average has been calculated as $(210 + 170 + 124 + 192) \div 4 = 696 \div 4 = 174$.

Calculate the rest of the four-point moving averages in the same way. The next two are $(170 + 124 + 192 + 222) \div 4 = 177$ and $(124 + 192 + 222 + 186) \div 4 = 181$.
Repeating this process gives the complete table.

Time period	1	2	3	4	5	6	7	8	9
Moving average	174	177	181	184	188	192	194	196	199

b The crosses are the original values and the dots are the moving averages.

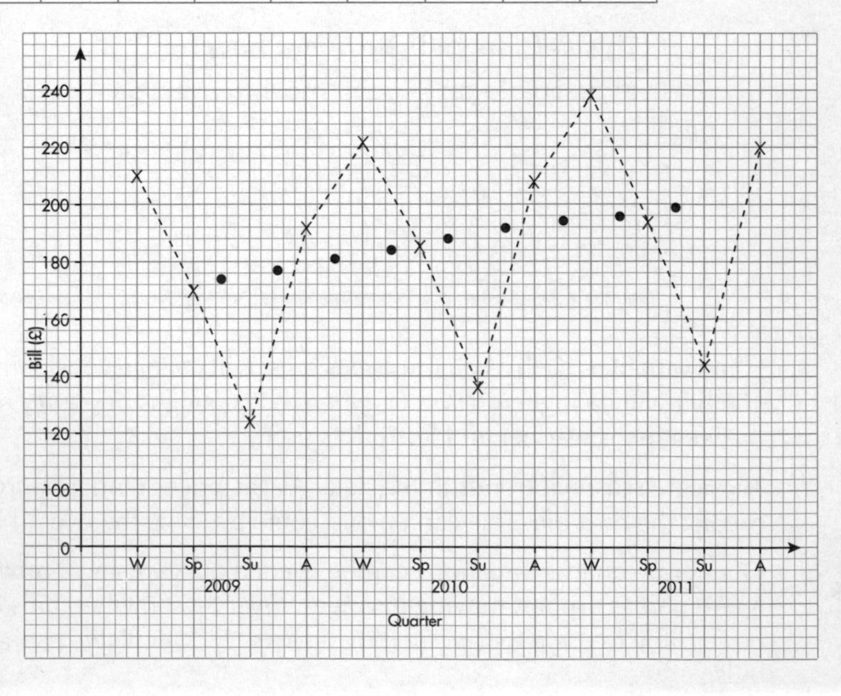

You can see that the original data has a similar pattern each year and the moving averages show a steady increase. It would be easy to put a line of best fit through these to predict the next moving average. These graphs are called time series as they show the trend of data over time. The lines joining the points cannot be used to find the bills for intermediate times. In that respect they have no meaning but they are used to show a trend in the data.

Note that the moving average is plotted at the midpoint of the consecutive values used to calculate it, so the first is plotted between spring and summer 2009 and the second is plotted between summer and autumn 2009.

EXERCISE 6B

1 The table shows the daily sales of milk at a local corner shop for four weeks.

Sun	Mon	Tue	Wed	Thu	Fri	Sat
12	8	6	9	4	11	15
11	7	7	6	3	15	14
14	9	7	7	5	12	15
11	12	8	7	4	14	19

Make a table showing the seven-point moving average, and draw a graph to show the trend of milk sales over the month.

2 John opens a shop selling specialist comic books. The shop is open on Wednesday, Thursday morning, Friday and Saturday each week. John records the numbers of customers in the first three weeks he is open.

John wants to know how his business is performing.

a Why would using a four-point moving average be appropriate?

b Work out the four-point moving averages for the data.

c Plot the raw data and the four-point moving averages on the same graph.

Week	Day	Number of customers
1	Wednesday	18
	Thursday	12
	Friday	22
	Saturday	28
2	Wednesday	22
	Thursday	14
	Friday	25
	Saturday	30
3	Wednesday	24
	Thursday	18
	Friday	28
	Saturday	34

AU **3** A college records how many students attend adult education classes each term, for a period of three years. The table shows some of the data and some of the three-point moving averages.

Term	Summer 2008	Autumn 2008	Spring 2009	Summer 2009	Autumn 2009	Spring 2010	Summer 2010	Autumn 2010	Spring 2011
Number	480	310	380	513	340	410	540	364	
Three-point moving average		390	401	411	421			445	

a Calculate the two missing three-point moving averages.

b Calculate the missing value for spring 2011.

B

PS 4 The table shows the numbers of candidates who sat a GCSE mathematics examination in June and November over a period of four years.

Date	Jun 08	Nov 08	Jun 09	Nov 09	Jun 10	Nov 10	Jun 11	Nov 11
Number of entries	76 000	12 600	74 800	11 400	73 600	10 200	72 400	9000

a Explain why a two-point moving average would be appropriate.

b Calculate the first two-point moving average.

c **i** The rest of the two-point moving averages are 43 700, 43 100, 42 500, 41 900, 41 300, 40 700. Look at the sequence of moving averages. If the pattern continues what will the next moving average be?

ii Show that this value predicts the number of entries for June 2012 to be 71 200.

AU 5 A teacher gives her class a 10-question multiplication tables test (M) on alternate Monday mornings and a 10-question spelling test (S) in the alternate Monday mornings between. These are the average scores over a term of 14 weeks.

Week	1 (M)	2(S)	3(M)	4(S)	5(M)	6(S)	7(M)
Average	6.2	3.7	6.5	3.8	6.6	3.4	6.9

Week	8(S)	9(M)	10(S)	11(M)	12(S)	13(M)	14(S)
Average	3.5	7.0	3.7	7.3	3.8	7.5	3.6

a Work out the two-point moving averages for this data.

b The teacher says that the moving averages show that her class is getting better at their tables and their spelling. Is the teacher correct? Give a reason for your answer.

PS 6 The mean of four numbers, 4.9, 6.8, 4.5 and x, is 5.6. What is the value of x?

PS 7 The mean of three numbers, 23, 45 and a, is 31 and the mean of three numbers, 45, a and b is 39. Work out the values of a and b.

A

PS 8 The table shows Trevor's gas bills for the first six months of 2010. The values for April and May are missing.

Month	Jan	Feb	Mar	Apr	May	Jun
Amount (£)	35.20	28.40	25.60			19.00

The first four-point moving average is £28.00 and the second is £24.25. Work out the third four-point moving average.

PS 9 The table shows Ahmed's oil bills for the first six months of 2010. The values for January, May and June are missing but the bill in June was half the bill in January.

Month	Jan	Feb	Mar	Apr	May	Jun
Amount (£)	$2x$	38.80	33.60	28.40		x

The first four-point moving average is £36.65. The third four-point moving average is £27.525. Work out the bill for May.

6.3 Using moving averages

This section will show you how to:
- use a moving average to predict future values

Key words
moving average
trend line

You have seen that **moving averages** can be used to identify trends in data. The next step is to use them to make predictions.

EXAMPLE 4

A van rental firm keeps a record of how many vans were hired in each month of a year. This data is shown in the table. Using a four-point moving average, predict the number of vans the firm will rent out during the following January.

Months	Jan	Feb	Mar	Apr	May	Jun	Jul	Aug	Sep	Oct	Nov	Dec
Vans	9	22	37	14	18	24	42	17	20	27	48	20

SOLUTION

First, plot the raw data. The resulting line graph shows a normal variation of business for the hire firm, but does not reveal the general trend of business. Is the firm's business improving, declining or remaining the same?

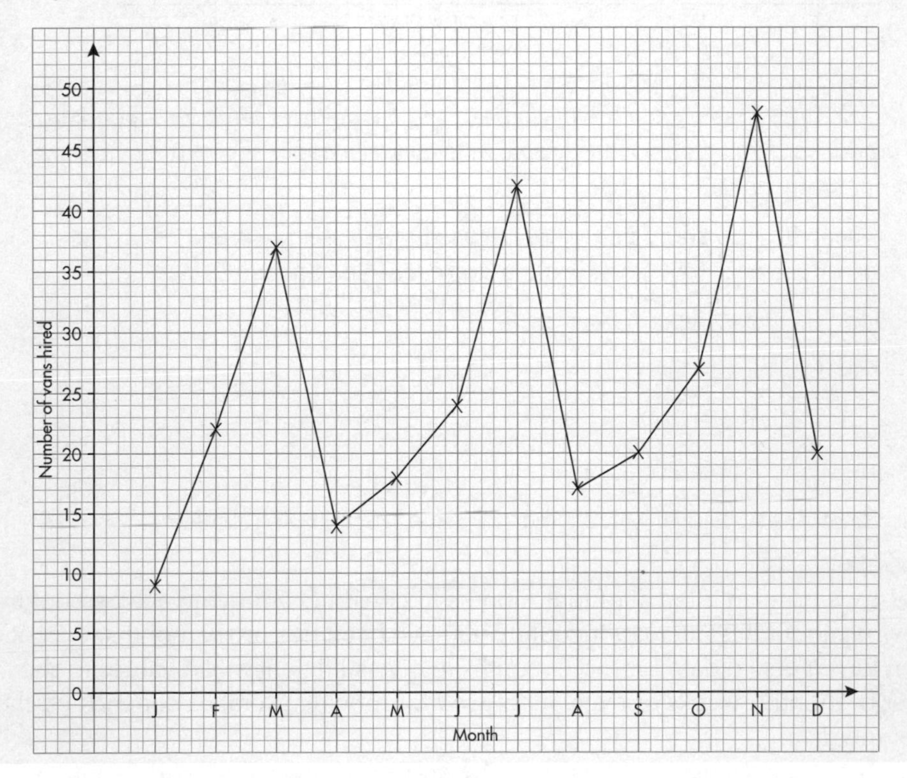

continued

Mean for January, February, March and April = $(9 + 22 + 37 + 14) \div 4 = 20.5$

Mean for February, March, April and May = $(22 + 37 + 14 + 18) \div 4 = 22.75$

Mean for March, April, May and June = $(37 + 14 + 18 + 24) \div 4 = 23.25$

And so on, giving 24.5, 25.25, 25.75, 26.5, 28, 28.75 as the remaining averages.

You can show the general trend by first calculating the four-point moving average.

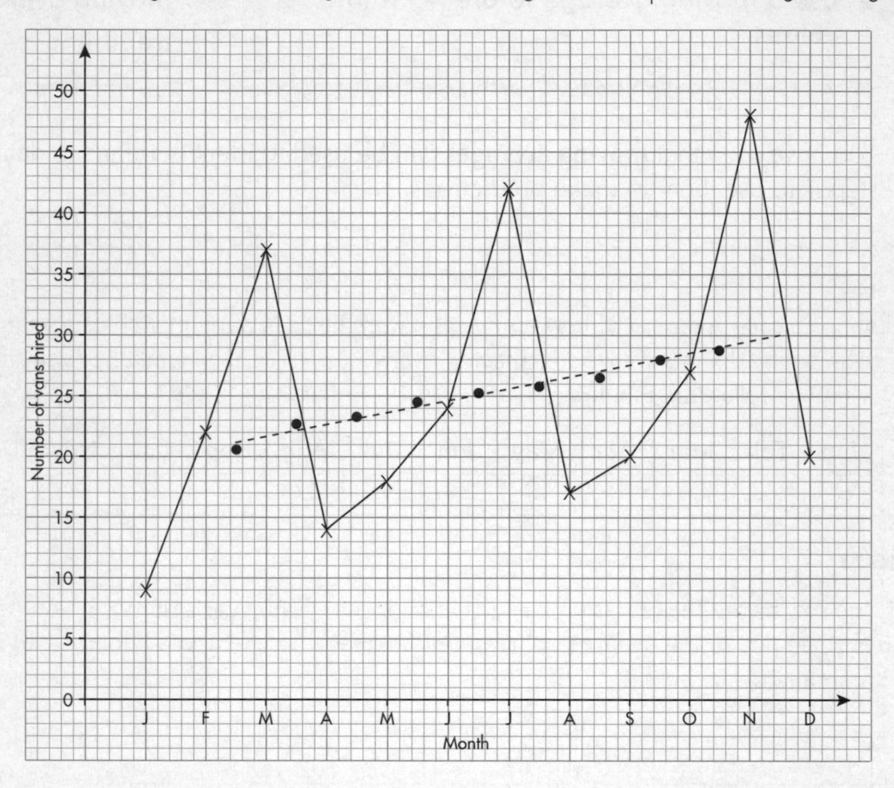

Then plot, on the first graph, each mean value at the midpoint of the corresponding four-month span. This produces a much smoother graph, which, in this case, shows a slight upward trend. In other words, business is improving.

Draw a line of best fit (the **trend line**) through the moving averages and read from it the predicted value of the next four-point moving average. This is about 30.

Let the value for the next January be x, then:

$(27 + 48 + 20 + x) \div 4 = 30$

$$95 + x = 120$$

$$x = 25$$

So we can predict that the firm will rent out 25 vans the following January.

In Example 4 we used an interval of four months to construct a moving average but there is nothing special about this interval. It could well have been five or six months, except that we should then have needed data for more months to give sufficient mean values to show a trend. The number of months, weeks or even years used for moving averages depends on the likely variation of the data.

You would not expect to use less than two or more than seven items of data at a time.

EXERCISE 6C 

1 The table shows the telephone bills for a family over four years.

	2007	2008	2009	2010
First quarter	82	87	98	88
Second quarter	87	88	95	91
Third quarter	67	72	87	78
Fourth quarter	84	81	97	87

a Plot a line graph showing the amounts paid each month.

b Plot a four-point moving average.

c Comment on the trend shown and give a possible reason for it.

d Use the trend line of the moving averages to predict the bill for the first quarter of 2011.

2 A factory making computer components has the following sales figures (in hundreds) for electric fans.

	Jan	Feb	Mar	Apr	May	Jun	Jul	Aug	Sep	Oct	Nov	Dec
2009	12	13	12	14	13	3	15	12	14	13	14	12
2010	13	14	12	14	13	14	13	13	15	15	15	14

a Plot a line graph of the sales and add a three-point moving average.

b Comment on the trend in the sales.

c Use the trend line of the moving averages to predict the number of electric fan sales in January 2011.

3 The table shows the total sales of video recorders (VCRs) and DVD players from 2004 to 2010 for an electrical store in the USA.

	2004	2005	2006	2007	2008	2009	2010
VCRs (thousands)	3.4	3.8	3.9	3.2	2.8	2.5	2.3
DVDs (thousands)	0.2	0.8	0.9	1.5	1.9	2.8	3.7

a Plot a line graph showing the sales for each product over these years.

b On the same diagram, plot the three-point moving average of each product.

c Comment on the trends seen in the sales of video recorders and DVD players.

d Use the trend line of the moving averages to predict the number of video recorders and DVD players sold in 2011.

4 A college records how many students used a tutorial service each term, over a period of three years. The table shows some of the data and some of the three-point moving averages.

Term	Summer 2008	Autumn 2008	Spring 2009	Summer 2009	Autumn 2009	Spring 2010	Summer 2010	Autumn 2010	Spring 2011
Number	64	40	52	70	46	58	76	58	70
Three-point moving average		52	54	56	58			66	68

a Calculate the two missing three-point moving averages.

b Calculate the missing value for spring 2011.

c i The time series graph for the original data is shown in the graph. Copy the graph and plot the value of the moving average on the same axes.

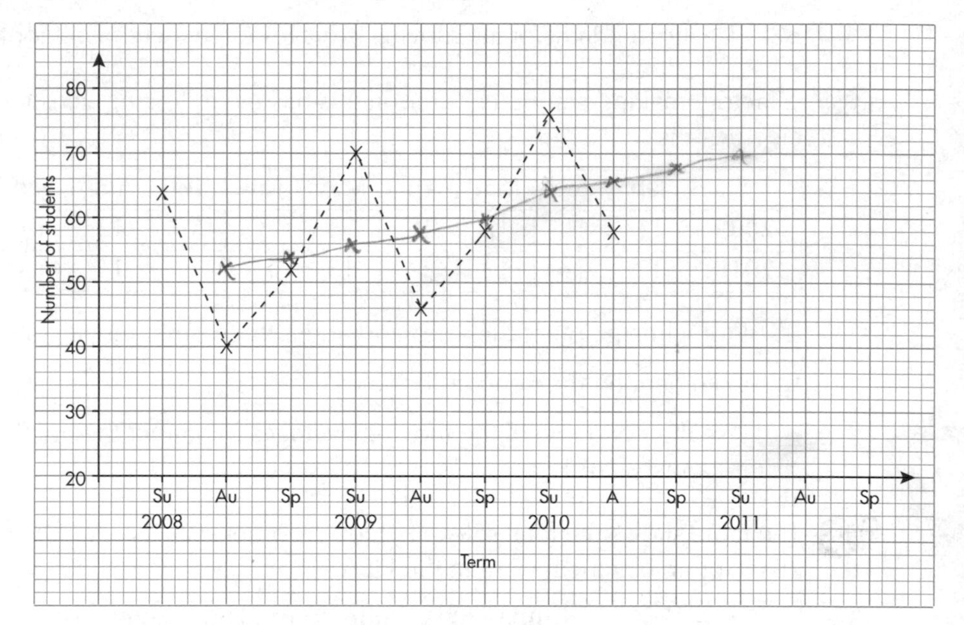

ii Use the trend of the moving averages to predict the number of students using the tutorial service in summer 2011.

5 The table shows the numbers of candidates who sat a GCSE English examination in June and November over a period of four years, with some of the two-point moving averages.

Date	Jun 08	Nov 08	Jun 09	Nov 09	Jun 10	Nov 10	Jun 11	Nov 11
Number of entries	56 000	18 600	52 000	16 600	48 400	14 800	44 000	12 200
Two-point moving average	37 300	35 300	34 300	32 500				

a Calculate the three missing two-point moving averages.

b Plot the moving averages on a graph. Take the horizontal axis as the dates from June 2008 to June 2012 and the vertical axis as the number of entries, from 10 000 to 60 000.

c Draw a line of best fit through the moving averages.

d Estimate the next moving average and use this to predict the number of entries for June 2012.

6 The table shows the water bills for a household.

Date	Winter 2006	Summer 2007	Winter 2007	Summer 2008	Winter 2008	Summer 2009	Winter 2009	Summer 2010
Amount (£)	132	168	138	175	142	180	146	188

Using a two-point moving average, predict the water bill for winter 2010.

Draw the graph with the dates on the horizontal axis and the amounts on the vertical axis.

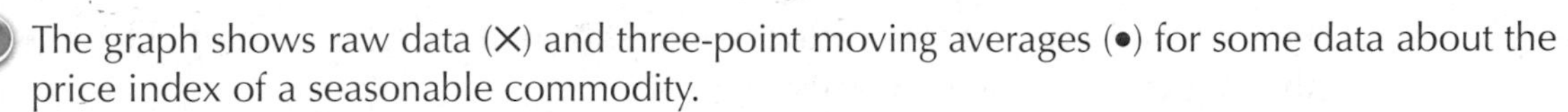

AU 7 The graph shows raw data (X) and three-point moving averages (•) for some data about the price index of a seasonable commodity.

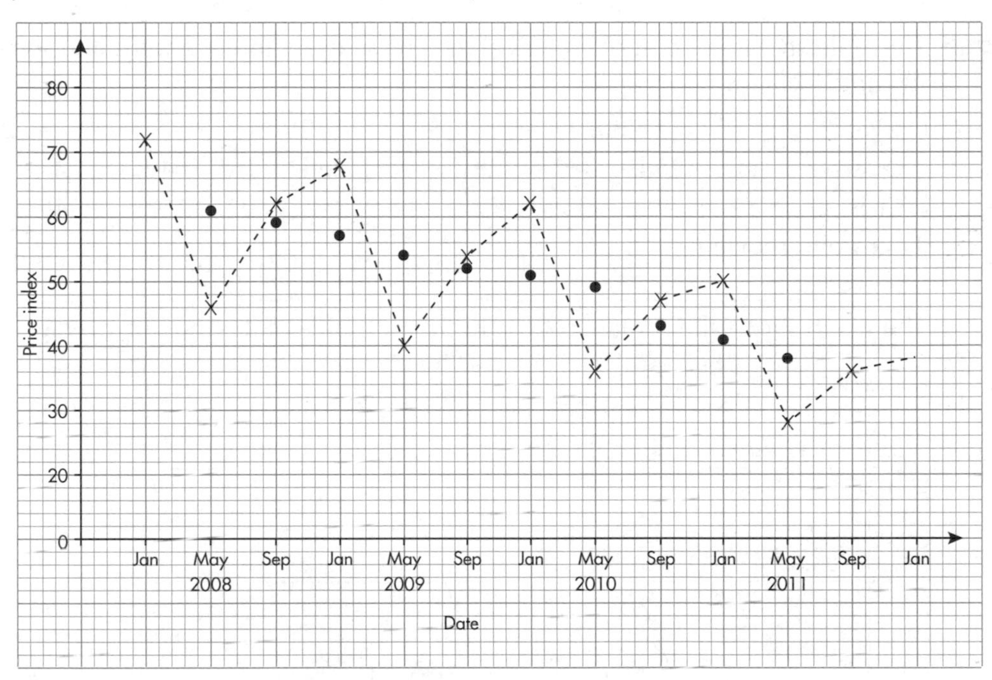

Use the graph to predict the price index of the commodity in January 2012.

8 The table shows Rob's electricity bills from March 2010 to June 2011. The entry for December 2010 is missing.

Date	March 2010	June 2010	Sept 2010	Dec 2010	March 2011	June 2011
Amount (£)	44.70	33.80	29.40		48.70	37.80

a The value of the first four-point moving average is £38.50.

Calculate the bill for December 2010.

b How can you tell from the table that the second four-point moving average will be higher that £38.50?

c The three four-point moving averages are £38.50, £39.50 and £40.50.

Continue the pattern of the moving averages and predict the bill for December 2011.

9 The table shows quarterly electricity bills over a four year-period.

	2007	2008	2009	2010
First quarter (£)	123.39	119.95	127.39	132.59
Second quarter (£)	108.56	113.16	117.76	119.76
Third quarter (£)	87.98	77.98	102.58	114.08
Fourth quarter (£)	112.47	127.07	126.67	130.87

Plot the line graph of the electricity bills shown in the table and, on the same axes, plot the four-point moving averages.

10 The table shows the cost of the gas at the end of every three months and some four-point moving averages

Year	**2008**				**2009**				**2010**		
Quarter	**1st**	**2nd**	**3rd**	**4th**	**1st**	**2nd**	**3rd**	**4th**	**1st**	**2nd**	**3rd**
Cost (£)	86	90	93	99	94	94	97	103	94	98	101
Four-point moving average		92	94	95	96	97	97	98	99		

The graph shows the actual cost of the gas and some of the moving averages.

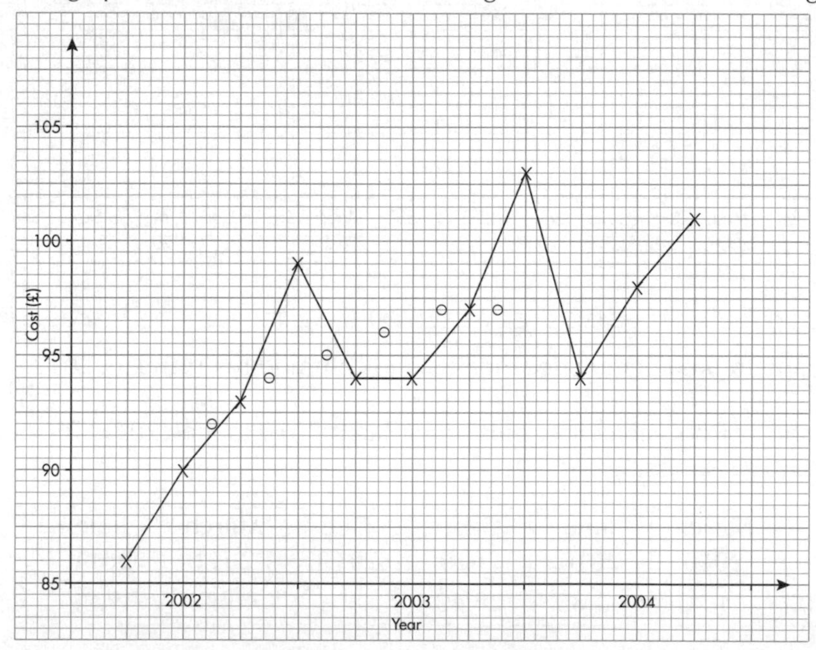

On a copy of the graph, plot the last two four-point moving averages.

PS 11 The table shows some raw data and three-point moving averages for this data. Some of the raw data has had coffee spilt on it. Work out the missing values.

Raw data	12	8		15		25				21
Three-point moving average		14	15		17	18	20	21	22	

GRADE BOOSTER

B You can calculate an n-point moving average

A You can use a moving average to predict future values

A You can interpret and draw a time-series graph

What you should know now

- How to calculate and use a moving average to predict future values
- How to use a time–series graph

EXAMINATION QUESTIONS

1 The retail price index (RPI) is used to measure how the cost of living changes over time. The values of the RPI for various years are shown in the table and on the graph.

Year	2000	2001	2002	2003	2004	2005	2006
RPI	170	173	176	181	187	192	197

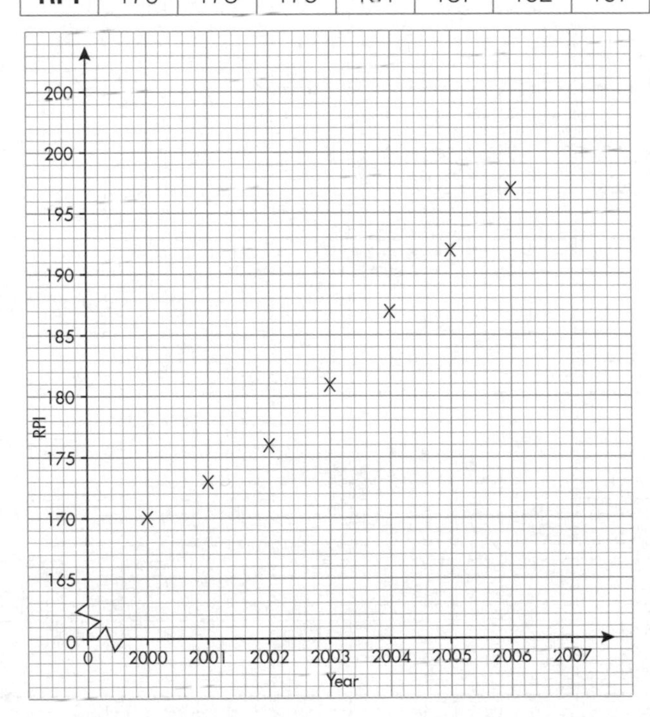

a Use the graph to estimate the RPI for 2007.

b The RPI for 1987 was 100. State the percentage increase in the RPI from 1987 to 2005.

c Calculate the percentage increase in the RPI from 2000 to 2006.

November 07, Specification A, Intermediate 2, Question 14 (amended)

2 The spreadsheet shows information about a £10 000 loan, taken over 12 months at an annual interest rate of 5.8%. The table shows the interest each month, the amount paid off the principal each month and the outstanding balance each month.

	Loan information		Summary	
1	Loan amount	10 000.00	Rate (per period)	0.483%
2	Annual interest rate	5.80%	Number of payments	12
3	Term of loan in years	1	Total payments	10 316.94
4	First payment date	01/01/2009	Total interest	316.94
5	Payment frequency	Monthly	Est. interest savings	0.00
6	Compound period	Monthly		
7	Payment type	End of period		
8	Monthly payment	859.75		

No.	Due date	Payment	Additional payment	Interest	Principal	Balance
						10 000.00
1	1/1/09	859.75		48.33	811.42	9188.58
2	1/2/09	859.75		44.41	815.34	8373.24
3	1/3/09	859.75		40.47	819.28	7553.96
4	1/4/09	859.75		36.51	823.24	6703.72
5	1/5/09	859.75		32.53	827.22	5903.50
6	1/6/09	859.75		28.53	831.22	5072.28
7	1/7/09	859.75		24.52	835.23	4237.05
8	1/8/09	859.75		20.48	839.27	3397.78
9	1/9/09	859.75		16.42	843.33	2554.45
10	1/10/09	859.75		12.35	847.40	1707.05
11	1/11/09	859.75		8.25	851.50	855.55
12	1/12/09	859.69		4.14	855.55	0.00

a Copy this graph and plot the interest for each month.

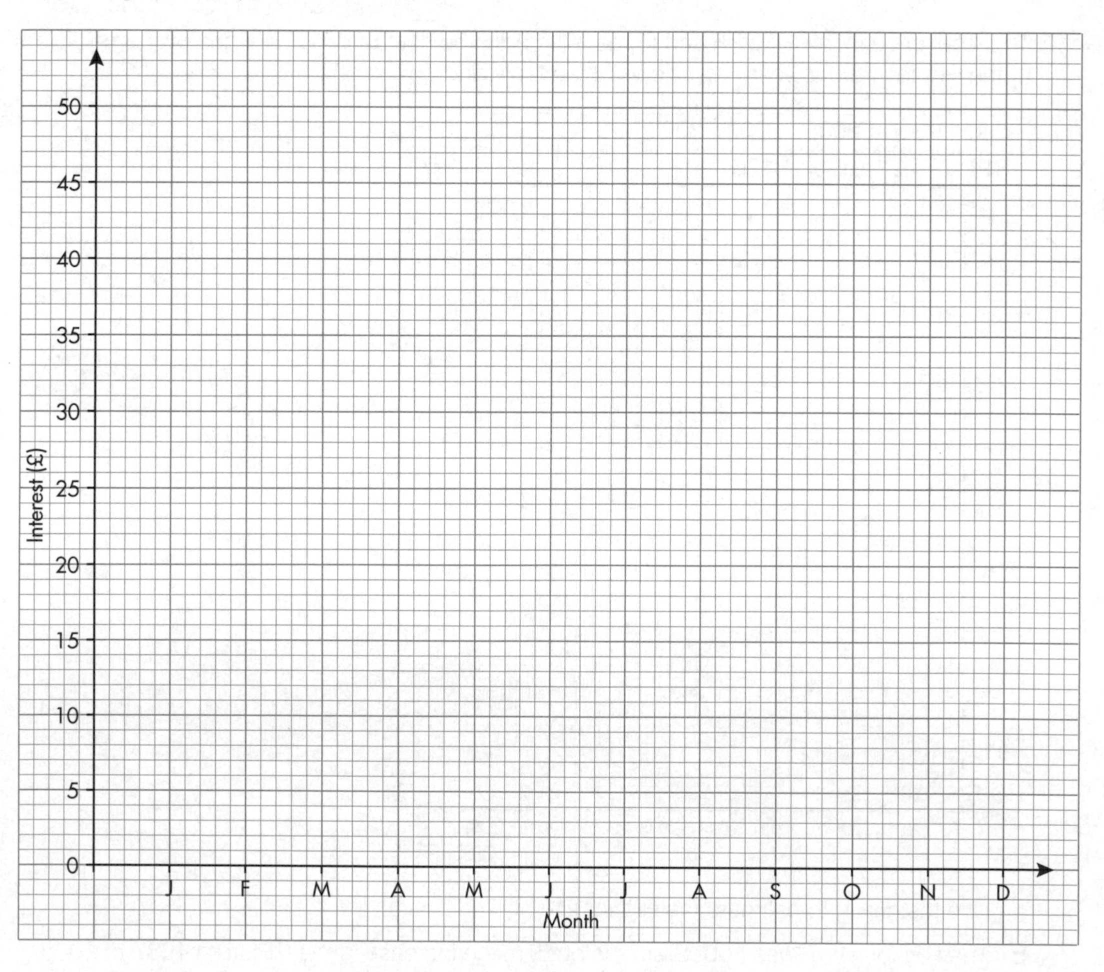

continued

D C B A

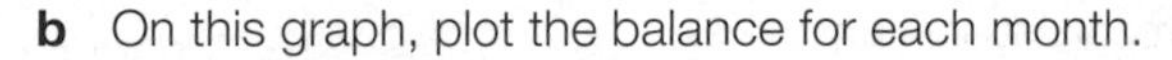

b On this graph, plot the balance for each month.

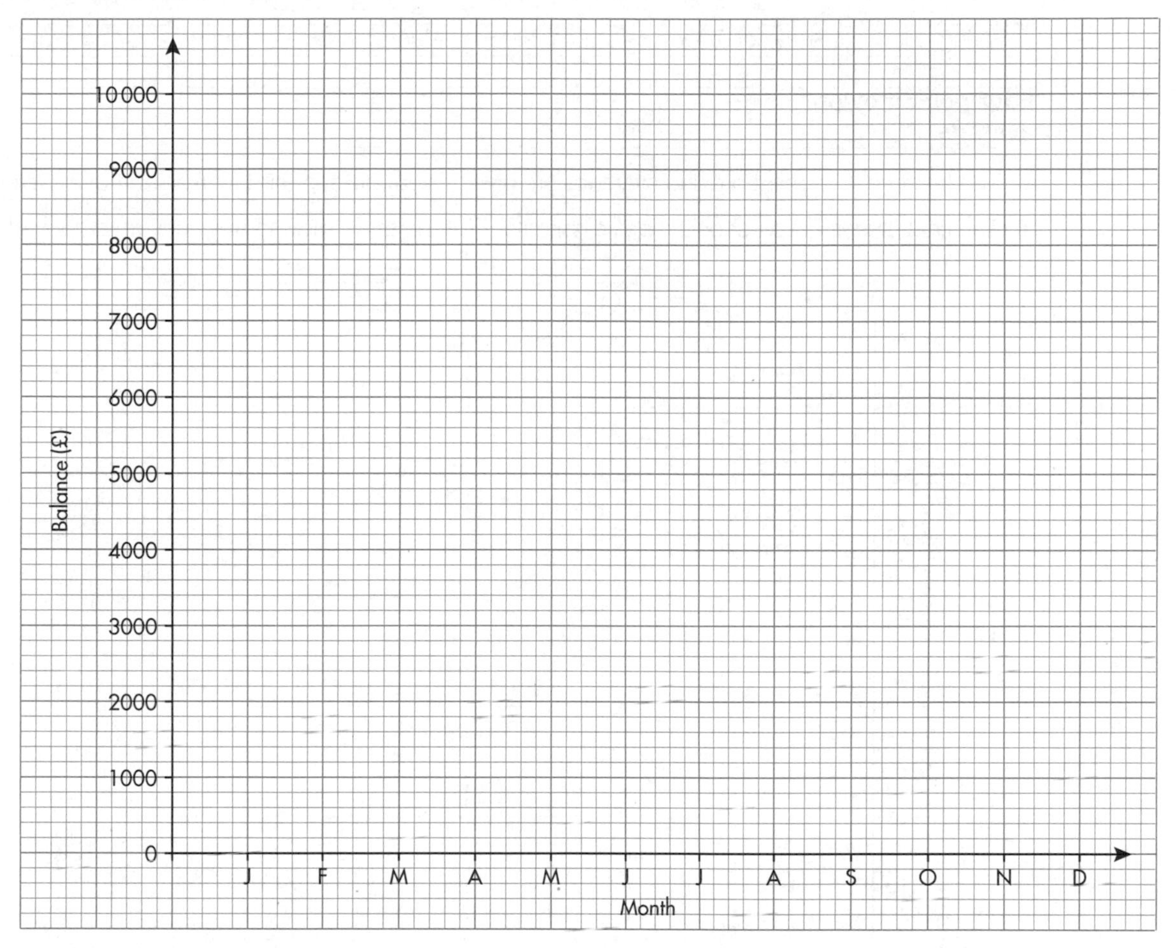

3 The table shows the numbers of candidates who sat a GCSE French examination in June and November from June 2005 until November 2008.

Date	Jun 05	Nov 05	Jun 06	Nov 06	Jun 07	Nov 07	Jun 08	Nov 08
Number	82 300	4700	79 800	5200	76 400	5400	72 100	6100

a The data is used to predict the entries for 2009. Explain why a 2-point moving average would be appropriate.

b Calculate the first 2-point moving average for the data.

June 2009, Specification A, Higher paper 2, Question 18

4 A youth club is open on Monday, Wednesday and Friday evenings each week.

The table shows the numbers of people at the youth club each evening over two weeks. One of the values in the table is missing.

Week 1			**Week 2**		
Monday	Wednesday	Friday	Monday	Wednesday	Friday
42	37	62	39		66

a Show that the first three-point moving average is 47.

b The third three-point moving average is 51. Calculate the number of people at the youth club on Wednesday of Week 2.

November 2007, Specification B, Module 1, Question 4

5 The table shows the numbers of sunny days recorded in Sunnyville.

Spring 2006	Summer 2006	Autumn 2006	Winter 2006	Spring 2007	Summer 2007
45	72	38	29	53	84

Four-point moving average	46	48	

a The first two four-point moving averages are given. Work out the third four-point moving average.

b The time–series graph shows the original data. Plot the three moving averages on a copy of the graph.

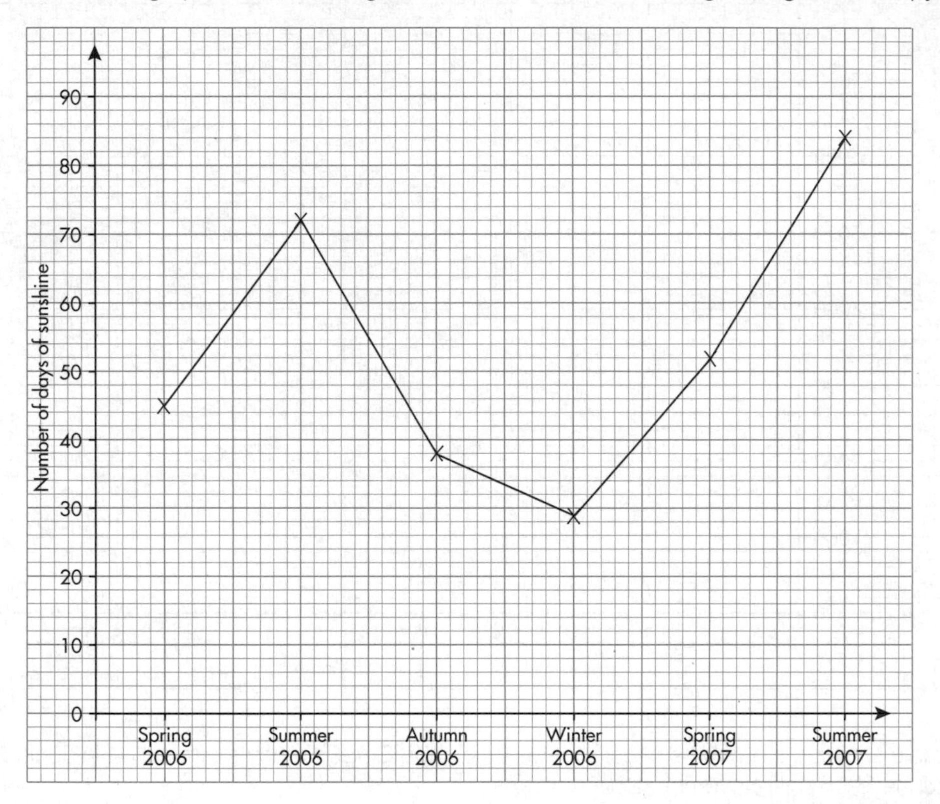

June 2008, Specification B, Module 1, Question 3

6 The table shows the numbers of students at a tutorial college each term since autumn 2002. The table also shows the three-point moving averages for this data except for spring 2003 and summer 2005.

	Autumn 2002	Spring 2003	Summer 2003	Autumn 2003	Spring 2004	Summer 2004	Autumn 2004	Spring 2005	Summer 2005	Autumn 2005
Number of students	48	30	81	54	39	93	69	57	114	
3-point moving average			55	58	62	67	73	80		

a Calculate the three-point moving average for spring 2003. You **must** show your working.

b i By continuing the number sequence for the moving averages, predict the three-point moving average for summer 2005.

ii Show how the college predicted that the number of students in autumn 2005 would be 93.

November 2005, Specification A, Intermediate paper 2, Question 4

D C B A

Worked Examination Questions

PS A **1** A Scottish regional park opens in the Summer (S) and Winter (W). It is closed in the Spring and Autumn for conservation reasons. The table shows the number of visitors to the park (in thousands) since it opened in Summer 2002.

Year	2002		2003		2004		2005		2006	
Season	S	W	S	W	S	W	S	W	S	W
Number of visitors (1000s)	12.7	8.6	13.2	9.5	14.8	10.3	16.3	12.0	18.1	14.2

Use a 2-point moving average and the grid below to predict the number of visitors in Summer 2007. You must show your working.

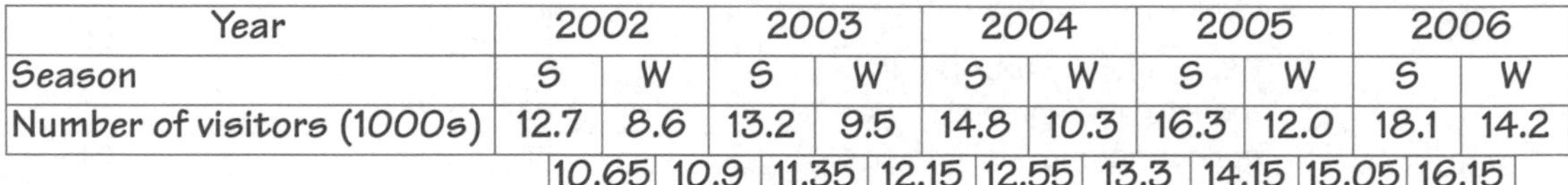

Year	2002		2003		2004		2005		2006	
Season	S	W	S	W	S	W	S	W	S	W
Number of visitors (1000s)	12.7	8.6	13.2	9.5	14.8	10.3	16.3	12.0	18.1	14.2
	10.65	10.9	11.35	12.15	12.55	13.3	14.15	15.05	16.15	

Calculate the 2-point moving averages. This gains **1 method mark** and **1 accuracy mark**.

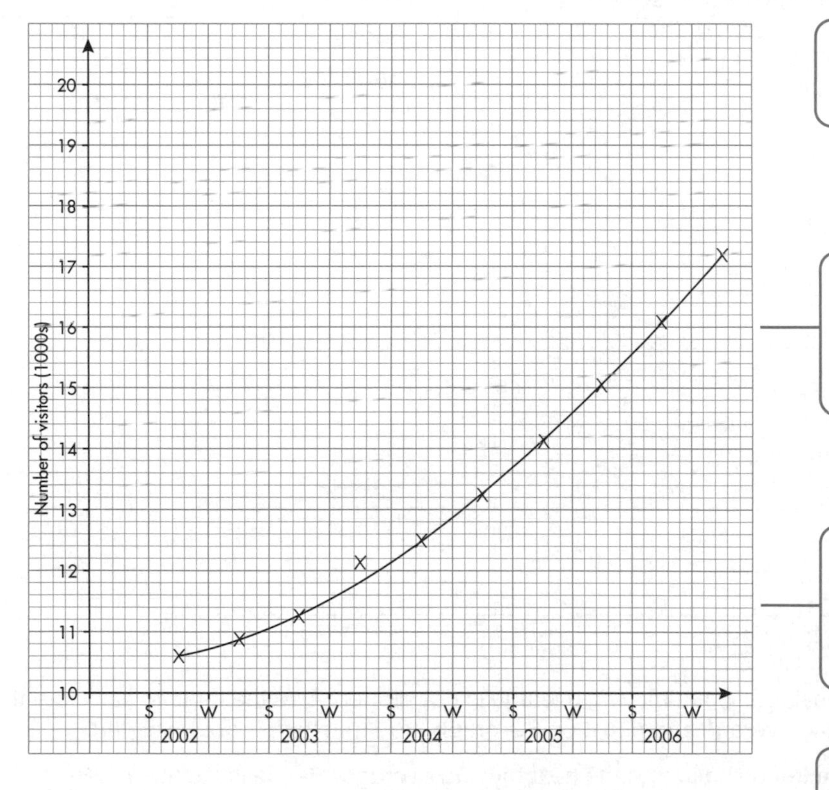

Plot the moving averages on the graph. Remember to plot halfway between the values used to calculate the averages. This gains **1 mark**.

Extend the trend line and read off the next value. This is 17.2 here, but as this is a **follow-through mark** whatever value you read off will be fine.

= 17.2

$x = 2 \times 17.2 - 14.2 = 20.2$

Let the value for summer 2007 be x and set up the equation, using the value from the graph. This gains **1 method mark**.

So number of visitors in summer 2007 = 20 200

Solve the equation to get the value for summer 2007. This gains **1 accuracy mark**.

Total: 6 marks

June 2007, Specification A, Higher paper 2, Question 22

Worked Examination Questions

AU A **2** **a** What is the purpose of moving averages?

b Julie is a hairdresser. Her shop is open Tuesday, Wednesday, Friday and Saturday each week. The numbers of customers at her shop each day over a period of 3 weeks are shown in the table.

Week	Day	Number of customers
1	Tuesday	15
	Wednesday	10
	Friday	20
	Saturday	25
2	Tuesday	21
	Wednesday	14
	Friday	22
	Saturday	27
3	Tuesday	21
	Wednesday	20
	Friday	26
	Saturday	33

Julie wants to know how her business is performing.

Explain why a four-point moving average would be appropriate for Julie to use.

c The graph shows the raw data (x) and some of the four-point moving averages (•).

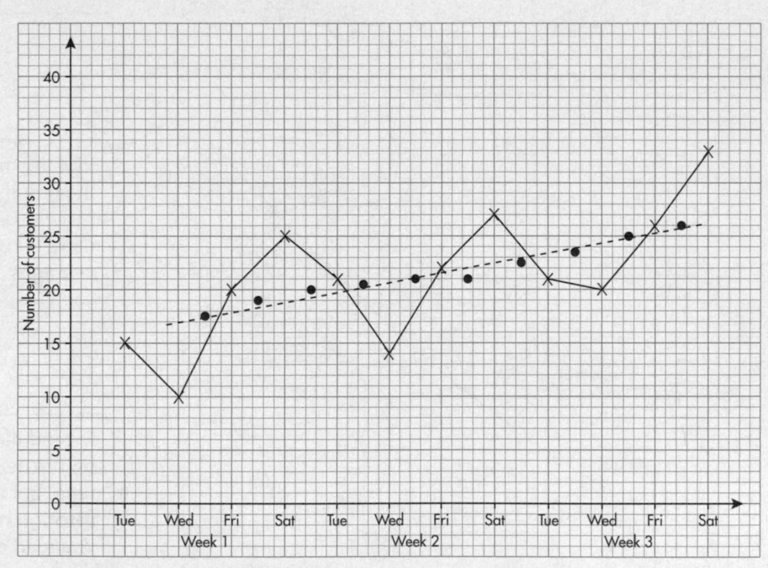

i The last four-point moving average is not plotted. Use the table to calculate this moving average and plot it on the graph. You must show your working.

ii Use a trend line to find the next moving average. Hence calculate a prediction of the number of customers on the Tuesday of week 4. You must show your working.

a *To smooth out variations in data and show a trend over time.*

Just give a brief explanation. This gains **1 mark**.

b *There are four days each week when the shop is open.*

It is usually obvious why an n-part moving average is appropriate. This explanation gains **1 mark**.

continued

c i $\frac{21 + 20 + 26 + 33}{4} = 25$

Use the last four values in the table and calculate the average. This gains **1 method mark** and **1 accuracy mark**.

ii $\frac{20 + 26 + 33 + x}{4} = 26$

$x = 25$

You gain **1 mark** for reading the next moving average from the graph and **1 method mark** for setting up the equation.

You gain **1 accuracy mark** for correctly solving the equation.

Total: 7 marks

June 2009, Specification B, Module 1, Question 5

D **3** The table shows the percentage of students with mobile phones in each year in an 11–18 comprehensive school.

Year	7	8	9	10	11	12	13
% who have a mobile phone	40	45	49	71	85	90	96

a Plot the data on a graph. Take the horizontal axis to be the year and the vertical axis to be the percentage of students with phones.

b In which year do the largest number of students have a mobile phone for the first time?

c There are 250 students in Year 9. How many of them have a mobile phone?

d Only 2 students in Year 13 do **not have** a mobile phone. How many students are there in Year 13?

a

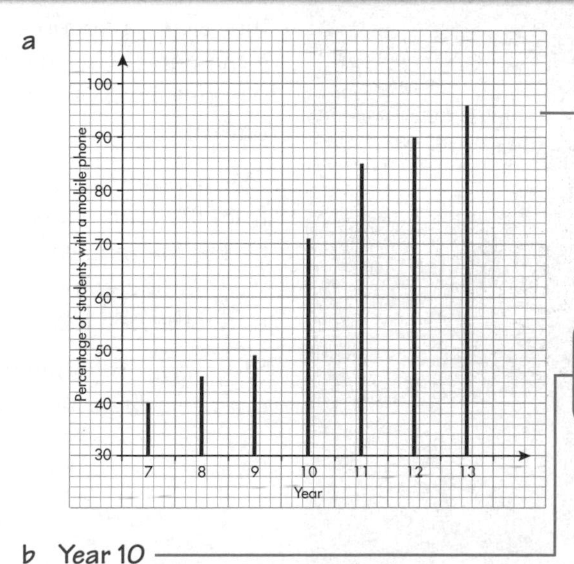

A set of axes will always be drawn for you in an examination. Plot the points. You can draw this as a vertical line graph or a set of crosses joined with dashed lines as the intermediate values have no meaning. You earn **1 mark** for this.

b Year 10

Work out between what years there is the greatest increase in the percentage. This is between years 9 and 10. You earn **1 mark** for this.

c 122–123

Work out 49% of 250. You will get **1 method mark** for an appropriate calculation such as 0.49×250 and **1 mark** for a correct answer.

d 50

If 4% is 2 students then 100% is 50 students. You will get **1 method mark** for 4% = 2 students and **1 accuracy mark** for the answer.

Total: 6 marks

Why this chapter matters

What are your chances of winning the lottery?
What are your chances of being hit by a car, when crossing the road?

We can answer the first question easily, as the chance of winning the lottery is 14 million to one.

We cannot answer the second question as there are too many **variables**.

- Did you look both ways?
- Is it a quiet road or a busy main road?
- Are you crossing where there are lots of vehicles parked?
- Are you talking on your mobile phone and not paying attention?

Using statistics helps us to predict the **risk** associated with certain events. By paying attention, looking both ways and crossing where the road is clear, you can reduce to zero the risk of being hit by a car.

Actuaries use data related to life expectancy, how likely you are to get hit by a car, how many people catch life-threatening diseases and cancers, and so on, to work out how much to charge a person for a life insurance policy.

Gamblers take a risk every time they place a bet on a horse-race. If the horse wins, the risk was worth it; if not, the risk didn't pay off.

Our lives are governed by risk from the moment we get out of bed to the moment we get back in. Then there are the monsters under the bed to worry about!

Risk plays an important part in our everyday lives.

Chapter

Statistics: Risk

The grades given in this FOUNDATION and HIGHER chapter are target grades.

1 The risk factor and comparing risk

This chapter will show you ...

C how to estimate the risk of an event

B how to compare the risks of certain events

A how to compare and decide on the risk of complex events

Visual overview

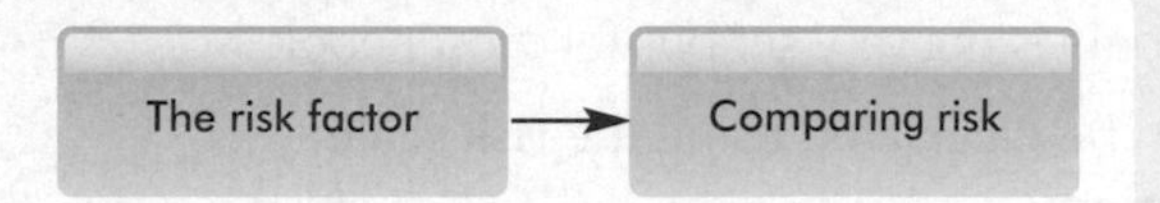

What you should already know

- How to work out the probability of events **(KS3 level 5, GCSE grade E)**
- How to calculate with ratios **(KS3 level 6, GCSE grade D)**
- How to calculate the relative frequency of an event **(KS3 level 7, GCSE grade C)**

Quick check

1 When an ordinary six-sided fair dice is thrown, what is the probability that the score will be:

a a prime number

b a square number

c a factor of 6

d 7?

2 If a bag contains red and blue balls in the ratio 3 : 5, which of the following statements is definitely true (T), may be true (M) or is definitely false (F)?

a There are 16 balls in the bag.

b The probability of taking a red ball from the bag is $\frac{3}{5}$.

c The probability of taking a blue ball from the bag is 0.625.

d If the number of red balls in the bag is doubled but the number of blue balls stays same, the ratio of red balls to blue balls is now 6 : 5.

3 A home-made four-sided spinner was spun 100 times. The results are recorded below.

Score	1	2	3	4
Frequency	15	24	28	33

a Calculate the relative frequency of each score.

b Is the spinner biased? Give a reason for your answer.

7.1 The risk factor and comparing risk

This section will show you how to:
- calculate the risk associated with events
- compare the risk of different events

Key words
hazard probability
risk risk factor
risk scale

Most everyday activities have a **risk factor** associated with them. This depends on two things: the **probability** of the event happening and the **risk** or **hazard**, the catastrophic outcome if it does. For example, flying is a very safe activity but if anything goes wrong then the outcome is usually very severe. If you walk in the country in the summer, the chances of being stung by nettles is quite high, but the outcome is not very serious for most people.

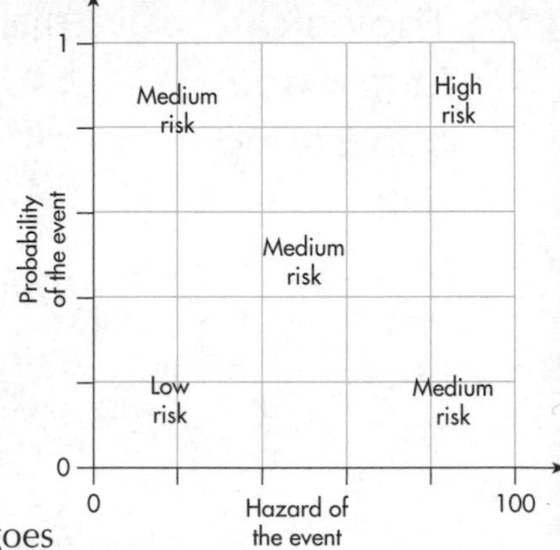

The risk factor can be calculated using the **risk scale** shown in this diagram.

The vertical scale shows the probability of the event, which goes from 0 to 1. The horizontal scale is the hazard. This is not a standard figure and, in this case, is given on a scale of 0 to 100. 'No hazard' would be shown as 0, 'total hazard' would be 100.

The risk factor of an event is the probability multiplied by the index associated with the hazard.

risk factor of an event = P(event) × index factor of the hazard

Risk can actually be a good thing. If they didn't understand risk, many more people would walk across a road without looking. By estimating risk, we make judgements in many everyday situations; health and safety rules are based on risk. When students are taken out of school on an activity or a trip, teachers have to complete a risk assessment of the likely hazards that might be met.

Sometimes it is useful to take biased measurements. Measurements can vary because of human or mechanical error, which could lead to a disaster. For example, an old bridge can carry a load of 40 tonnes. Some lorries can weigh as much as 42 tonnes. If a 42-tonne wagon tried to cross the bridge would probably collapse. To avoid this, the local council put up a sign saying 'Weight limit 35 tonnes'. The under-estimate of 5 tonnes is to make absolutely sure that the bridge is safe.

EXAMPLE 1

The grid shows the risk factors of five events: A, B, C, D and E. Associate the five activities and outcomes described below with each event on the grid.

1: A person who smokes 60 cigarettes a day getting a smoking-related illness.
2: A person playing golf being struck by lightning on the golf course.
3: A person sharing a room with someone with a cold catching a cold as well.
4: A person hammering a nail into a wall hitting their thumb with the hammer.
5: A person cutting grass, sneezing due to pollen.

1
0
Probability of the event
A B
C
D E
0
Hazard of the event
100

continued

SOLUTION

Activity and outcome 1: A person who smokes a lot has a very high probability of getting a smoking-related disease and if they do then they are very likely to die from it, so this is high probability and high hazard, and is event **B**.

Activity and outcome 2: A person playing golf is very unlikely to get hit by lightning but if they do the outcome is likely to be very serious so this is low probability but high hazard, and is event **E**.

Activity and outcome 3: A person sharing a room with someone who has a cold is very likely to catch a cold but these are not serious, so this is high probability but low hazard, and is event **A**.

Activity and outcome 4: A person hammering nails is quite likely to hit their thumb and this could be quite painful so this is medium probability and medium hazard, and is event **C**.

Activity and outcome 5: A person cutting the grass is only likely to sneeze if they are allergic or have hay fever, so this is low probability and low hazard, and is event **D**.

EXAMPLE 2

Every week, Matt plays badminton for 4 hours and rugby for 90 minutes.
Here is some information about the two sports.

	For every 100 000 hours played	
Badminton	0.001 fatalities	14 non-fatal injuries
Rugby	0.0025 fatalities	290 non-fatal injuries

a Is he more likely to have a fatal accident when playing badminton or rugby?

b How much more likely is he to have a non-fatal accident when playing rugby than when playing badminton?

SOLUTION

a In 4 hours of badminton there will be $4 \div 100\,000 \times 0.001 = 4 \times 10^{-8}$ fatal accidents.

In 90 minutes of rugby there will be $1.5 \div 100\,000 \times 0.0025 = 3.75 \times 10^{-8}$ fatal accidents.

Hence Matt is more likely to have a fatal accident playing badminton.

b In 4 hours of badminton there will be $4 \div 100\,000 \times 14 = 5.6 \times 10^{-4}$ non-fatal injuries.

In 90 minutes of rugby there will be $1.5 \div 100\,000 \times 290 = 4.35 \times 10^{-4}$ non-fatal injuries.

Hence Matt is $5.6 \div 4.35 = 1.3$ times as likely to get injured playing rugby as badminton.

EXERCISE 7A

C

1 This table shows the probabilities and hazards associated with asteroid impacts on the Earth.

Diameter	Probability	Hazard	Number of fatalities
Less than 50 m	0.2	Burns up in atmosphere before impact.	0
50 m	0.08	Will survive atmosphere and hit ground, causing widespread localised damage.	400
500 m	0.0002	Impacts with the energy of a small nuclear bomb. Can cause giant tidal waves (tsunami) if impact is in ocean.	100 000
1 km	0.000 002	Devastates local regions. Impact energy of hundreds of nuclear bombs.	1 000 000
2 km	0.000 000 3	Massive destruction. Debris in atmosphere causes global climate change. Starvation due to lack of plant growth.	5 000 000
15 km	0.000 000 01	Global extinction of most plants and animals. Last impact, 65 million years ago, caused extinction of dinosaurs.	50 000 000

Using the number of fatalities as the hazard factor, calculate the risk factor of each size of asteroid.

2 In each case below:

i say if the estimate should be higher, lower or exactly the same as the quoted figure.

ii give an estimate of the hazard of the event, in the context of the situation, on a scale of 1–10, if the estimate is wrong and the event goes wrong. For example, in **a**, if the driver did not brake quickly enough and did not stop before the crossing.

a A car travelling at 30 mph needs 90 feet to come to a complete stop. How far before a pedestrian crossing should the driver of a car travelling at 30 mph start to brake?

b An area of 10 m^2 of a kitchen wall is to be tiled. 25 tiles cover 1 m^2 so 250 tiles would cover 10 m^2. How many tiles should be bought?

c A man wants to fix a window in a frame that is 20 cm by 30 cm. Should he buy a pane of glass that is slightly bigger than the frame or slightly smaller?

d A TV licence costs £150 per year. Mary saves for this by spending £3 on special stamps each week. How many stamps will she need to buy?

e The area of wall to be wallpapered is the same as 2.1 rolls of wallpaper. How many rolls of wallpaper will be needed?

f A recipe for bread needs 150 g of flour plus some flour for kneading. How much flour should be used?

g A running club is organising a race. 200 people have entered. There will be no entries on the day of the race. Each runner gets a T-shirt. How many T-shirts should the club order?

h The same running club is organising another race. 150 people have entered before the deadline but entries will be allowed on the day of the race. Each runner gets a T-shirt. How many T-shirts should the club order?

3 This table shows some hazards associated with golf and caving. Assume that the average golfer plays for 3 hours a week and that the average caver goes caving for an average of 1 hour a week.

Sport	Number of regular participants	Deaths per 100 000 hours of participation	Injuries per 100 000 hours of participation
Golf	3.6 million	0.0006	2
Caving	20 000	0.157	2.5

a How many deaths per year would you expect among golfers?

b Work out the ratio:

deaths caused by golf : deaths caused by caving

giving your answer in the form $n : 1$.

c Work out the ratio:

injuries caused by golf : injuries caused by caving

giving your answer in the form $n : 1$.

4 The risk diagram shows five events: A, B, C, D and E. Associate the five activities and outcomes described below with each event in the diagram.

1: A motorcycle stunt rider trying for a new record jumping buses.

2: An international athlete having a heart attack.

3: An electricity cut causing all the food in a freezer to be spoilt.

4: A traffic jam in London causing someone to miss a meeting for a £50 000 contract.

5: A person catching a train at rush hour and not getting a seat.

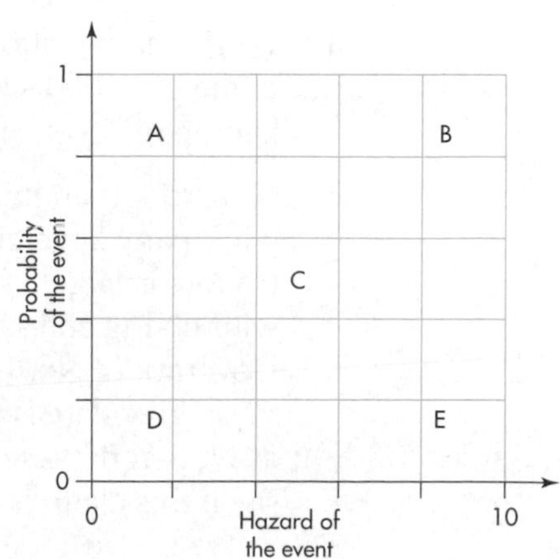

C

5 The table shows the numbers of murders in 2009 in eight countries.

Country	Number of murders	Population (millions)
India	37 170	1140
Russia	28 904	142
United States	16 204	308
United Kingdom	1201	61.4
Germany	914	82.1
Colombia	26 539	43.7
Japan	637	127.7
Ireland	38	4.4

Work out the number of murders per 1000 population for each country and rank the countries in order of danger of death from murder.

B

6 This table shows the annual risk of death for different age groups and genders.

Age group	Annual risk of death
Men aged 55–64	1 in 74
Women aged 55–64	1 in 120
Men aged 35–44	1 in 590
Women aged 35–44	1 in 910
Boys aged 5–14	1 in 5000
Girls aged 5–14	1 in 6700

a Give a reason why, in each age group, males have a greater risk of death than females do.

b Compare the ratio of the risk for males to females in each age category. Give your ratios in the form $1 : n$.

c Does the risk of death for males increase or decrease, compared to females, as people get older? Give a reason for your answer.

7 The table shows the number of different types of crime for two US states, New York and North Dakota in 2008.

	New York	North Dakota
Area	140 000 km^2	181 000 km^2
Population	19 490 297	641 481
Violent	77 585	1068
Property	388 533	12 152
Robbery	31 778	72
Aggravated assault	42 170	761
Burglary	65 735	2106
Larceny theft	297 684	9164
Vehicle theft	25 114	882

a Work out the ratio of each crime per 1000 people in each state.

b Work out the ratio of each crime per 1000 square kilometres in each state.

c New York is an industrial state with many large cities. North Dakota is largely farmland with no big cities. The Governor of North Dakota says: 'This state is twice as safe as New York.' Use the data to say if this claim is true.

8 In these graphs the solid line shows the distance and time needed to stop a car under emergency braking from 60 mph. The dotted line shows the risk of death if hit by a car travelling at this speed.

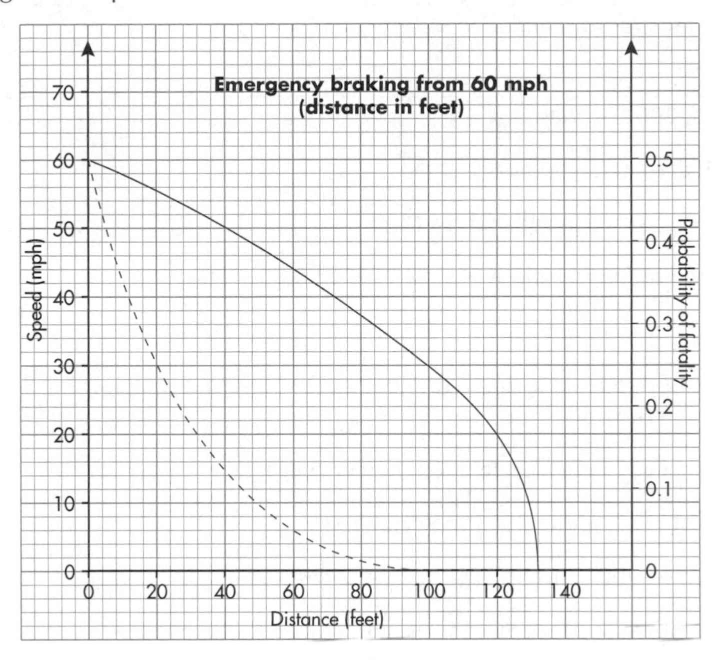

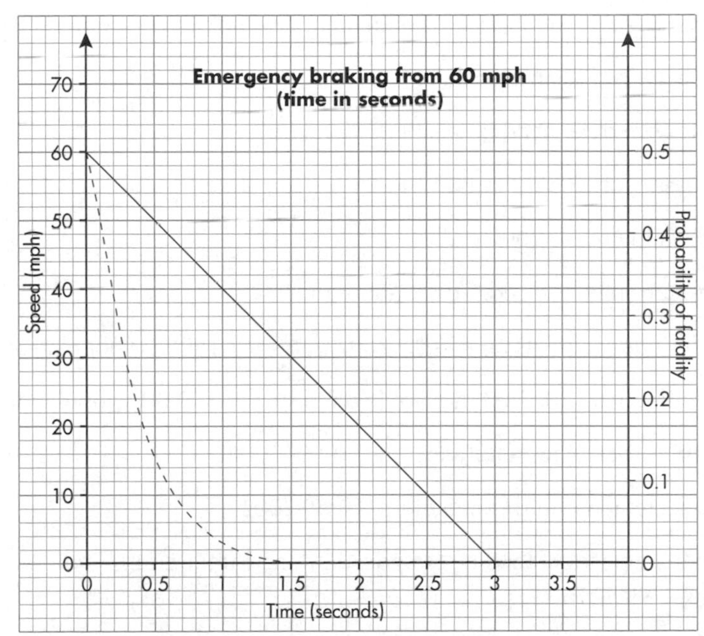

a How far has a car travelled before the speed has reduced to 20 mph?

b How long does it take for the speed to reduce to 20 mph?

c **i** At what speed does the chance of death become less than 0.05?

ii How far has the car travelled when it reaches this speed?

iii How long does it take to reach this speed?

d How long after braking does the chance of death reduce to 0.25?

9 These tables from the UK government statistics website show some economic and other data about two English regions: London and Yorkshire and the Humber.

1.13 Key statistics for London		
	London	**UK**
Population, 2002[1] (thousands)	7355	59 229
Percentage aged under 16[1]	19.6	19.9
Percentage pension age and over[1]	14.0	18.4
Standardised mortality ratio (UK = 100), 202	98	100
Infant mortality rate,[2] 2002	5.6	5.3
Percentage of pupils achieving 5 or more grades A*–C at GCSE level or equivalent, 2001/02	50.6	52.5
Economic activity rate,[6] spring 2003 (percentages)	75.6	78.8
Employment rate,[6] spring 2003 (percentages)	70.3	74.7
Unemployment rate,[6] spring 2003 (percentages)	7.1	5.1
Average gross weekly earnings: males in full-time employment, April 2002 (£)	704.8	511.3
Average gross weekly earnings: females in full-time employment, April 2002 (£)	503.6	382.1
Gross value added, 2001 (£ million)	140 354	874 227
Gross value added per head index, 2001 (UK = 100)	133.2	100.0
Total business sites, 2002 (thousands)	384.9	2538.1
Average dwelling price, 2002 (£)[3]	241 080	145 320
Motor cars currently licensed,[4] 2002 (thousands)	2473	25 782
Fatal and serious accidents on roads,[5] 2002 (rates per 100 000 population)	71	59
Recorded crime rate, 2002/03 (notifiable offences per 100 000 population)[3]	15 175	11 327
Average gross weekly household income, 1999–2002	676	510
Average weekly household expenditure, 1999–2002	463.0	379.7
Households in receipt of Income Support/Working Families Tax Credit,[5] 2001/02 (percentage)	17	17

1 Population figures for 2002 are mid-year population estimates and include provisional results from the Manchester matching exercise. Pension age is men aged 65 and over and women aged 60 and over.

2 Deaths of infants under one year of age per 1000 live births.

3 Figure labelled as the United Kingdom relate to Great Britain.

4 Totals for the United Kingdom include vehicles where the country of the registered vehicle is unknown, that are under disposal or from counties unknown within Great Britain.

5 Figure labelled as the United Kingdom relates to Great Britain.

6 Seasonally adjusted data for people of working age, men aged 16 to 64 and women aged 16 to 59.

7 Data combined for years, 1999/2000, 2000/01 and 2001/02.

A

1.5 Key statistics for Yorkshire and the Humber		
	Yorkshire and the Humber	UK
Population, 2002[1] (thousands)	4983	59 229
Percentage aged under 16[1]	20.2	19.9
Percentage pension age and over[1]	18.7	18.4
Standardised mortality ratio (UK = 100), 202	101	100
Infant mortality rate,[2] 2002	6.2	5.3
Percentage of pupils achieving 5 or more grades A*–C at GCSE level or equivalent, 2001/02	45.6	52.5
Economic activity rate,[6] spring 2003 (percentages)	78.4	78.8
Employment rate,[6] spring 2003 (percentages)	74.1	74.7
Unemployment rate,[6] spring 2003 (percentages)	5.5	5.1
Average gross weekly earnings: males in full-time employment, April 2002 (£)	447.1	511.3
Average gross weekly earnings: females in full-time employment, April 2002 (£)	345.0	382.1
Gross value added, 2001 (£ million)	61 929	874 227
Gross value added per head index, 2001 (UK = 100)	86.4	100.0
Total business sites, 2002 (thousands)	187.0	2538.1
Average dwelling price, 2002 (£)[3]	92 157	145 320
Motor cars currently licensed,[4] 2002 (thousands)	2000	25 782
Fatal and serious accidents on roads,[5] 2002 (rates per 100 000 population)	64	59
Recorded crime rate, 2002/03 (notifiable offences per 100 000 population)[3]	13 597	11 327
Average gross weekly household income, 1999–2002[7]	444	510
Average weekly household expenditure, 1999–2002[7]	340.5	379.7
Households in receipt of Income Support/Working Families Tax Credit,[5] 2001/02 (percentage)	19	17

Jonathan works in Yorkshire. His wife stays at home to look after their two young children. He earns £500 a week. He gets an offer of a job in London that will pay £600 a week.

a By reference to the data in the tables, give two reasons why Jonathan should take the job in London.

b By reference to the data in the tables, give two reasons why Jonathan should not take the job in London.

A

10 These pie charts show the risks associated with two different forms of investment – a high-risk dot-com start-up or a low-risk blue chip established company.

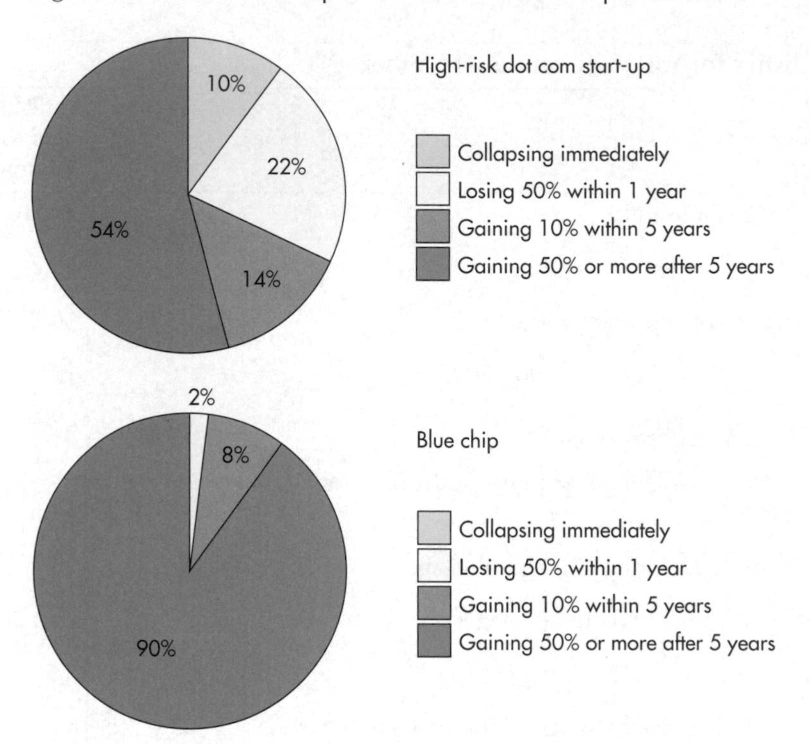

a Assuming £2000 is available for each type of investment, calculate the average profit for each.

b Give a reason why someone with £2000 might choose to invest in a dot-com start-up company.

c Give a reason why someone with £2000 might choose to invest in a blue chip company.

GRADE BOOSTER

C You can calculate the risk factor of an event from the probability and the index factor of the hazard of the event

B You can compare the risks associated with two events

A* You can compare the risks associated with events with more than one variable

What you should know now

- How to calculate a risk factor
- How to compare the risks of events

1 The risk diagram shows five events:

A, B, C, D and E.

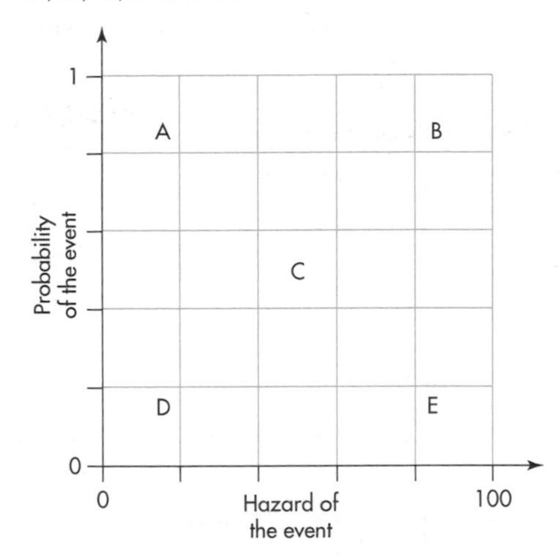

a Explain why an aeroplane crash would be associated with event E.

b Associate the activities and outcomes described below with the other four events.

1: Hanging washing out to dry in the autumn and it raining during the day.

2: Waving a sparkler on bonfire night and getting a burn on the arm from a spark.

3: A sudden downpour in summer causing a cricket match to be cancelled.

4: An outbreak of dysentery after severe flooding in a remote part of a developing country.

2 Marie is a professional horse event rider. She trains every day, Mondays to Fridays, and takes part in events on Sundays. Each training day she swims, runs and cycles for 1 hour each and rides her horse for 3 hours. On event days she rides her horse for 2 hours.

	Risks associated per 100 000 hours			
	Horse riding	**Swimming**	**Running**	**Cycling**
Fatal accidents	0.01	0.012	0.8	1.5
Injuries	3	2	10	32

a Rate the activities in order of risk of a fatality in any week.

b Rate the activities in order of risk of an injury in any week.

c Marie decides to cycle for only 30 minutes a day and increase her run to 1 hour and 30 minutes. Does this change the order of risk for a fatality or an injury? Show your working.

3 It is estimated that a tsunami (tidal wave) will hit a coastal village in Malaysia with a probability of 0.003 and will kill about 700 people.

It is also estimated that a typhoon (strong wind) will hit the same village with a probability of 0.07 and will kill about 20 people.

a Calculate the risk factor of each event, taking the number of likely deaths as the index factor of the hazard.

b How many more times as great is the risk factor of the tsunami than the typhoon?

4 Beryl lives on the outskirts of London and works in the centre. If she is late for work she loses money. If she has an accident and cannot work she does not get paid at all.

She can travel to work by bus, train, car or motorcycle. The table shows the daily cost, the probability of being late and the probability of an accident for each means of transport.

Means of transport	Cost per day	Probability of being late	Probability of an accident
Bus	£10	0.4	0.009
Train	£12	0.2	0.005
Car	£20	0.3	0.01
Motorcycle	£15	0.1	0.08

a Taking the cost per day as the hazard index, work out the risk factor of being late with each means of transport.

b Which means of transport would you advise Beryl to use? Give reasons to justify your choice.

5 This table shows the risk factor associated with tornados in seven US states. The risk factor is defined as the probability of death or serious injury from a tornado times the population of the state.

State	Risk factor	Population (millions)
Indiana	4.25	6.5
Florida	12.75	18.5
Texas	17	25
Wisconsin	20.75	5.6
Colorado	35	5
Maine	40	1.3
Alaska	50	0.7

Calculate the probability of death or serious injury for each state and rank them in order from the least dangerous to the most dangerous.

6 A drug company conducts an experiment with a 'flu vaccine on a group of 200 people.

150 of the group are injected with the vaccine.

The chance of getting 'flu without the vaccine is estimated as 12%.

The chance of getting 'flu with the vaccine is estimated as 3%.

One of the group gets 'flu.

How much more likely is it to be a person who has not had the vaccine than one who has had the vaccine?

Worked Examination Questions

C **1** This table shows the chances of dying from various causes in a year in the UK.

Cancer	1 in 360
External causes	1 in 3070
Road accidents	1 in 15 700
Death by being struck by lightning	1 in 15 000 000

The population of the UK is 61 million.

a Work out how many people will die each year from each of the causes listed above.

b How many more times likely is it that a person will die from cancer than from being struck by lightning?

a Cancer: 61 000 000 ÷ 360 ≈ 170 000 to the nearest ten thousand.

External causes: 20 000 to the nearest thousand.

Road accidents: 3900 to the nearest hundred.

Lightning: 4 to the nearest whole number.

Divide 61 million by the '1 in…' chance for each cause. This will earn **1 method mark** and **2 marks** for all answers correct, or 1 mark if any answers are wrong. Round the answers to a sensible figure.

b 15 000 000 ÷ 360 = 41 666.67
so approximately 41 500 or 42 000.

Divide the chance of dying from being struck by lightning by the chance of dying from cancer to earn **1 mark**.

Total: 4 marks

C **2** A micromort is the one in a million chance of an activity causing death.

It is estimated that a sky-diver has 0.3 micromorts per sky-dive.

It is estimated that a car driver has a 1 micromort per 250 miles travelled.

Mr Black drives a 400 miles round trip to the airfield where he does three sky-dives.

How many times more likely is he to have a fatality driving to the airfield than sky-diving while there?

Total micromorts for 3 sky-dives = 3 × 0.3 = 0.9

Total micromorts for driving = 400 ÷ 250 × 1 = 1.6

1.6 ÷ 0.9 = 1.77… ≈ 1.8

Calculating the total micromorts for either sky-diving or driving will earn **1 method mark**. Getting both answers correct will earn **1 accuracy mark**.

So the driving is 1.8 times as risky as the sky-diving.

Divide one total by the other to get a comparison, to earn **1 mark**

Total: 3 marks

Why this chapter matters

In everyday life you will need to solve mathematical problems in all sorts of situations. You have already used Venn diagrams to solve probability problems (Chapter 1). Another use for Venn diagrams is in solving numerical problems involving two events, when the outcomes are not mutually exclusive, which means that both outcomes can happen. A Venn diagram is a useful way of representing a problem visually.

Here are some shapes.

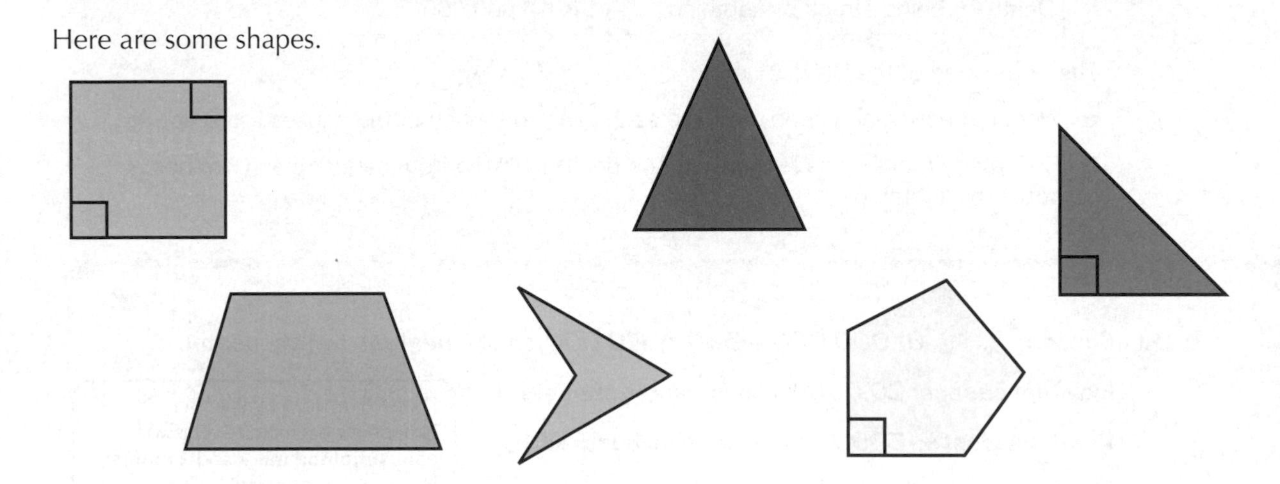

These shapes can be sorted according to whether they have right angles or are quadrilaterals, or both or neither.

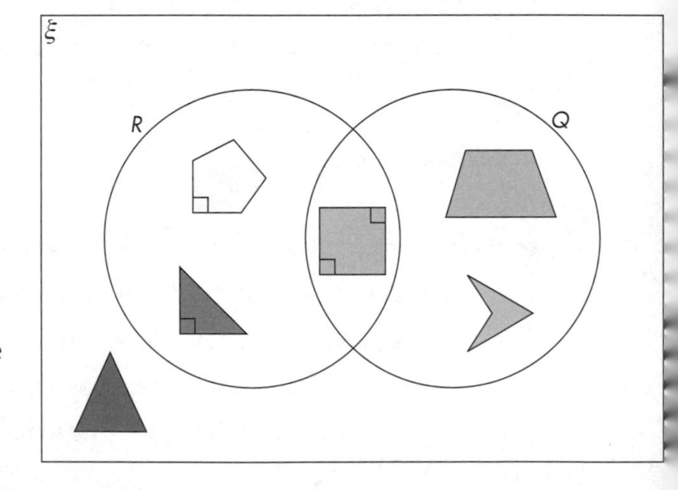

You can see that:

- only the square both has right angles and is a quadrilateral
- the trapezium and the arrowhead are quadrilaterals but do not have right angles
- the right-angled triangle and the pentagon have a right angle but are not quadrilaterals
- the isosceles triangle shown does not have a right angle and is not a quadrilateral.

Suppose that in a box of 100 shapes, 34 have right angles, 47 are quadrilaterals, and 30 are neither right-angled nor quadrilaterals.

Counting the shapes in each area of the diagram, we can use numbers to represent the same information.

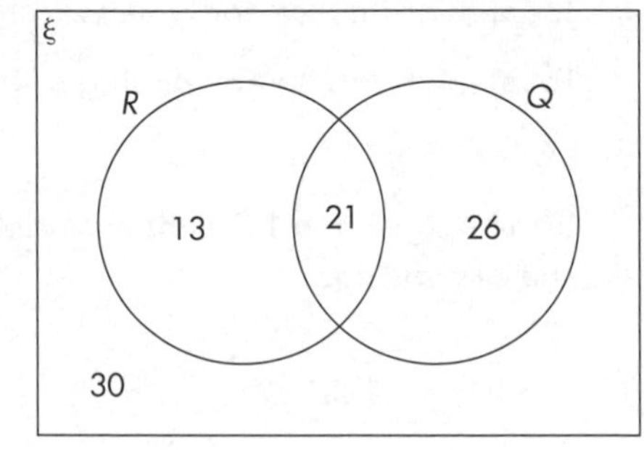

This shows that 21 shapes both have right angles and are quadrilaterals. Can you see how to work this out? This chapter will show you how to use Venn diagrams to solve problems like this one.

Number: Understand and use Venn diagrams to solve problems

The grades given in this FOUNDATION and HIGHER chapter are target grades.

1 Using Venn diagrams to solve problems

This chapter will show you …

C how to understand and use Venn diagrams to solve problems

Visual overview

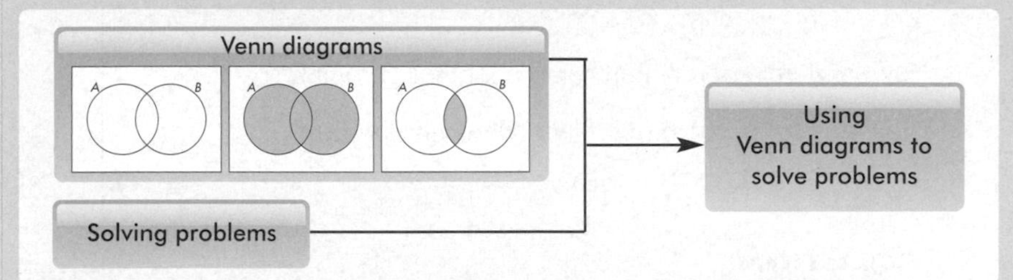

What you should already know

- How to use set notation with probabilities **(GCSE Grade D)**
- The meaning of union (∪) and intersection (∩) **(GCSE Grade D)**
- How to use Venn diagrams to calculate probabilities **(GCSE Grade D)**

Quick check

1 $\xi = \{1, 2, 3, 4, 5, 6, 7, 8, 9, 10\}$

$A = \{1, 3, 4, 5, 8, 9, 10\}$

A number is chosen at random from the universal set.

Write down:

a $P(A)$ **b** $P(A')$.

2 $\xi = \{1, 2, 3, 4, 5, 6, 7, 8, 9, 10\}$

$A = \{1, 2, 3\}$

$B = \{2, 3, 4, 5, 6\}$

a Show this information in a Venn diagram.

b Use the Venn diagram to work out:

i $P(A)$ **ii** $P(A \cup B)$ **iii** $P(A \cap B)$

8.1 Using Venn diagrams to solve problems

This section will show you how to:

- use Venn diagrams to solve numerical problems

Key words

Venn diagram

Venn diagrams can be used to solve problems that can be expressed in terms of overlapping sets.

EXAMPLE 1

In a class of 30 students, 16 study French (F) and 23 study Spanish (S).

Six students study neither French nor Spanish.

Use this information to fill in the Venn diagram.

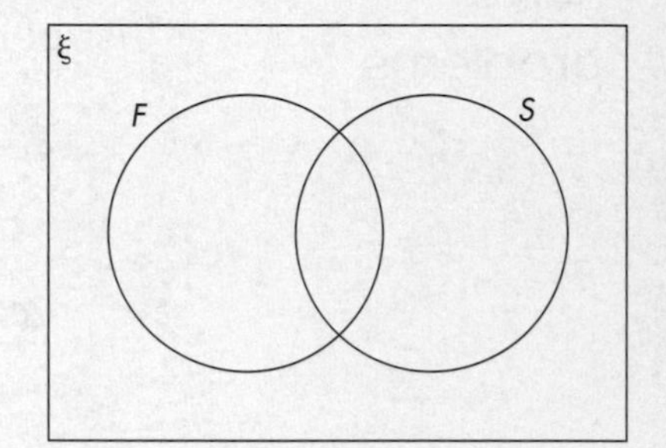

SOLUTION

Method 1: Using a logical argument

As six students do not study French or Spanish, this leaves $30 - 6 = 24$ students.

$16 + 23 = 39$, which means that $39 - 24 = 15$ students have been counted twice, as they study both French and Spanish. This means that $16 - 15 = 1$ studies only French and $23 - 15 = 8$ study only Spanish.

Check: $1 + 15 + 8 + 6 = 30$

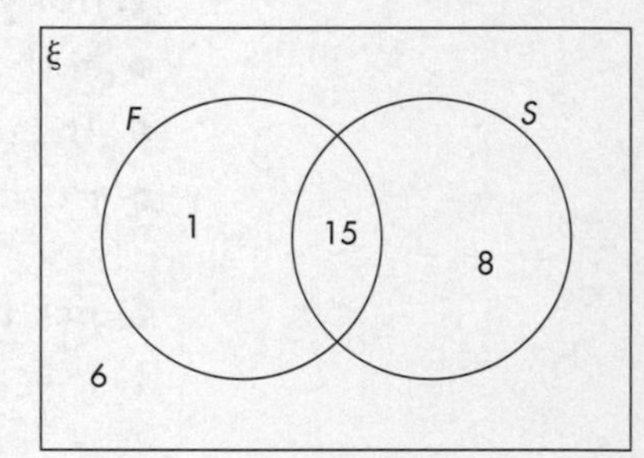

Method 2: Using algebra

Sometimes it is easier to use algebra to solve the problem.

Let the number who study both $= x$

Put in the 6 students who study neither.

The number who study French only will be $16 - x$

The number who study Spanish only will be $23 - x$

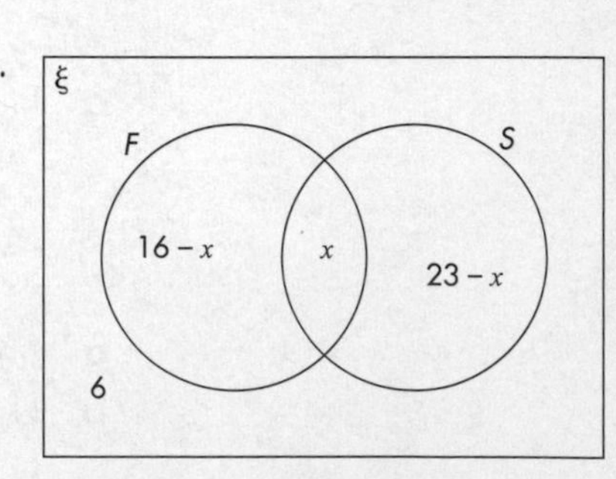

Setting up an equation for the total number of students gives:

$16 - x + x + 23 - x + 6 = 30$

Solving the equation:

$45 - x = 30$

$x = 15$

So $16 - x = 1$ and

$23 - x = 8$

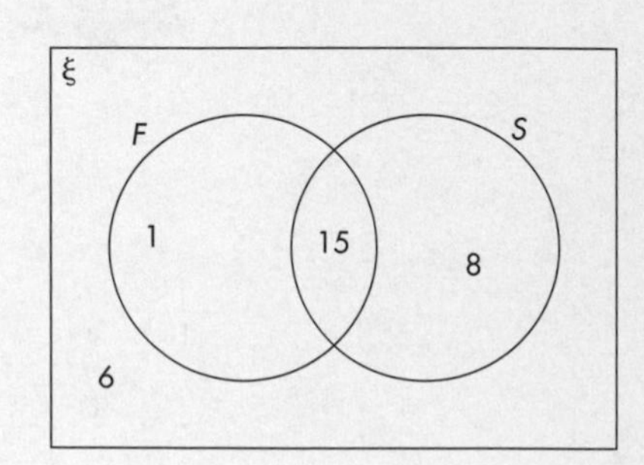

EXERCISE 8A 

For each question in this exercise, draw a Venn diagram unless one is given. Use appropriate letters for the sets.

1 In an athletics club, there are 180 members.

All members are runners (R), field athletes (F) or both.

130 are runners (R).

80 are field athletes (F).

a Copy and complete the Venn diagram.

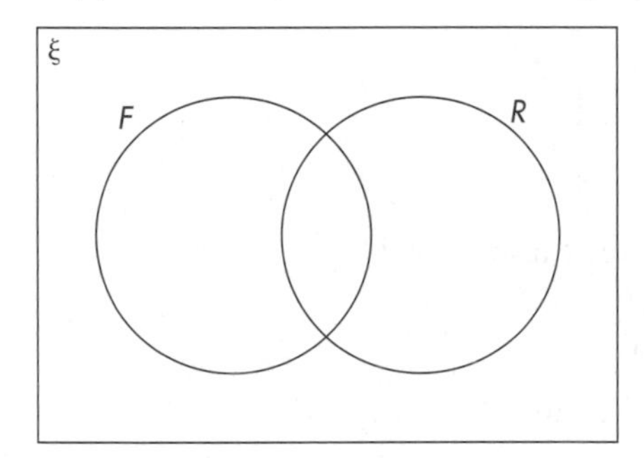

b How many athletes are both runners and field athletes?

2 At a barbeque there are 40 guests.

25 are wearing coats. 10 are wearing scarves. 8 are wearing neither.

How many are wearing both a coat and a scarf?

3 In a fishing competition there are 38 competitors.

22 catch trout. 15 catch bream. 7 catch nothing.

How many catch both trout and bream?

4 In a survey of 100 shoppers, 54 had bought food, 21 had bought clothes, 38 had bought neither food nor clothes.

How many had bought food but had not bought clothes?

5 In Year 11 there are 240 students.

225 take both English and mathematics examinations. 235 took the English examination. 230 took the mathematics examination.

How many did not take either examination?

6 50 students go a school activities trip.

35 do orienteering. 30 do climbing. 2 do not do either orienteering or climbing.

How many do both?

C

7 30 people visit a charity shop.

12 buy books. 8 buy toys. 15 buy nothing.

How many buy either books or toys but not both?

8 In a class of 30 students, 18 like watching films (*F*), 21 like sport (*S*). Four like neither.

a Copy and complete the Venn diagram.

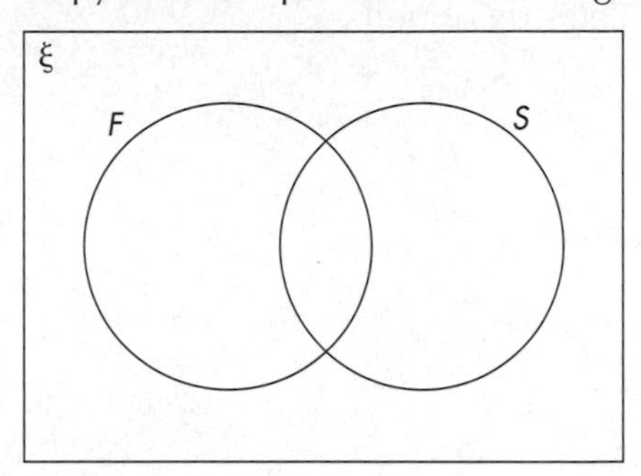

b How many like both watching films and sport?

9 There are 40 people at a leisure centre.

12 are doing both swimming (*S*) and weight training (*W*).

Altogether 26 people are swimming.

Three are doing neither swimming nor weight training.

a Use the information to copy and complete the Venn diagram.

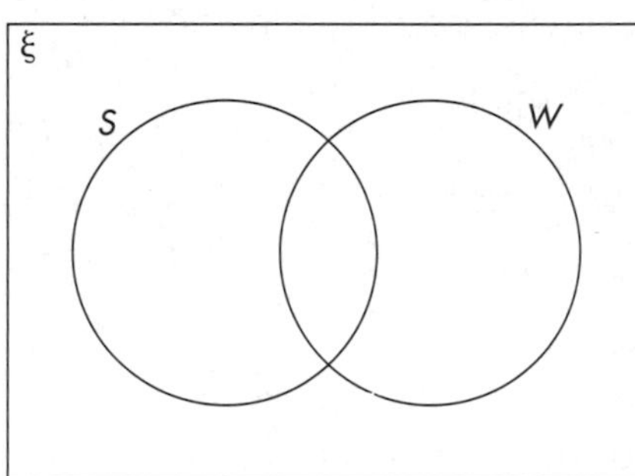

b How many people did weight training altogether?

10 In a school of 1500 students, one-fifth do not buy food for lunch.

720 of the remaining students buy only hot food (*H*).

260 of the remaining students buy only cold food (*C*).

a How many buy a mixture of hot and cold food?

b How many buy some hot food?

AU 11 In a youth club there are 90 members.

16 do not like discos and do not like sport. 14 like discos but do not like sport.

a What is the greatest possible number who like both discos and sports?

b What is the least possible number who like both discos and sports?

B

PS 12 In a mathematics test, 25 have a calculator, 17 have a ruler, 12 have neither, 6 have both.

How many sat the test?

13 110 workers were asked whether they liked bacon (B) or eggs (E) or both.

The results are shown in the Venn diagram.

How many like both?

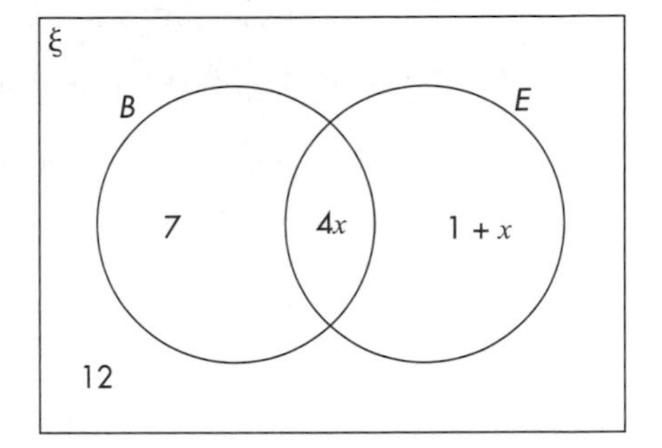

PS 14 In this Venn diagram the number of elements only in P is equal to the number only in Q.

Draw a Venn diagram to show the actual values in each part of the sets.

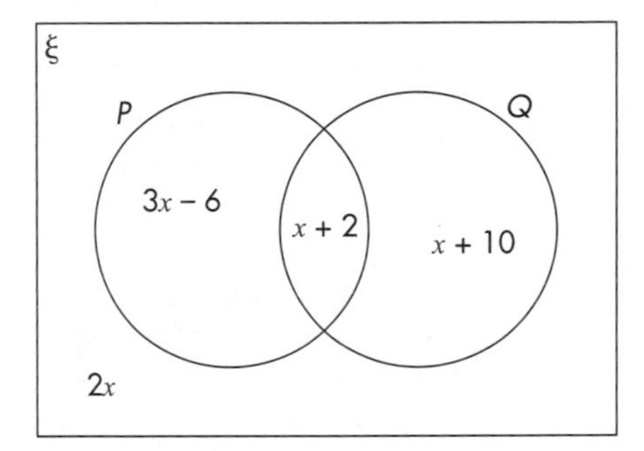

GRADE BOOSTER

- **C** You can complete a Venn diagram, given numerical information
- **C** You know the meaning of union (∪) and intersection (∩)
- **C** You can understand set notation
- **C** You can complete Venn diagrams and solve problems
- **B** You can use Venn diagrams to solve more complex numerical problems

What you should know now

- How to use set notation to describe events
- How to shade Venn diagrams to illustrate sets
- How to use Venn diagrams to solve problems

1 There are 32 students in a class.

18 like apples (A). 15 like bananas (B).
two students do **not** like apples or bananas.

a Use the information to copy and complete the Venn diagram.

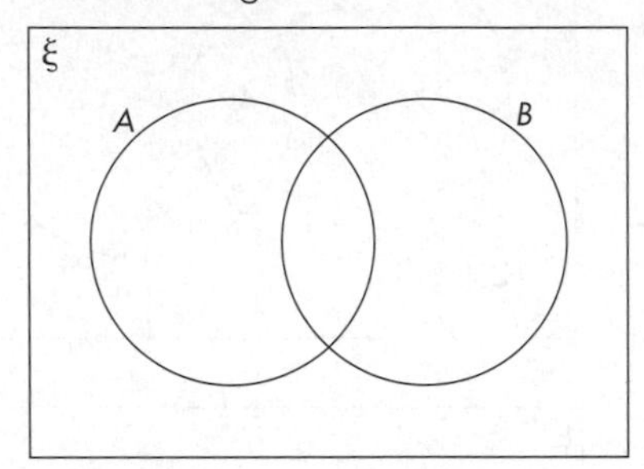

b How many students like only apples or only bananas?

2 In a village of 200 households, 160 have a car, 15 have a motorbike and 32 have no car and no motorbike.

How many of these households have both a car and a motorbike?

3 On a school trip, 100 students visit a park.

52 go rowing (R). 34 go walking (W).
15 neither go rowing nor go walking.

a Use the information to copy and complete the Venn diagram.

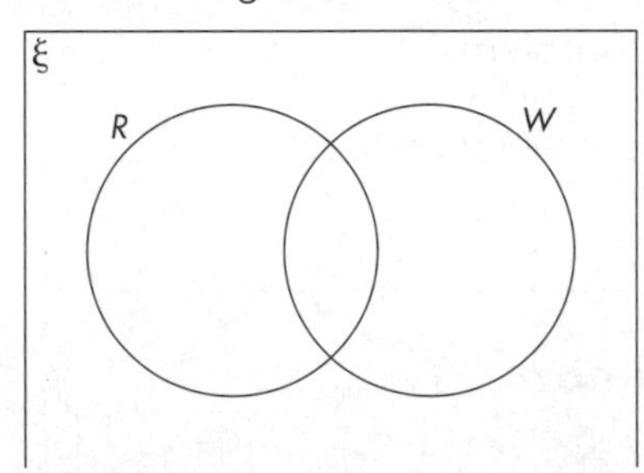

b How many people go both rowing and walking?

4 A flying club has 300 members.

175 of the members only fly two-seater planes (T). 64 of the members fly both two-seater planes and four-seater planes (F). 15 of the members do not fly two-seater or four-seater planes.

a How many of the members fly only four-seater planes?

b 25 new members join the club.

12 of these fly only two-seater planes.
Seven of these fly only four-seater planes.
The rest fly both.

Draw a Venn diagram to show the information for **all** members.

PS 5 At a food van, out of 200 people, twice as many people have burgers as have hotdogs as part of their meal.

10 people have both and 30 have neither.
How many have a burger but not a hotdog?

AU 6 100 teenagers were asked which TV channel they usually watched, BBC (B), ITV (T), both or neither.

The results are shown in the Venn diagram.

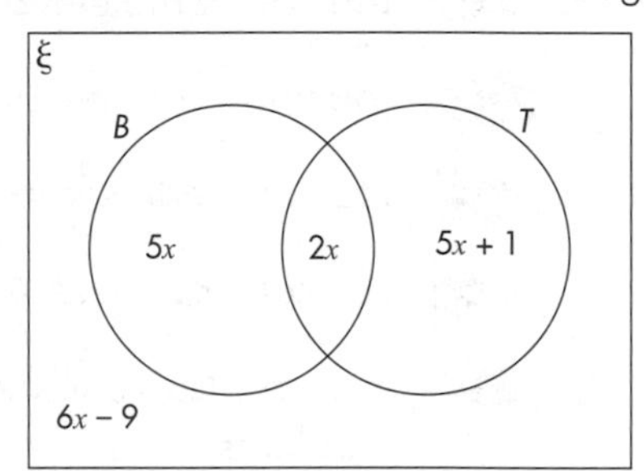

How many usually watched neither?

PS 7 In this Venn diagram the number of elements that are only in P is double the number of elements that are only in Q.

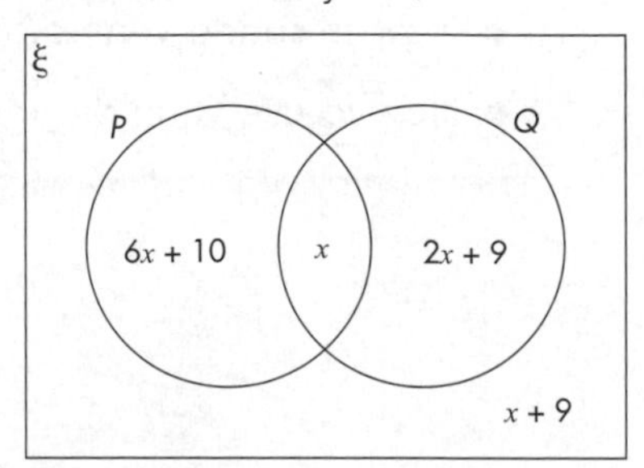

Draw a Venn diagram to show the actual values in each part of the sets.

Worked Examination Questions

C **1** A survey about the breakfasts of 45 people is carried out.

32 say they eat cereal (C).

16 say they have fruit (F).

Four do **not** eat breakfast.

Use a Venn diagram to find how many have cereal but not fruit.

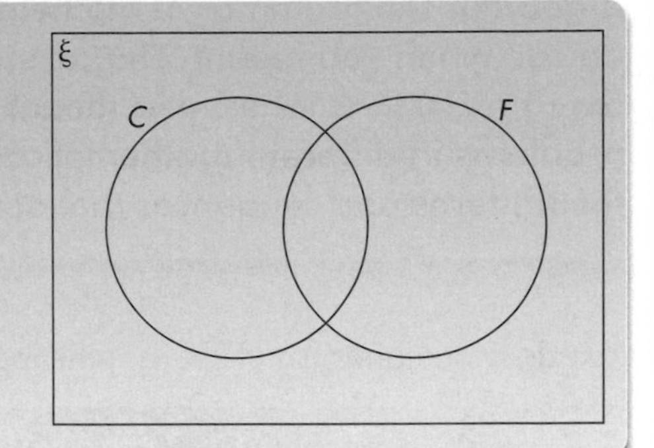

Since 4 do not eat breakfast, put 4 outside the circles. This leaves 41 people who eat breakfast.

16 + 32 = 48, so 48 − 41 leaves 7 who have both cereal and fruit.

Completing the Venn diagram gives:

So 32 − 7 = 25 have cereal but not fruit.

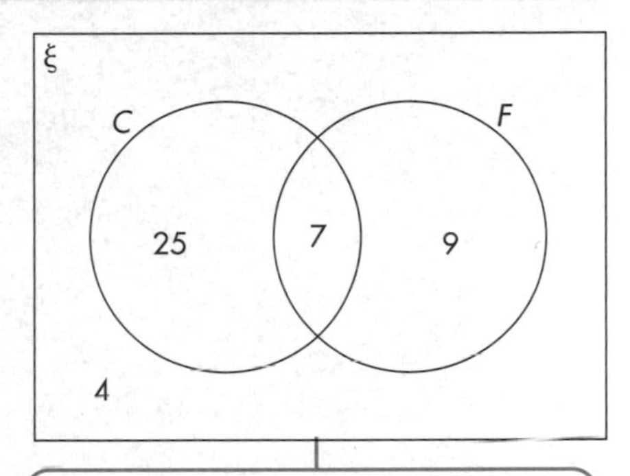

You will gain **1 mark** for the final correct answer.

You will gain **2 marks** for a completed Venn diagram or 1 mark for having the 4 and the 7 in the correct places.

Total: 3 marks

AU B **2** 100 people are asked whether they usually shop at supermarkets (S) or local shops (L). The Venn diagram represents the results.

How many only use local shops?

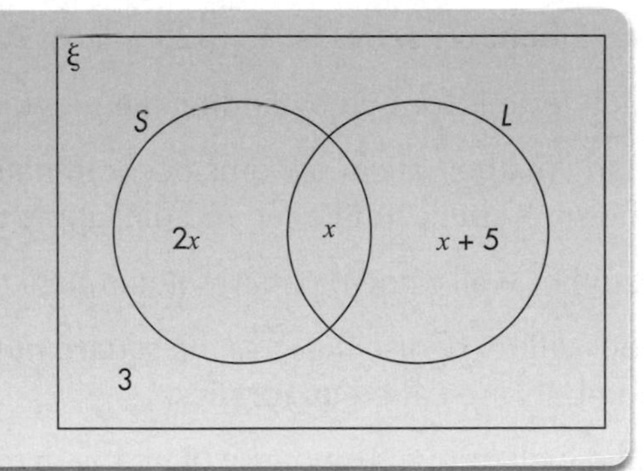

The entries on the Venn diagram total 100.

So $2x + x + x + 5 + 3 = 100$ — You will gain **1 method mark** for setting up the equation.

$4x + 8 = 100$

$4x = 92$ — You will gain **1 mark** for collecting like terms.

$x = 23$ — You will gain **1 mark** for solving the equation correctly.

So the number using only local shops is

$x + 5 = 23 + 5$

$= 28$ — You will gain **1 mark** for the final correct answer.

Total: 4 marks

Why this chapter matters

Algebra is the grammar of mathematics. You would not write 'mat the sat on cat' when you meant 'The cat sat on the mat' and you would not write $5x(+) = 10 + 3$ when you meant $5(x + 3) = 10$. Algebra lets us write problems in different mathematical notations and to investigate some of the patterns and sequences that occur in nature, such as quadratic series.

What do pine cones, snail shells, pineapples, and sunflowers have in common?

The answer is they all have a common mathematical pattern, called the **Fibonacci spiral**.

The **Fibonacci series** is: 1, 1, 2, 3, 5, 8, 13, 21,…

Each term is formed by adding the previous two terms.

Many mathematical patterns occur in nature but the Fibonacci series is the most well known. The snail shell, sunflower and pineapple all have a spiral pattern based on the Fibonacci series.

Another well-known pattern of numbers is 1, 4, 9, 16, 25, 36,…

You will recognise these as the **square numbers**. Functions and equations based on the square numbers are called **quadratics**.

One well-known use of quadratics is in the calculation of the stopping distances of cars.

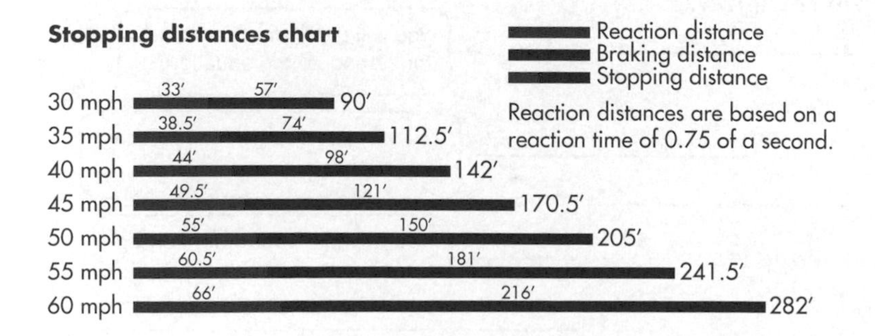

Finding rules to fit this data enables investigators to find out how fast cars involved in accidents had been moving, and to improve the safety of road travel.

Chapter

Algebra: Algebraic equivalence and quadratic sequences

The grades given in this FOUNDATION & HIGHER chapter are target grades.

1 Equivalence in numerical, algebraic and graphical representations

2 The nth term

3 Finding the nth term for quadratic sequences

This chapter will show you...

- **E** different ways to show the relationship between two variables
- **B** how to generate the terms of a quadratic sequence, given the nth term
- **A** how to find a quadratic formula to describe the nth term of a sequence

Visual overview

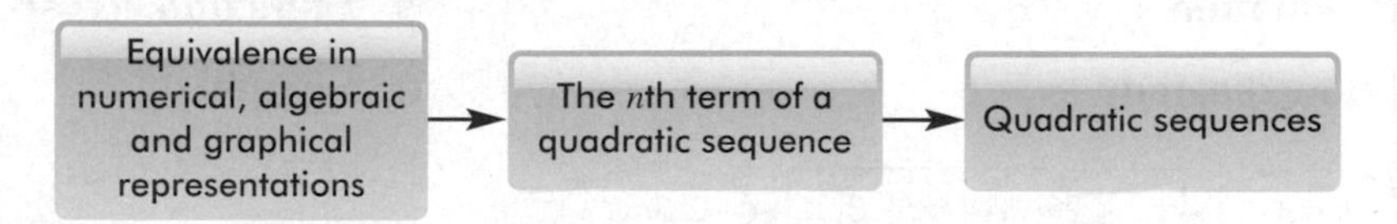

What you should already know

- How to solve simple linear equations **(KS3 grade 5, GCSE grade E)**
- How to generate the terms of a linear sequence, given the nth term **(KS3 Level 6, GCSE grade D)**
- How to find the nth term of a linear sequence **(KS3 level 7, GCSE grade C)**

Quick check

1 Solve these equations.

a $3x + 7 = 13$ **b** $\frac{x}{4} + 3 = 2$

2 Write down the first five terms of the linear sequences that have as an nth term:

a $3n - 1$ **b** $4n + 3$

c $6n - 5$

3 Find the nth term of each of the following linear sequences.

a 5, 8, 11, 14, 17,... **b** 20, 16, 12, 8, 4,...

c 4, 9, 14, 19, 24,...

9.1 Equivalence in numerical, algebraic and graphical representations

This section will show you:

- the different ways of showing the relationship between two variables

Key words
dependent
equation
flow diagram
function
independent
inverse
variable

There are many ways of showing the relationship between two variables. Although they all do the same, but each type of representation has its own uses and advantages. Moving between them can make solving problems easier. For example, the relationship 'multiply by 2 and add 3' can be shown as:

- an **equation** $y = 2x + 3$
- a **function** $f(x) = 2x + 3$
- a **flow diagram**
- a **graph**

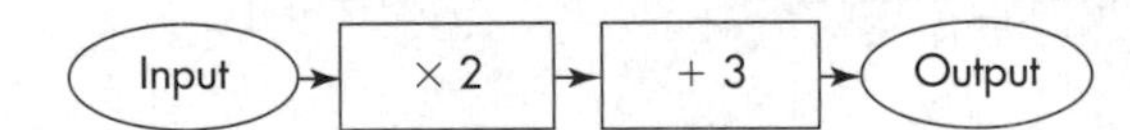

In all of the above cases, x (the input) is the **independent variable** and y (the output) is the **dependent variable**. That means that the value of y (the output) is defined by the value of x (the input).

Which to use

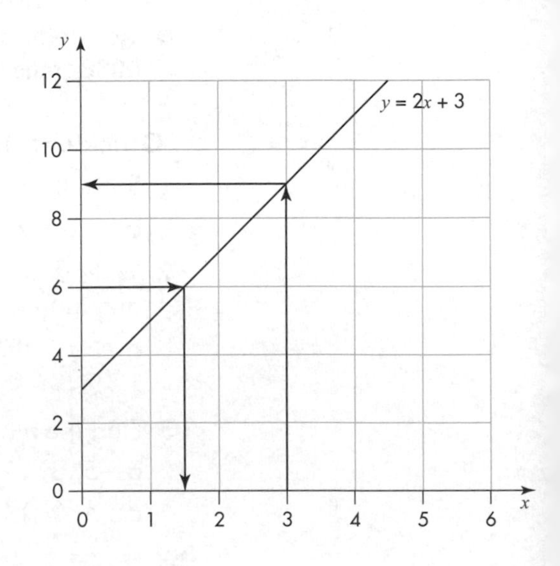

As the different forms are all equivalent, you can use whichever is appropriate for the problem.

Starting with the independent variable, for different values of x:

- **equation**, when $x = 4$, $y = 2 \times 4 + 3 = 11$
- **function**, when $x = 8$, $f(8) = 2 \times 8 + 3 = 19$
- **flow diagram**, when input = 3.5, output = $2 \times 3.5 + 3 = 10$
- **graph**, when $x = 3$, $y = 9$ (See lines on graph above.)

Working backwards from a dependent variable: The easiest way to work backwards is to start with the flow diagram (from above) and draw an **inverse flow diagram**.

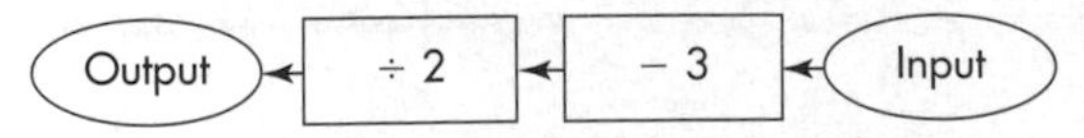

Note: You will **not** be expected to use function notation at Foundation tier.

Take a value for x, such as 5, and put it through the original flow diagram. The output will be 13.

Now put 13 through the inverse flow diagram. The answer will be 5.

The inverse function: Put the variable x through the inverse flow diagram. this gives an output of $\frac{x-3}{2}$, so the inverse function is $f(x) = \frac{x-3}{2}$.

Note: You will **not** be asked to find an inverse function at Foundation tier.

Self-inverse functions: The function $f(x) = 1 - x$ is self-inverse.

For example, if $x = 3$, then $f(3) = 1 - 3 = -2$ and $f(-2) = 1 - -2 = 1 + 2 = 3$.

Using a graph: When $y = 6$, $x = 1.5$ (See lines on graph on page 154.)

Solving an equation: If $2x + 3 = 8$ then $2x = 5$ and $x = 2.5$.

EXAMPLE 1

a Look at the two flow diagrams and show clearly that they both give the same output when the input is 13.

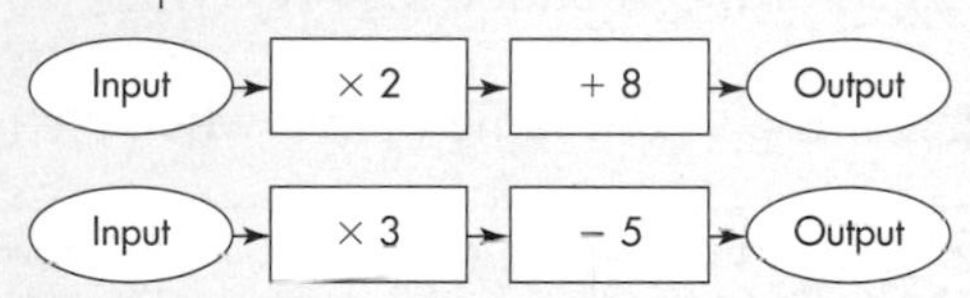

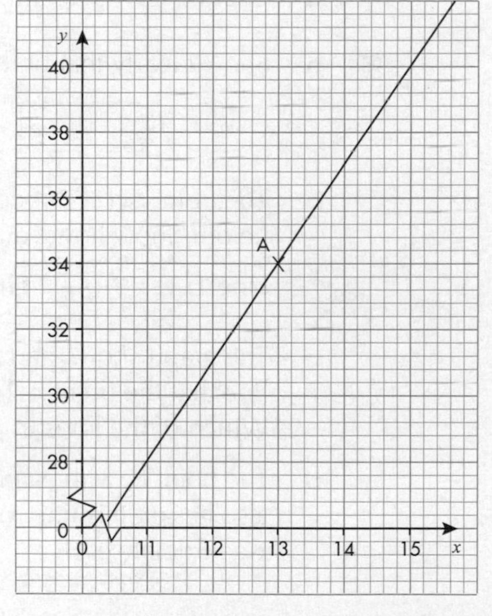

b The point A(13, 34) could lie on three of the following four lines. Which line could it not lie on? Show how you work out your answer.

Line 1: $y = 2x + 8$ **Line 2:** $y = 3x - 5$

Line 3: $y = x + 21$ **Line 4:** $y = \frac{3x - 10}{2}$

c The graph shows part of the actual line on which the point A(13, 34) lies. Which of the lines in part **b** is this?

d The point B, with y-coordinate 94, also lies on this line. What is the x-coordinate of B?

SOLUTION

a Put the input value of 13 through the first flow diagram: $13 \times 2 + 8 = 26 + 8 = 34$

Put the input value of 13 through the second flow diagram: $13 \times 3 - 5 = 39 - 5 = 34$

Hence both flow diagrams have the same output (34) for an input of 13.

b Looking at the equations of the lines, it is clear that lines 1 and 2 are equivalent to the flow diagrams in **a**, so only test lines 3 and 4.

Line 3: $y = 13 + 21 = 34$. Line 4: $y = (39 - 10) \div 2 = 14.5$, so point A could not be on line 4.

c Test another point in the flow diagrams in **a** or the equations in **b**.

Taking $x = 11$, $3 \times 11 - 5 = 28$, so the line is $y = 3x - 5$.

d Using the flow diagram in a reverse, put 94 through it backwards: $94 + 5 = 99$, $99 \div 3 = 33$

So the x-coordinate of point B is 33.

EXERCISE 9A

1 The sketch shows two points, A and B on a line.

a Which of the following is the equation of the line?

Line 1: $y = \frac{x}{2}$ **Line 2:** $y = 2x - 1$

Line 3: $y = 3x - 5$

b A point on the line has a y-coordinate of 7.
What is the x-coordinate of the point?

2 Study the flow diagram.

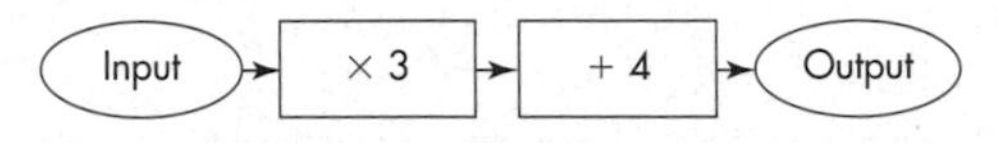

a Show clearly that the point $(-2, -2)$ lies on the line $y = 3x + 4$.

b Use the flow diagram, or otherwise, to solve the equation $3x + 4 = 13$.

3 **a** Show clearly that both these flow diagrams give the same output when the input is 18.

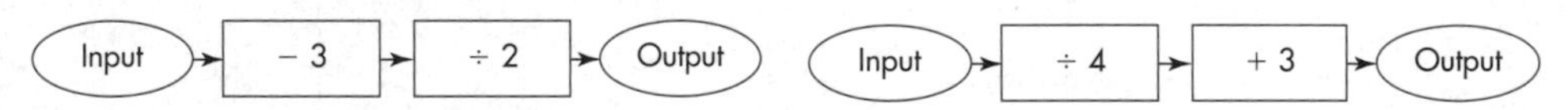

b The point $A(18, 7\frac{1}{2})$ could lie on three of the following lines. Which line could it not lie on? Show how you work out your answer.

Line 1: $y = \frac{x - 3}{2}$ **Line 2:** $y = \frac{2x + 9}{6}$ **Line 3:** $y = x - \frac{3}{2}$ **Line 4:** $y = \frac{x}{4} + 3$

c The graph shows part of the actual line that the point $A(18, 7\frac{1}{2})$ is on. Which of the lines in part **b** is this?

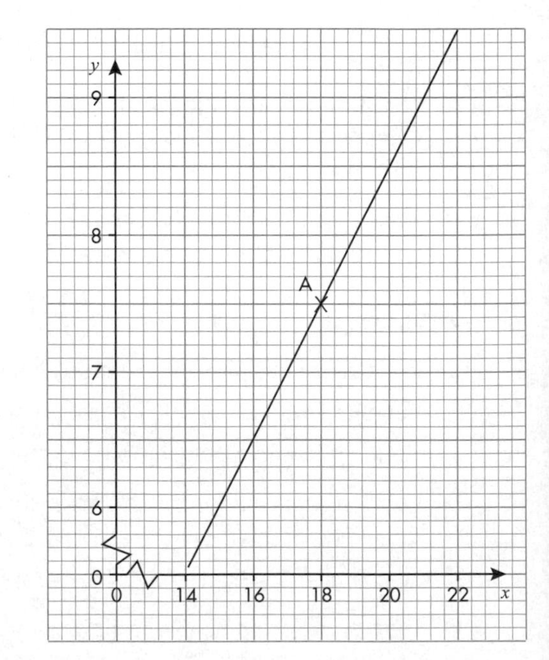

d The point B, with y-coordinate 103, also lies on this line. What is the x-coordinate of B?

4 The diagram shows a number machine.

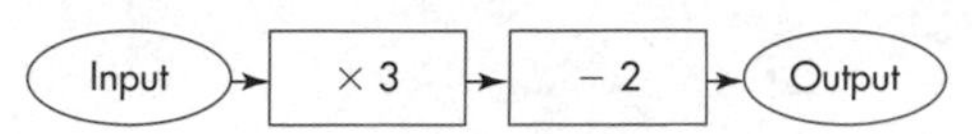

a What is the output when the input is 5?

b What is the input when the output is 7?

c The diagram shows part of the graph of $y = 3x - 2$ for $x = 21$ to 23.

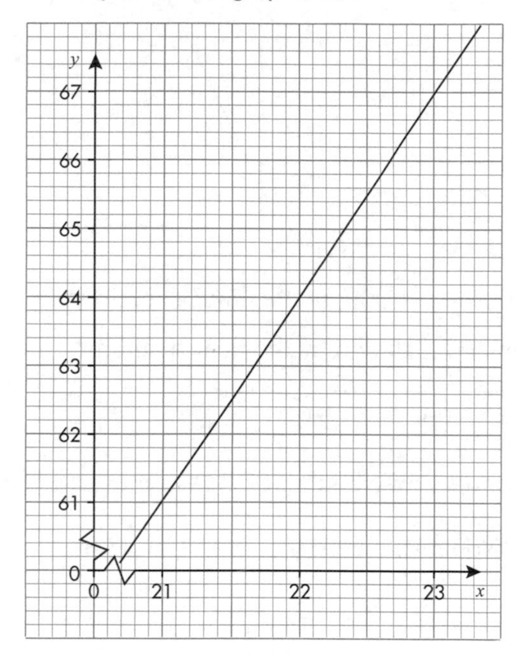

i Use the graph to write down the x-value when $y = 64$.

ii Does the point (25, 73) lie on the graph? Show how you decide.

5 The diagram shows crosses, circles and triangles.

a Which of the following relationships is true for the crosses?

$x - y = 6 \quad xy = 6 \quad x + y = 6$

b Write down the relationship that is true for the circles.

c Does the point (102, 100) lie on the same line as the triangles? Explain your answer.

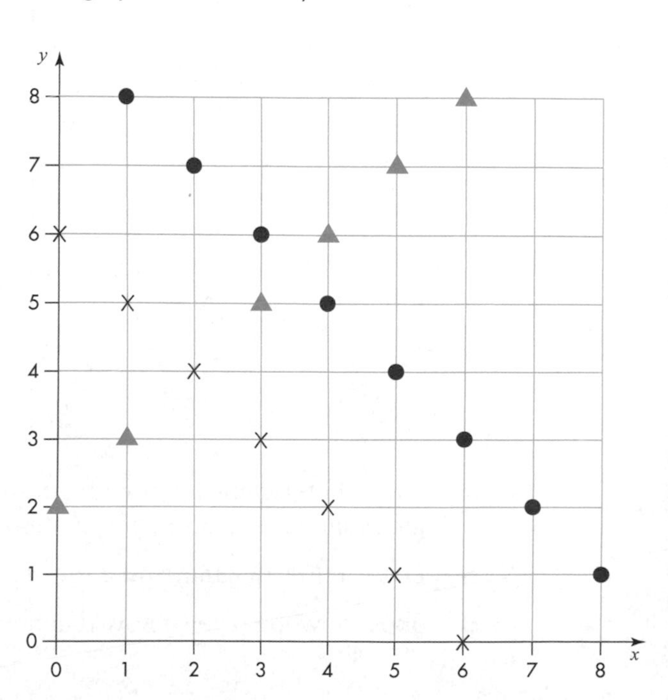

C

AU 6 The sketch shows the graph of $y = 2x + 1$.

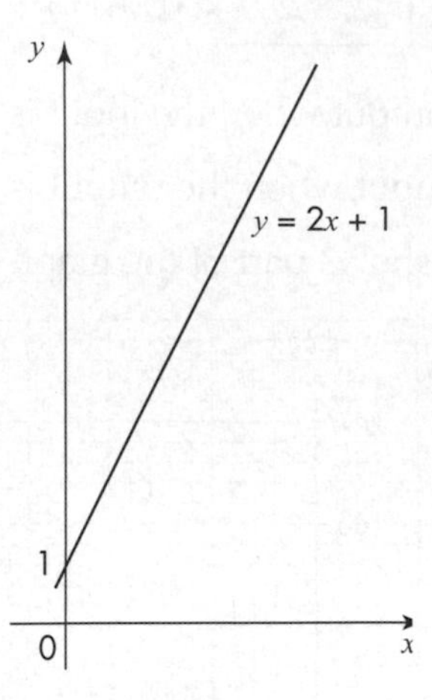

a Does the point $(-1, -3)$ lie on the line? Explain your answer.

b What relationship is shown by the flow diagram?

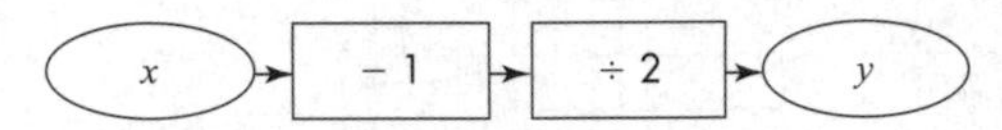

c Explain the connection with the relationship in part **b** and the graph $y = 2x + 1$.

AU 7 The graph shows two lines, line A: $y = 2x + 4$ and line B: $y = \frac{x}{2} - 2$.

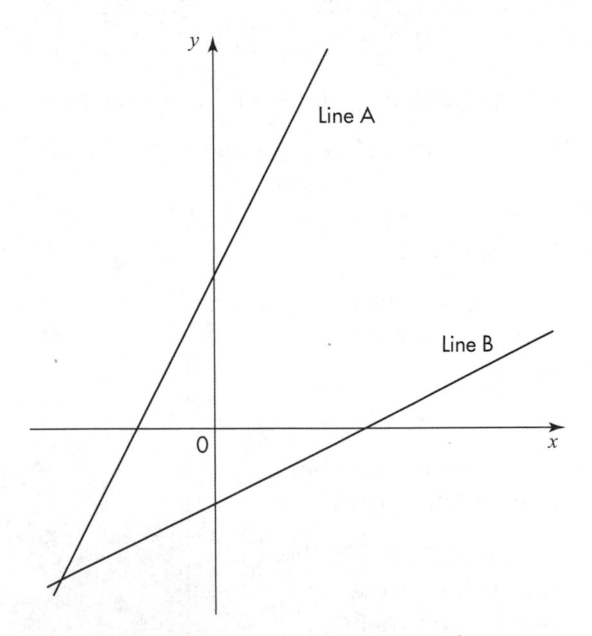

a Show that, for any point with coordinates of the form (x, y) that lies on line A, the point with reverse coordinates (y, x) will lie on line B.

b Draw a flow diagram and a reverse flow diagram for line A.

c Show how the reverse flow diagram in **b** relates to line B.

8 The flow diagram shows the equation $y = 6 - x$.

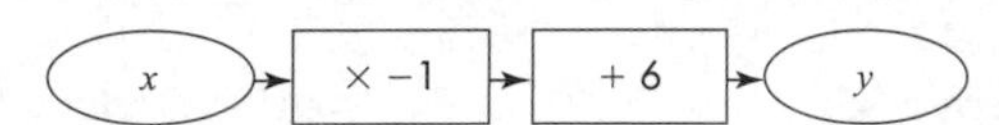

a Work out the value of y when $x = 8$.

b Use the inverse flow diagram to work out the value of x when $y = 10$.

c Show, by means of an example, that $y = 6 - x$ is self-inverse.

PS 9 The flow diagram works out the reciprocal of a number. Show, by means of two examples, that the reciprocal is a self-inverse function.

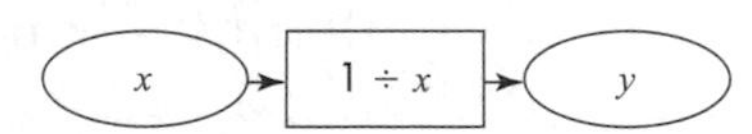

10 Show that these two flow diagrams show the same function.

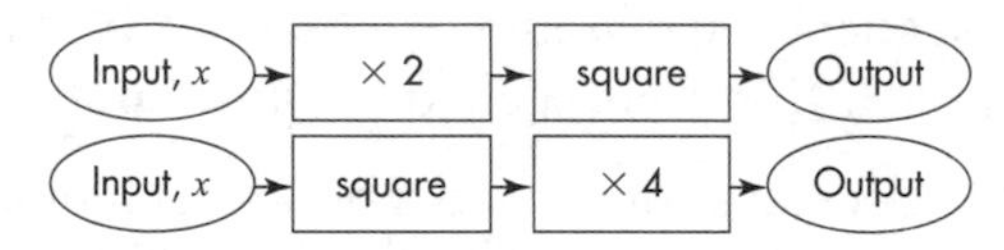

11 The graph shows a sketch of the lines $y = x + 2$ and $y = 2x - 1$.

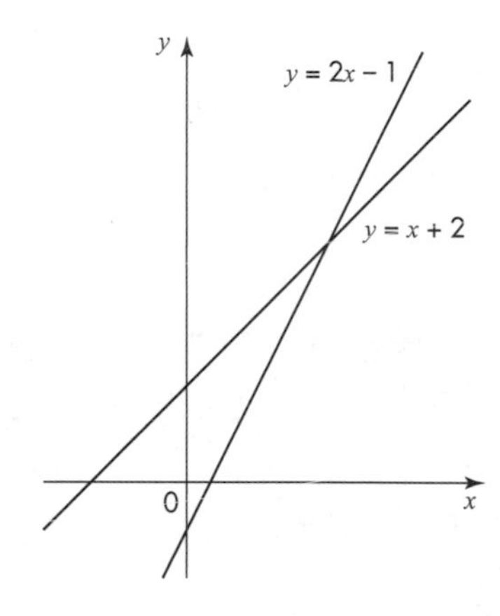

a Use algebra to find the point of intersection of the lines.

b Explain why you can use the flow diagrams below to check your answer to part **a**.

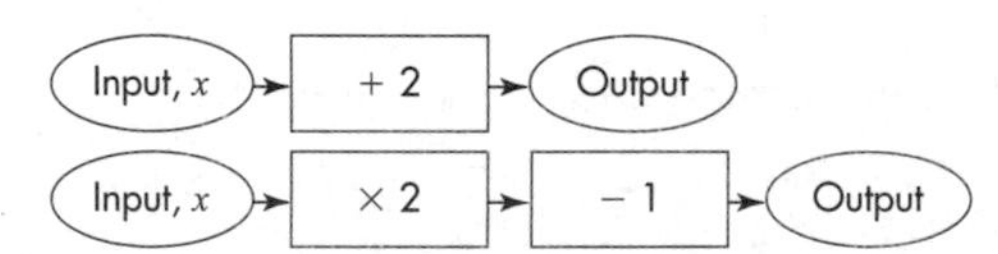

9.2 The *n*th term

This section will show you how to:

- generate the terms of a quadratic sequence, given the *n*th term

Key words
first difference
*n*th term
position-to-term rule
quadratic rule
quadratic sequence
second difference
term-to-term rule

Some problem-solving questions involve number sequences that are governed by a **quadratic rule**. Such a sequence is called a **quadratic sequence** and would be described by a quadratic expression for the ***n*th term**, such as $n^2 + 2n - 3$. This is also called the **position-to-term rule**.

In general, the **term-to-term rule** increases or decreases by a fixed amount each time, called the **first difference.** The differences of the first differences are called the **second differences**. If the second differences of a sequence are constant, it is a quadratic sequence.

EXAMPLE 2

Write down the first five terms of the following sequences, given that the *n*th term is:

a $n^2 + 5$ **b** $n^2 + 2n$ **c** $n^2 + 3n - 4$

SOLUTION

Substitute $n = 1, 2, 3, 4$ and 5 into the expressions to get the first five terms.

a 6, 9, 14, 21, 30,… **b** 3, 8, 15, 24, 35,… **c** 0, 6, 14, 24, 36,…

EXAMPLE 3

Work out the first and second differences, then state if these are quadratic sequences.

a 5, 7, 10, 14, 19, 25, 32,… **b** 1, 1, 2, 3, 5, 8, 13,…

SOLUTION

Work out the first differences and the second differences of each sequence.

a

	5	7	10	14	19	25	32
First differences	2	3	4	5	6	7	
Second differences	1	1	1	1	1		

The second difference is constant so this is a quadratic sequence.

b

	1	1	2	3	5	8	13
First differences	0	1	1	2	3	5	
Second differences	1	0	1	1	2		

The second difference is not constant so this is not a quadratic sequence.

EXERCISE 9B

1 In each of these sequences:

i write down the next two terms **ii** say how the sequence is building up.

a 1 4 8 13 19 26 ....

b 3 4 6 9 13 18 ....

c 9 14 20 27 35 44 ....

d 102 92 83 75 68 62 ....

2 Work out the first five terms of the sequences with nth term:

a $n^2 + 3$

b $2n^2$

c $n^2 + n$

d $n^2 + 2n + 1$

e $n^2 + 3n - 2$

f $2n^2 - 3n + 5$

3 **a** Write down the nth term of this linear sequence.

3 5 7 9 11 13 ...

b Write down the nth term of this linear sequence.

1 2 3 4 5 6 ...

c Hence write down the nth term of this sequence.

$1 \times 3, 2 \times 5, 3 \times 7, 4 \times 9, 5 \times 11, 6 \times 13, \ldots$

d Now write down the nth term of this sequence.

4, 11, 22, 37, 56, 79,...

4 This pattern of rectangles is made from squares.

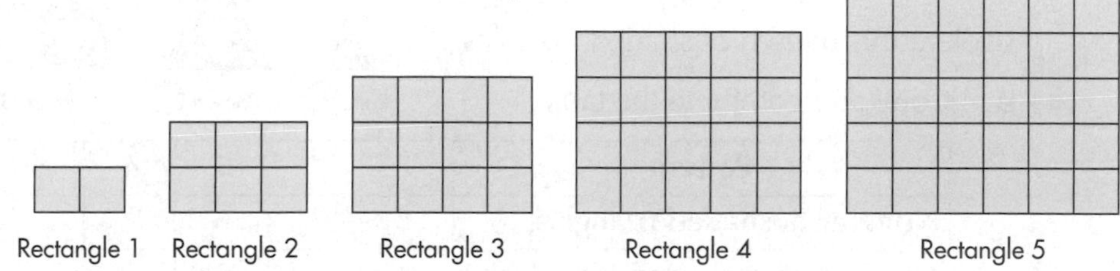

a The heights of the rectangles are the sequence 1, 2, 3,...

Write down the nth term of this sequence of heights.

b Write down the nth term of the sequence formed by the lengths of the rectangles.

c Hence, write down the nth term of the sequence formed by the areas of the rectangles.

d What is the area of the 99th rectangle in the pattern?

A

5 By finding the first and second differences, say if these are quadratic sequences.

a 6, 8, 11, 15, 20, 26, 33, 41,...

b 1, 3, 5, 8, 11, 14, 18, 22,...

c 2, 6, 12, 20, 30, 42, 56,...

d 1, 3, 9, 27, 81, 243,...

e 5, 9, 14, 20, 27, 35, 44,...

f 0, 2, 2, 4, 6, 10, 16, 26,...

AU 6 The sequence of squares is made from smaller white and coloured squares.

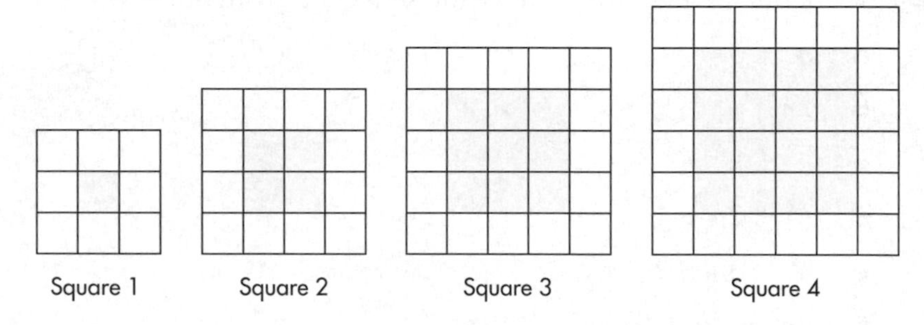

a The numbers of white squares make the sequence 8, 12, 16, 20,...

Write down the nth term of this sequence.

b The numbers of coloured squares make the sequence 1, 4, 9, 16,...

Write down the nth term of this sequence.

c Hence write down the nth term of the total number of smaller squares in each large square.

d Expand $(n + 2)^2$.

e Explain why the answers to parts **c** and **d** are the same.

7 The triangular numbers are 1, 3, 6, 10, 15, ...

The square numbers are 1, 4, 9, 16, 25, ...

Look at this pattern of shaded triangles.

a Copy and complete the table,

Pattern	1	2	3	4	5	6
Number of shaded triangles	1	3	6			
Number of unshaded triangles	0	1	3			
Total number triangles	1	4	9			

b A pattern in the sequence has 55 shaded triangles.

i How many unshaded triangles will be in this pattern?

ii How many triangles will there be altogether in this pattern?

9.3 Finding the nth term for quadratic sequences

This section will show you how to:

- find the nth term of a quadratic sequence

Key words
coefficient
nth term
quadratic sequence
second difference

There are several methods for finding the ***n*th term** of a **quadratic sequence** but they are all based on one simple rule:

If the **second difference** is constant then it is a quadratic sequence. The **coefficient** of n^2 is half the constant value of the second difference.

There are four ways of finding the nth term:

- spotting a simple rule
- breaking into factors
- subtracting the squared term
- extending the differences backwards.

You need to be able to recognise the sequence of square numbers: 1, 4, 9, 16, 25, 36, 49,... as 'simple rule' sequences are always based on n^2 alone. The question will often start with a hint that it is based on the sequence of square numbers.

EXAMPLE 4: SPOTTING A SIMPLE RULE

State the nth terms of each sequence.

a 1, 4, 9, 16, 25, 36, 49, 64,...

b 4, 7, 12, 19, 28, 39, 52, 67,...

c 3, 12, 27, 48, 75, 108, 147, 192,...

SOLUTION

a You should recognise this as the square numbers, so the nth term is simply n^2.

b Each term is 3 more than the corresponding square number in **a** so the nth term is $n^2 + 3$.

c Each term is 3 times as big as corresponding square number in **b** so the nth term is $3n^2$.

Even if you do not recognise a simple sequence based on n^2, any of the following methods will always work.

EXAMPLE 5: BREAKING INTO FACTORS

Find the nth term in the sequence 2, 6, 12, 20, 30,…

SOLUTION

	2		6		12		20		30
First differences		4		6		8		10	
Second differences			2		2		2		

The second differences are constant (2) so the sequence is quadratic and the coefficient of n^2 is 1. So the nth term includes $1n^2$ which is just written as n^2.

Split each term into factors to try to find a pattern for how the numbers have been formed. Constructing a table like this can help to sort out which factors to use.

Term	2	6	12	20	30
Factors	1 × 2	2 × 3	3 × 4	4 × 5	5 × 6

Now break down the factors to obtain:

$1 \times (1 + 1)$ $2 \times (2 + 1)$ $3 \times (3 + 1)$ $4 \times (4 + 1)$ $5 \times (5 + 1)$

It is now quite easy to see that the pattern is $n \times (n + 1)$.

So the nth term is $n(n + 1) = n^2 + n$

It may not always be possible to spot how to break terms into factors but either of the following methods will still always work.

EXAMPLE 6: SUBTRACTING THE SQUARED TERM

Find the nth term for the sequence of the triangular numbers 1, 3, 6, 10, 15,…

SOLUTION

	1		3		6		10		15
First differences		2		3		4		5	
Second differences			1		1		1		

The second difference has a constant value of 1, so the coefficient of n^2 is $\frac{1}{2}$, hence the nth term includes $\frac{1}{2}n^2$.

Now subtract this term from each term of the original sequence. It can sometimes be easier to use a table.

n	1	2	3	4	5
Original	1	3	6	10	15
$\frac{1}{2}n^2$	$\frac{1}{2}$	2	$4\frac{1}{2}$	8	$12\frac{1}{2}$
Difference	$\frac{1}{2}$	1	$1\frac{1}{2}$	2	$2\frac{1}{2}$

continued

The differences form a linear sequence $\frac{1}{2}, 1, 1\frac{1}{2}, 2, 2\frac{1}{2},\ldots$ which has an nth term of $\frac{1}{2}n$
Combining this with the squared term gives the nth term for the whole sequence of $\frac{1}{2}n^2 + \frac{1}{2}n$ and this can be written as $\frac{1}{2}n(n + 1)$, which is the usual formula for the nth triangular number.

The fourth method is possibly the easiest as it does not involve any difficult calculations. It works the sequence back to the term for $n = 0$, then the general quadratic $an^2 + bn + c$ when $n = 0$ gives the value of c. The first difference between the $n = 0$ and $n = 1$ terms eliminates c and is just $a \times 1^2 + b \times 1 = a + b$.

EXAMPLE 7: EXTENDING THE DIFFERENCES BACKWARDS

Find the nth term of the sequence 5, 15, 31, 53,...

SOLUTION

Set up a difference table.

n	1	2	3	4
nth term	5	15	31	53
First differences	10	16	22	
Second differences		6	6	

Now extend the table backwards to get the term for $n = 0$ and call the three lines of the table c, $a + b$ and $2a$.

n	0	1	2	3	4
c	1	5	15	31	53
$a + b$	4	10	16	22	
$2a$		6	6	6	

Taking the values for $n = 0$ gives $2a = 6 \Rightarrow a = 3$ $\quad a + b = 4 \Rightarrow b = 1$ $\quad c = 1$

Thus the nth term is $3n^2 + n + 1$.

EXERCISE 9C

1 For each of the sequences **a** to **e**:

i write down the next two terms **ii** find the nth term.

a 1, 4, 9, 16, 25,...

b 0, 3, 8, 15, 24,...

c 3, 6, 11, 18, 27,...

d 4, 7, 12, 19, 28,...

e −1, 2, 7, 14, 23,...

f 11, 14, 19, 26,...

B

A

2 For each of the sequences **a** to **e**:

i write down the next two terms **ii** find the nth term.

a 5, 10, 17, 26,... **b** 3, 8, 15, 24,...

c 9, 14, 21, 30,... **d** 10, 17, 26, 37,...

e 8, 15, 24, 35,...

3 Look at each of the following sequences to see whether the rule is linear, quadratic in n^2 alone or fully quadratic. Then write down:

i the nth term **ii** the 50th term.

a 5, 8, 13, 20, 29,... **b** 5, 8, 11, 14, 17,...

c 3, 8, 15, 24, 35,... **d** 5, 12, 21, 32, 45,...

e 3, 6, 11, 18, 27,... **f** 1, 6, 11, 16, 21,...

4 Find the nth terms of each of the following sequences, in the form $an^2 + bn + c$.

a 1, 4, 11, 22, 37,... **b** 2, 13, 30, 53, 82,...

c 4, 8, 13, 19, 26,... **d** 3, 9, 16, 24, 33, ...

e 8, 11, 15, 20, 26, ... **f** 4, 7, 11, 16, 22, ...

AU 5 Work out a formula for the surface area of a large cube made up of smaller centimetre cubes with a side of n centimetres.

AU 6 The diagram shows four houses of cards. A one-level house of cards (L1) takes two cards. A two-level house of cards (L2) takes seven cards. A three-level house of cards (L3) takes 15 cards.

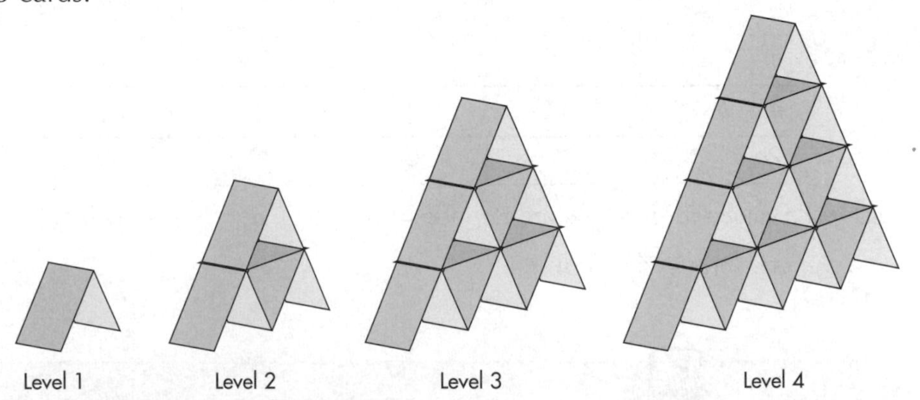

a How many cards are needed to make a four-level (L4) house of cards?

b Work out a formula for an n-level house of cards.

c The world record is a 75-level house of cards. How many cards were used to build this?

PS 7 A supermarket displays tins of beans by lining up n tins in a row and then putting $n - 1$ tins in a row on top of these, and then $n - 2$ tins in a row on top of these and so on until there is just 1 tin on top. The diagram shows a display that starts with four tins on the bottom.

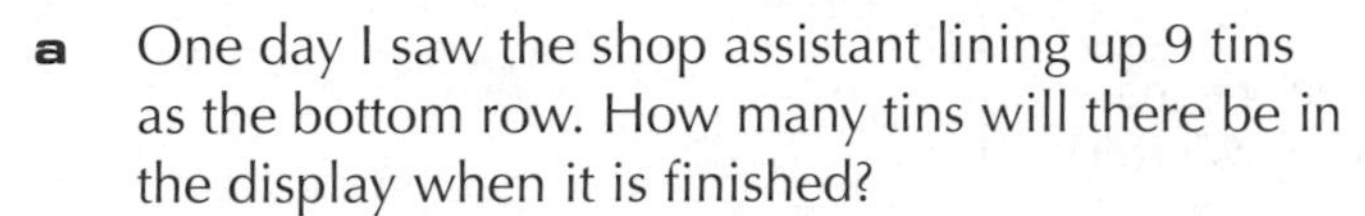

a One day I saw the shop assistant lining up 9 tins as the bottom row. How many tins will there be in the display when it is finished?

b Health and safety regulations state that the display cannot be more than 15 rows high. The supermarket has 100 tins of beans in stock. Can they make a display that is 15 rows high?

PS 8 Centimetre cubes are used to make patterns of cuboids. Work out an expression for the surface area of Cuboid n.

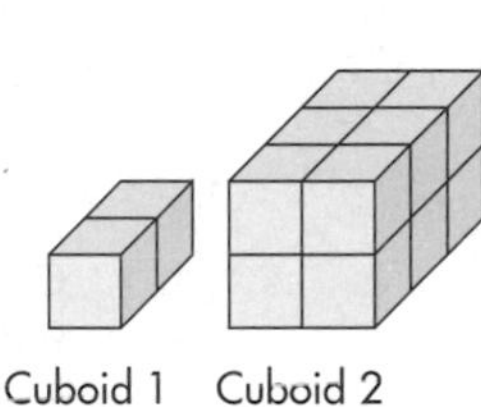

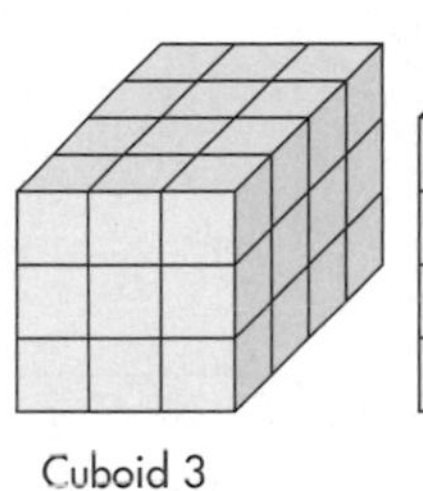

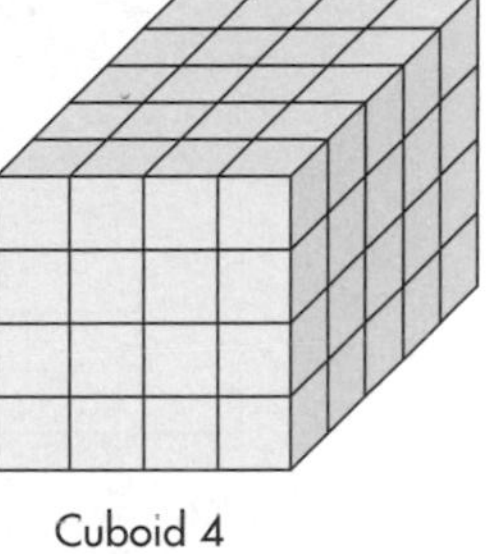

Cuboid 1 Cuboid 2 Cuboid 3 Cuboid 4

PS 9 The diagram shows the first four hexagonal numbers. Work out the 100th hexagonal number.

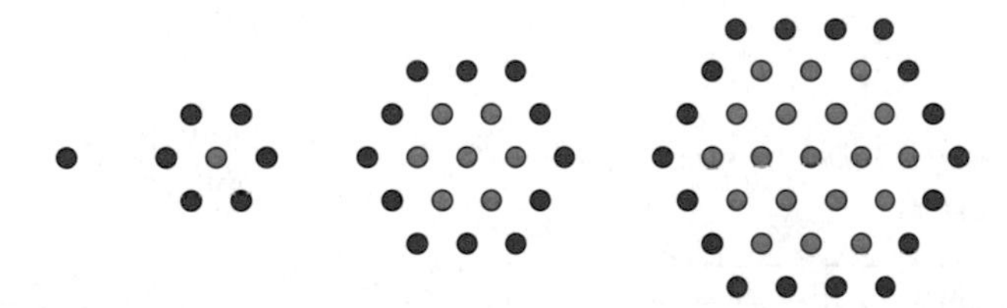

GRADE BOOSTER

E You can recognise the equivalence between different ways of showing the relationship between two variables

D You can use the relationship between two variables to choose the most appropriate method to solve problems

B You can generate the terms of a quadratic sequence, given the nth term

A You can find the nth term of a quadratic sequence

What you should know now

- How to generate the terms of a quadratic sequence, given the nth term
- How to find the nth term of a quadratic sequence

EXAMINATION QUESTIONS

1 **a** Here is a number machine. What is the output when the input is -3?

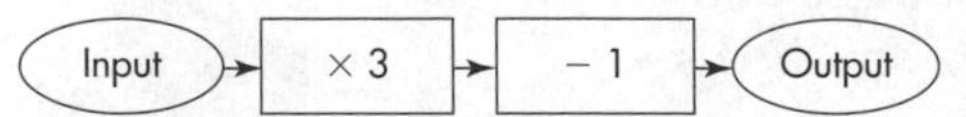

b Here is the graph of $y = 3x - 1$. Find the input value for the number machine that gives the same output value. Show clearly how you obtain your answer.

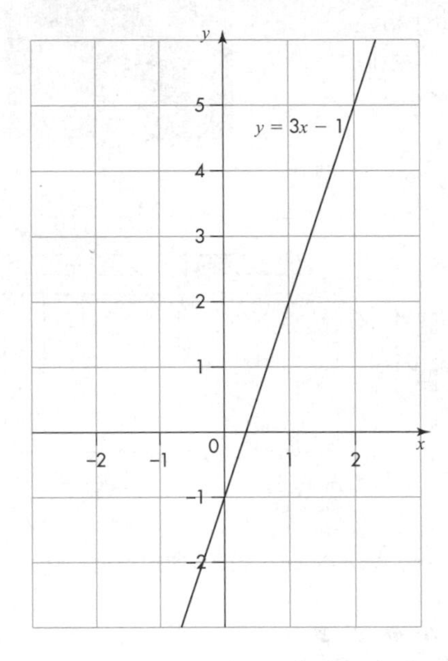

2 The middle column below shows the steps in a 'Think of a number' problem.

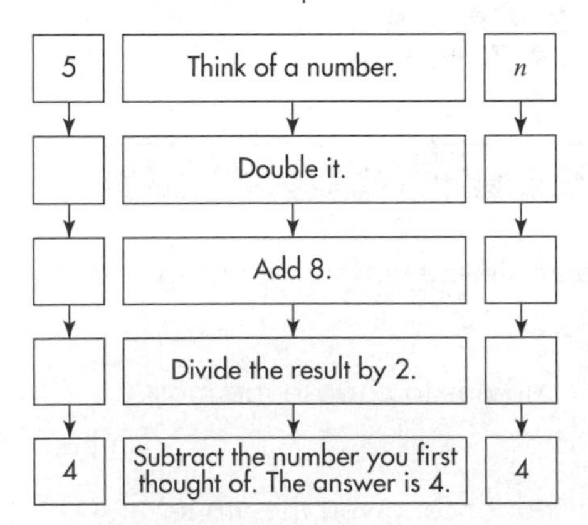

a Starting with a value of 5, work through the left-hand column, writing in all the missing numbers.

b Starting with a value of n, work through the right-hand column, writing in all the missing expressions.

3 Work out the first three terms of the sequence with an nth term of $2n^2 - n + 3$.

AU 4 The nth term of a sequence is:

$$\frac{2(n-1)(n-2)}{5}$$

Show that the third term is the first non-zero term and find its value.

5 Find the nth term of the quadratic sequence:

3, 7, 15, 27, 43, 63,…

PS 6 Mac is using small squares to make rectangle patterns.

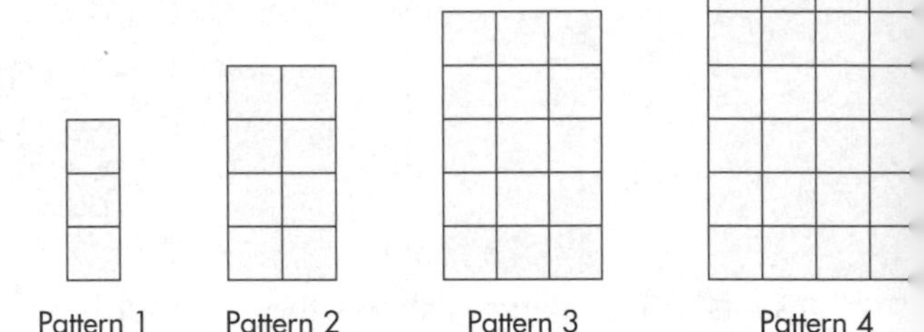

Pattern 1 Pattern 2 Pattern 3 Pattern 4

a Mac has 200 small squares. Can he make the 12th pattern?

b How many squares will there be in the 40th pattern?

7 These patterns of hexagons are formed with dots. How many dots will there be in the 20th pattern of hexagons?

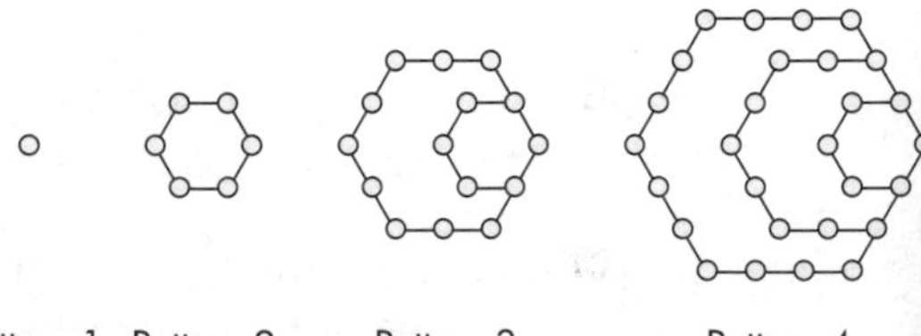

Pattern 1 Pattern 2 Pattern 3 Pattern 4

Worked Examination Questions

A **1** Work out the nth term of this quadratic sequence.

5 9 14 20 27 35 ...

First differences

Second differences

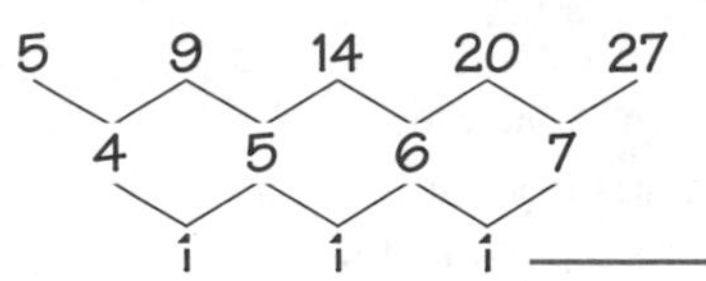

Work out the second differences. This tells you that the coefficient of n^2 is $\frac{1}{2}$ and this gains **1 mark**.

The linear sequence is

$4\frac{1}{2}$ 7 $9\frac{1}{2}$ 12 $14\frac{1}{2}$

Subtract $\frac{1}{2}n^2$ from each of the original terms. This gives a linear sequence, and earns **1 mark**.

The nth term of the linear sequence is $2\frac{1}{2}n + 2$

Write down the nth term of the linear sequence, to gain **1 mark**.

The nth term of the quadratic sequence is $\frac{1}{2}n^2 + 2\frac{1}{2}n + 2$

Put both expressions together to get the final nth term, to gain **1 mark**.

Total: 4 marks

PS **A** **2** Patterns are made with pentagons with increasing numbers of dots along each side. The diagram shows the first four patterns.

How many dots will there be in the 50th pattern?

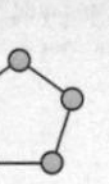

Pattern 1

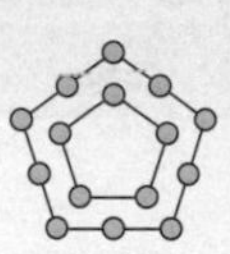

Pattern 2

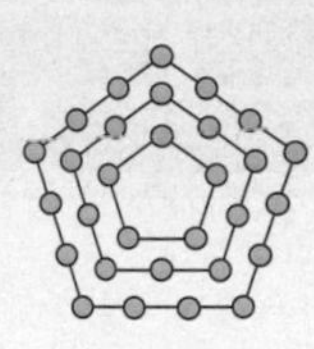

Pattern 3

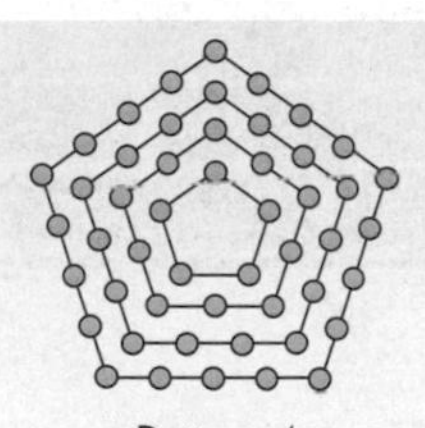

Pattern 4

The sequence of dots is

5, 15, 30, 50,...

Write down the number of dots in each pattern, to gain **1 mark**.

n	0	1	2	3	4
c	0	5	15	30	50
$a + b$	5	10	15	20	
$2a$	5	5	5		

Fill in the table, using the differences, and extend the table backwards, to gain **1 mark**.

$a = 2\frac{1}{2}$, $b = 2\frac{1}{2}$ and $c = 0$,

so the nth term is $2\frac{1}{2}n^2 + 2\frac{1}{2}n$

Write down the values of a, b and c to get the nth term, to gain **1 mark**.

$2\frac{1}{2} \times 50^2 + 2\frac{1}{2} \times 50 = 6375$

Substitute 50 into the expression for the nth term to get the number of dots in the 50th pattern, and to gain **1 mark**.

Total: 4 marks

Why this chapter matters

From the earliest of time, humankind has used simple shapes to create patterns. Some may be simple, others are exquisitely complicated.

Look around you, at walls and pavements, and you will see that bricks and paving slabs are arranged in patterns. Normally, walls of houses do not have any gaps, except for windows and doors. Pavements do not normally have any gaps except for, perhaps, a flowerbed or an inspection cover. Patterns that are continuous and do not have any gaps are called **tessellations**.

Craftsmen and artists can be very creative in producing tessellations, as in the design on the floor of St Mark's Basilica in Venice. Many tessellations and similar designs can be seen in Islamic art, such as in the Alhambra, a palace that was built by the Moors in southern Spain.

A famous artist whose work was based on tessellations was the Dutch artist M.C. Escher (1898–1972), who produced many amazing drawings after being inspired by a visit to the Alhambra. You could look up M.C. Escher on the internet to see some of his drawings.

This is a detail from the floor of a church.

Chapter

Geometry: Tessellations and tiling patterns

The grades given in this FOUNDATION chapter are target grades.

This chapter will show you ...

- **E** what a tessellation is
- **D** how to draw a simple tessellation
- **C** how to explain why a pattern tessellates

Visual overview

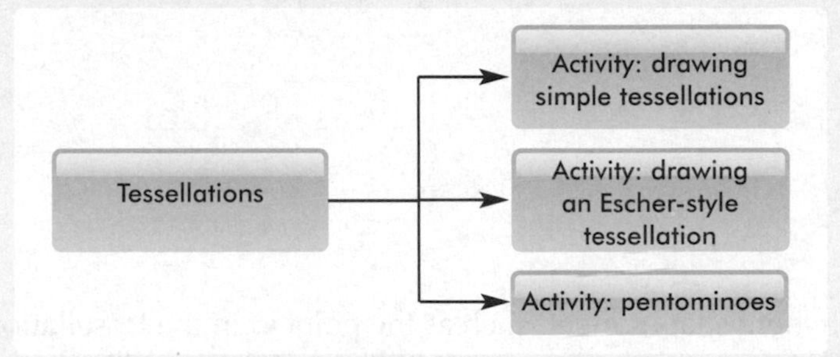

What you should already know

- That there are 360° in a full turn **(KS3 level 4, GCSE grade F)**
- The names and shapes of the standard quadrilaterals **(KS3 level 5, GCSE grade E)**
- The names of polygons up to a 12-sided polygon **(KS3 level 5, GCSE grade E)**
- How to calculate the interior angles of a regular polygon **(KS3 level 7, GCSE grade C)**

Quick check

1 Find the values of the angles marked with letters in these diagrams.

a 105°, 100°, x

b $4x$, x

2 Calculate the interior angle of a regular pentagon.

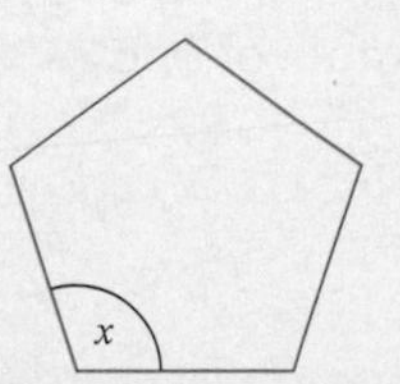

10.1 Tessellations and tiling patterns

This section will show you:
- what a tessellation is
- how to draw a tessellation
- what defines a tessellation

Key words
gap
node
plane
tessellation
vertex

A **tessellation** is a regular arrangement of shapes that completely covers the **plane**, or the page, without leaving any **gaps**.

The number of tessellations is infinite. A grid arrangement of squares is probably the simplest tessellation.

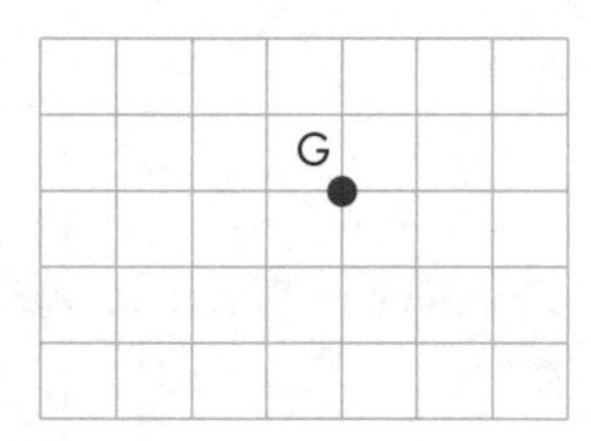

Any point where shapes meet, such as the point G in the tessellation above, is called a **node** or a **vertex**. The angles around any node in the tessellation will total 360°, so there are no gaps. In the tessellation above there are four angles of 90° at every node. In the example below, which shows part of a tessellation made up of octagons and squares, the angles at P are 135°, 135° and 90° and 135 + 135 + 90 = 360.

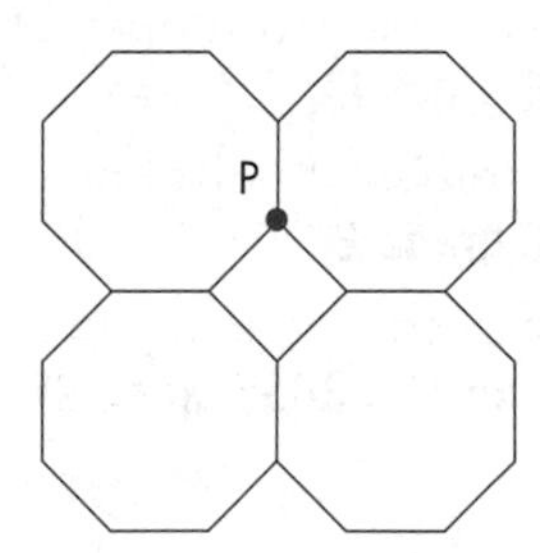

Many regular shapes, such as squares, rectangles and hexagons will tessellate with themselves. Other regular shapes will tessellate when combined with other shapes, for example, octagons will tessellate when combined with squares.

Before looking in more detail at some tessellations made from regular shapes, work through the activities that follow, to make some tessellations, including your own Escher-style pattern.

There are some shapes that will never tessellate. The circle is the best example of one of these.

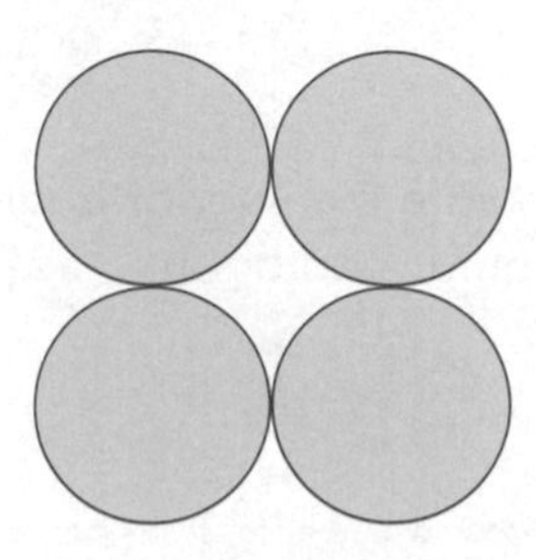

10.2 Drawing a simple tessellation

You will need:

- triangular dotty paper
- square dotty paper.

ACTIVITY A

Mark a page of triangular dotty paper with triangles, like this.

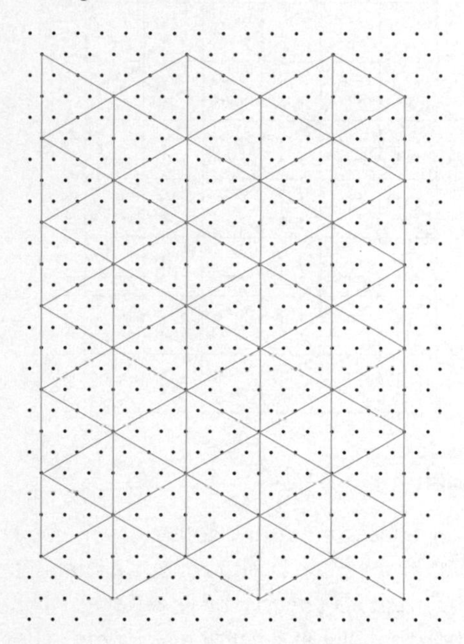

STEP 1:

Draw a curve on half of one side of a triangle.

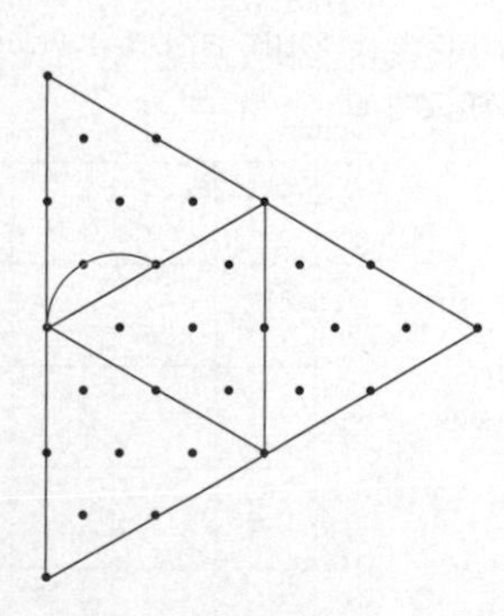

STEP 2:

Rotate this and draw it again on the other half.

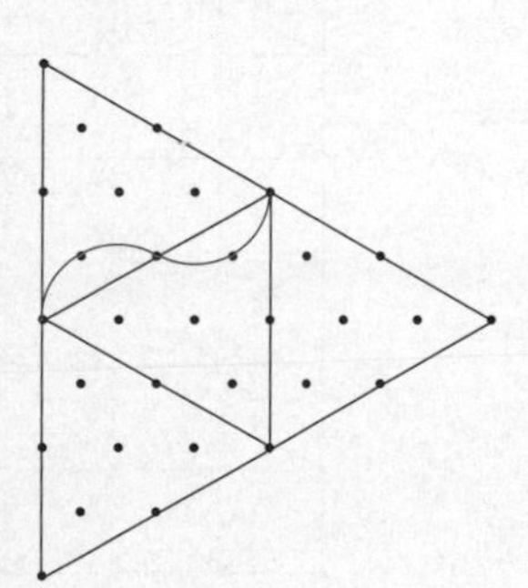

STEP 3:

Repeat this design on the other sides of the triangle.

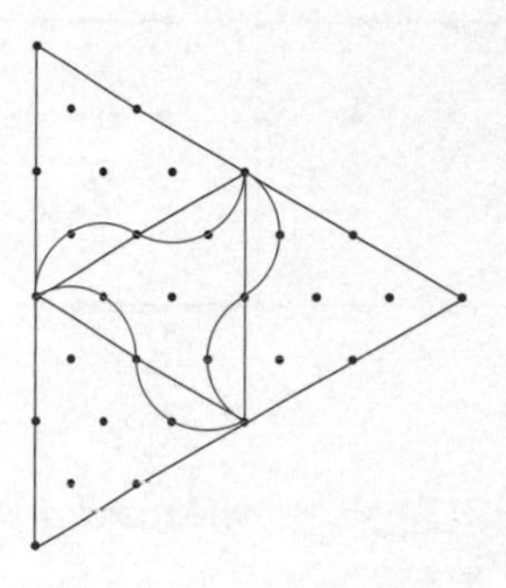

STEP 4:

Repeat the design on all the other triangles.

ACTIVITY B

Mark a page of square dotty paper into squares of side three units.

STEP 1:

Draw a design in the top left square of a group of four.

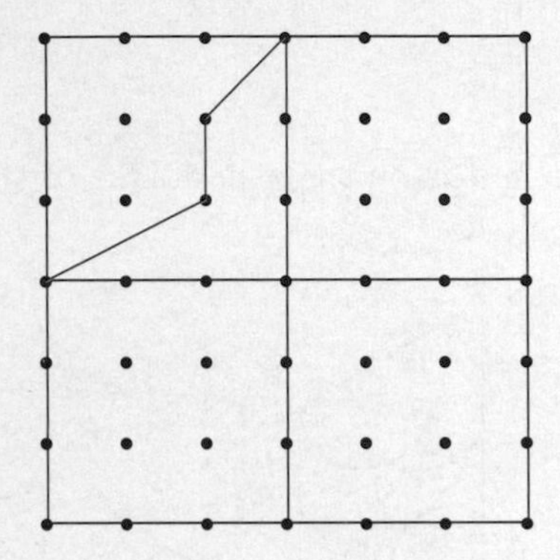

STEP 2:

Rotate this 90° anticlockwise and draw it in the next square.

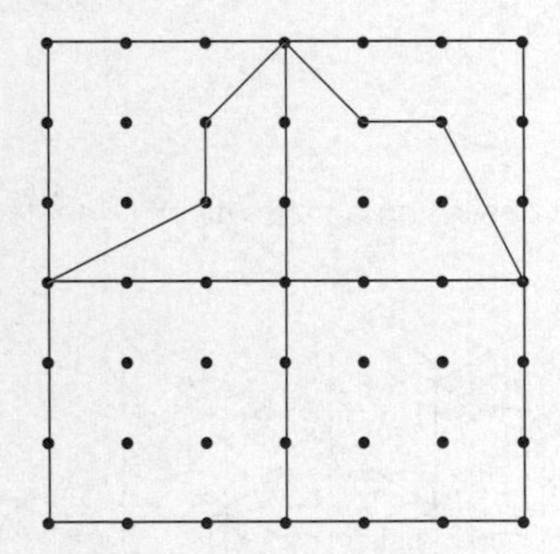

STEP 3:

Reflect this design, using the bottom of these squares as the mirror line.

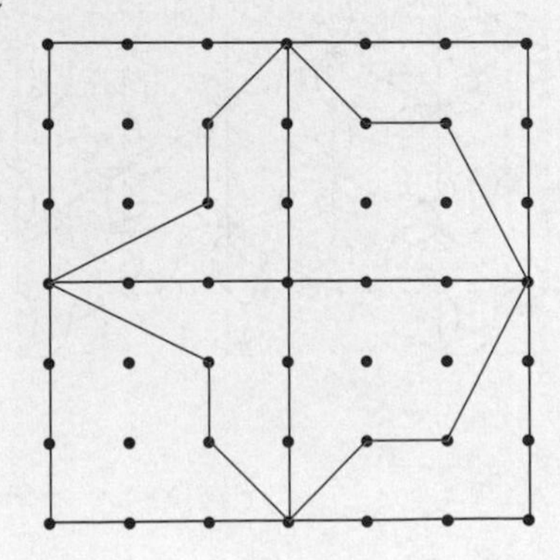

STEP 4A:

Repeat this horizontally and vertically. The same design should appear within the horizontal patterns.

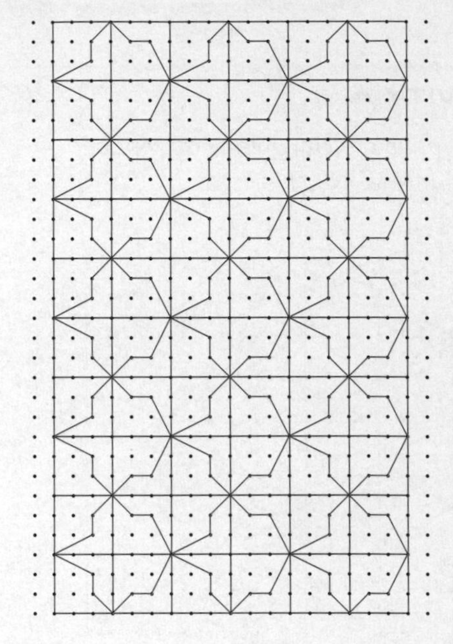

STEP 4B:

On another sheet of square dotty paper, copy your design from Step 3 and reflect it, using the right-hand edge as the mirror line. Repeat the designs horizontally and vertically. Some other shapes should appear within the horizontal patterns.

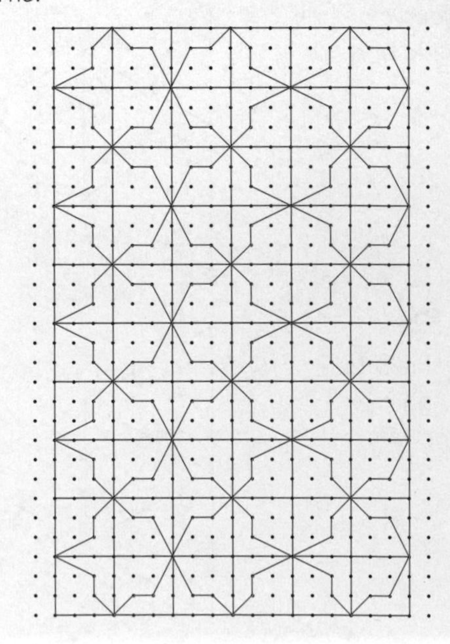

10.3 Drawing an Escher-style tessellation

Note: This activity is best done with a computer drawing program, although it can be done on paper. The instructions in italics refer to drawing facilities within Microsoft Word™.

ACTIVITY C

STEP 1:

Draw any quadrilateral, for example, a kite.

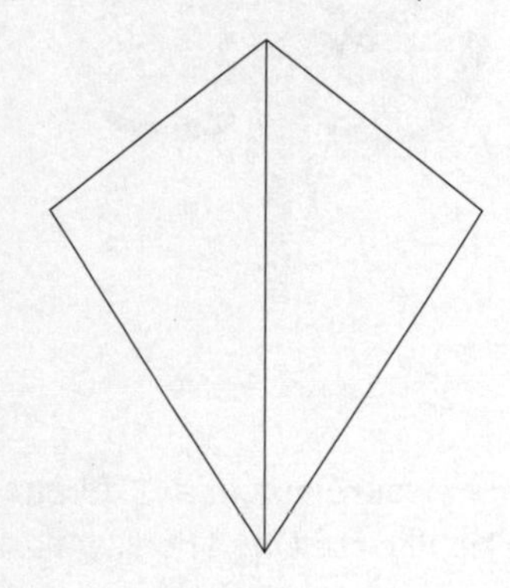

STEP 2:

Mark the midpoints of each side.

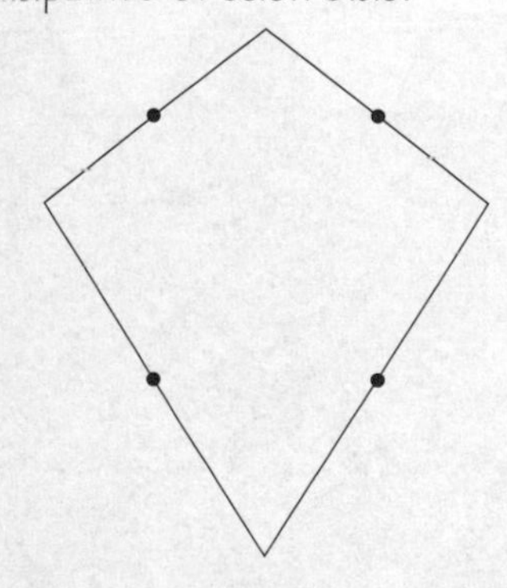

STEP 3:

Now draw any shape, such as a curve or line, on half of two adjacent unequal sides.

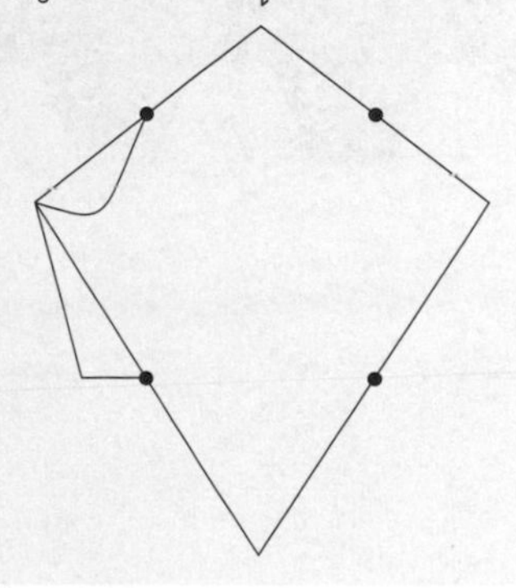

STEP 4:

Using the midpoints of the lines as centres, rotate the shapes to the other halves of the sides. [*Duplicate, Flip Vertical, Flip Horizontal*]

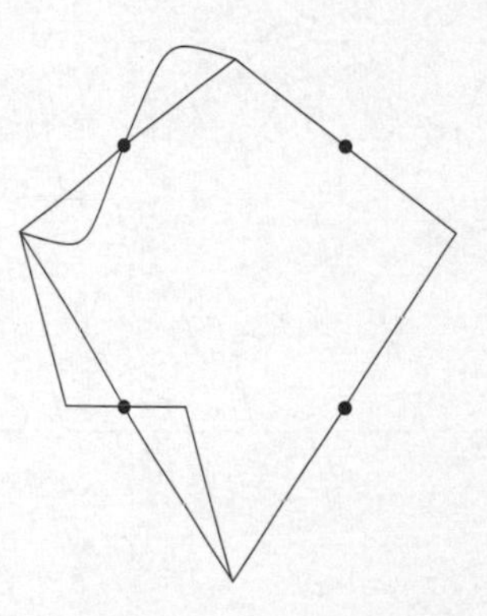

STEP 5:

Reflect these onto the other two sides. [*Group, Flip Horizontal*]

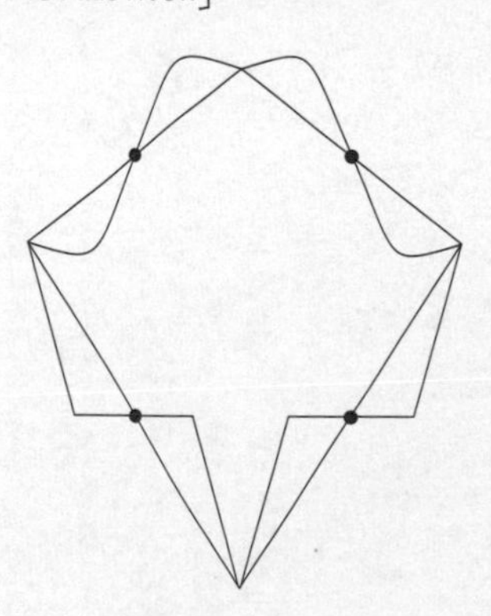

continued

STEP 6:

Remove the original kite and midpoints. [*Delete*]

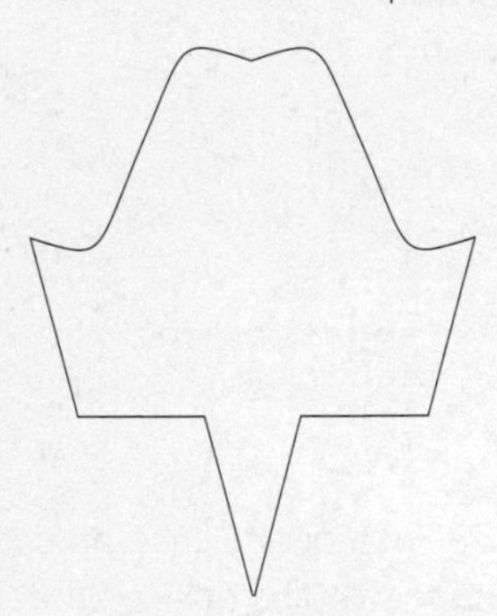

You now have a shape that will tessellate.

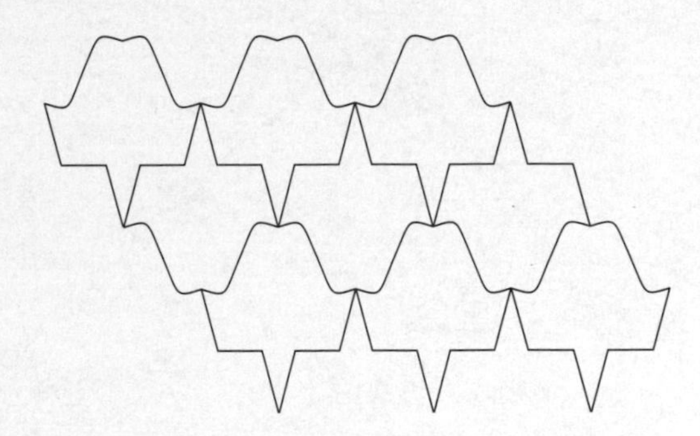

To make it into an Escher-style tessellation, draw something inside the original shape. You could use clip art from the internet or scan in a picture, for example:

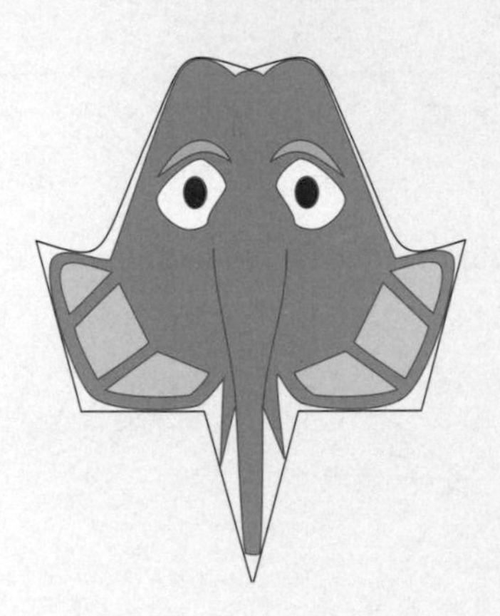

Copy the same picture but in a different colour and flip it vertically. Putting the two together gives an Escher-style tessellation.

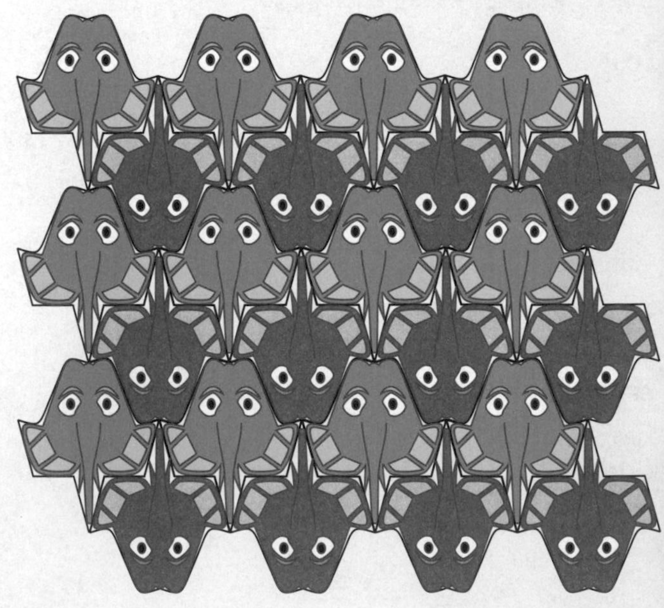

10.4 Pentominoes

Note: This is best done as a group activity. You will need lots of squared paper.

Pentominoes are shapes made from five squares.

There are 12 different pentomino shapes, assuming that rotations and reflections count as the same as the original. Six of the 12 pentominoes are shown below. Each of these pentominoes has been given a letter that closely resembles its shape.

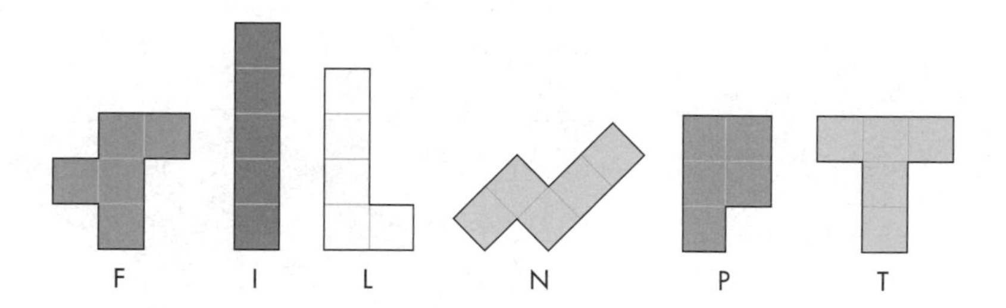

ACTIVITY D

Find the other six pentomino shapes.

Hint: The letters of the other six are U, V, W, X, Y and Z.

ACTIVITY E

Draw a poster showing which of the pentominoes have line and/or rotational symmetry.

ACTIVITY F

All of the pentominoes will tessellate with themselves. Draw diagrams to show how each pentomino will tessellate.

ACTIVITY G

The 12 pentominoes have a total area of 60 squares.

a There are 2339 ways of fitting the 12 pentominoes into a 6×10 rectangle. Can you find just one of these? Draw and colour in your solution.

b There are 368 ways of fitting the 12 pentominoes into a 4×15 rectangle. Can you find just one of these? Draw and colour in your solution.

c There are just two ways of fitting the 12 pentominoes into a 3×20 rectangle. Can you find just one of these? Draw and colour in your solution.

10.5 Tessellations

This section will show you:
- recognise and draw tessellations

Key words
polygon
tessellate

Shapes that tessellate with themselves

Triangles

All triangles will **tessellate**.

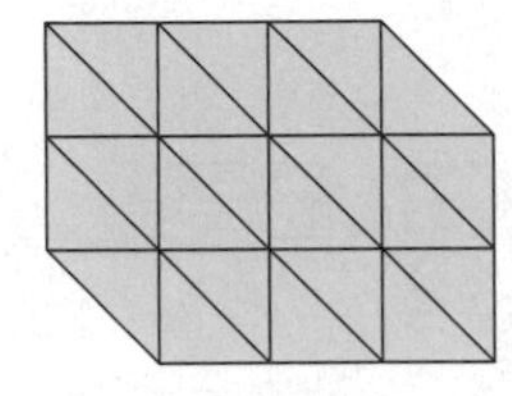

Right-angled

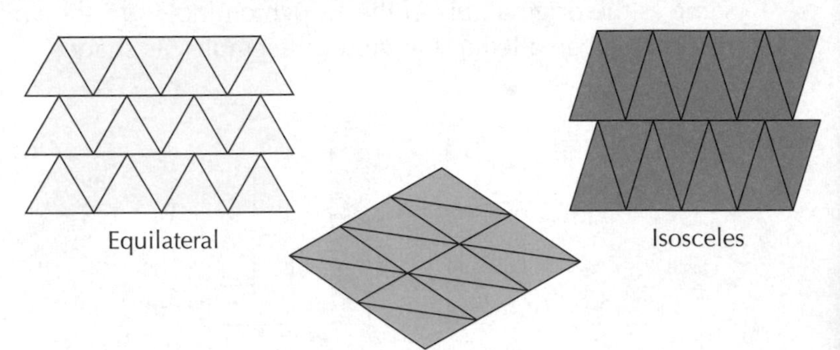

Equilateral

Scalene

Isosceles

Quadrilaterals

All quadrilaterals will tessellate.

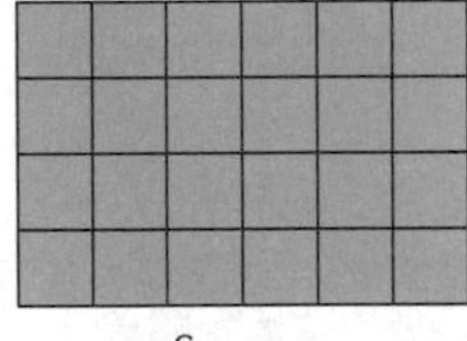

Square

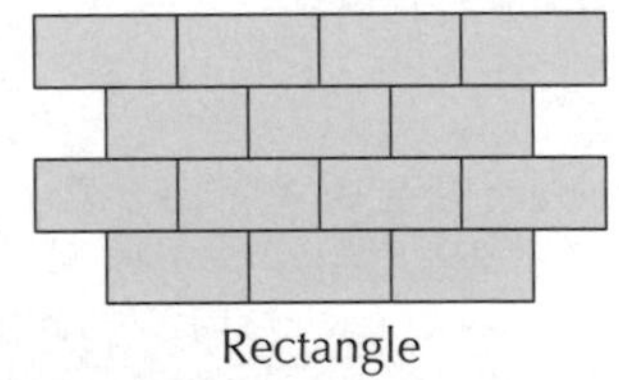

Rectangle

Parallelogram

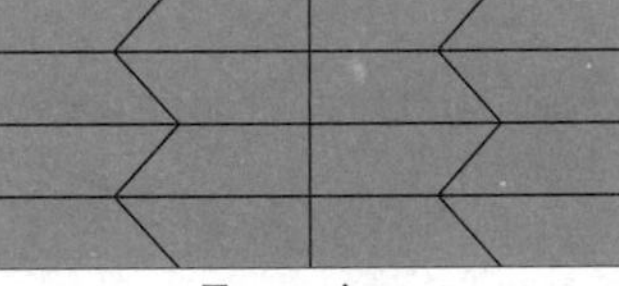

Trapezium

Rhombus

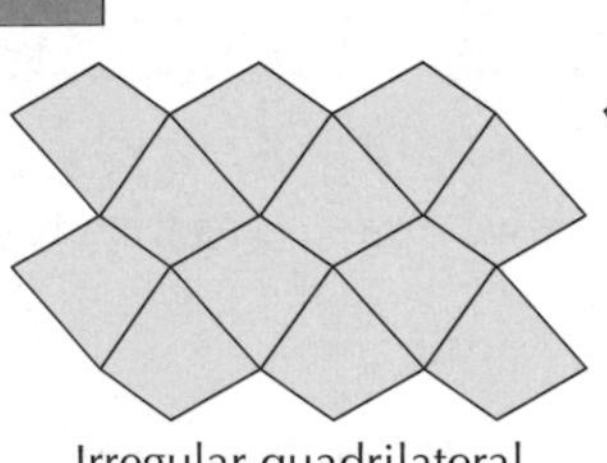

Irregular quadrilateral

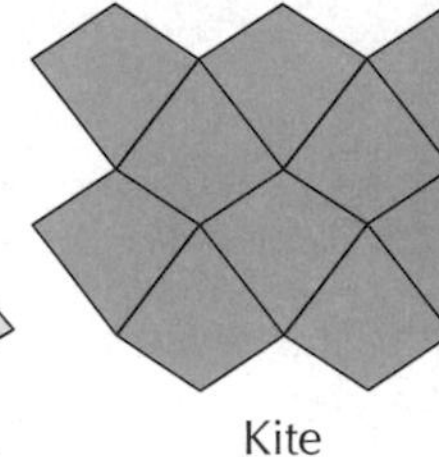

Kite

Other polygons

Other **polygons** will tessellate, depending on how they are drawn. For example:

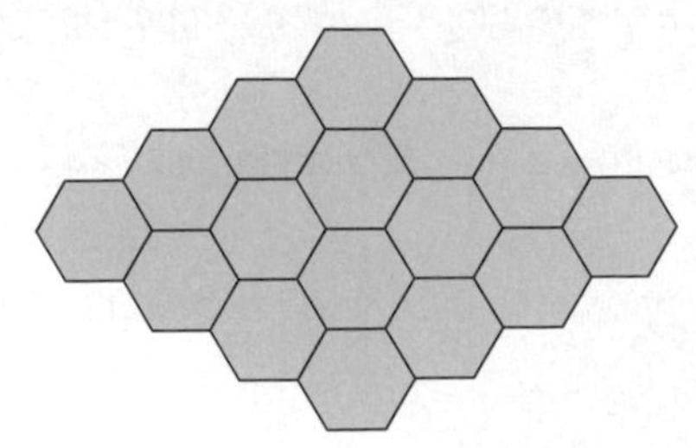

Regular hexagon

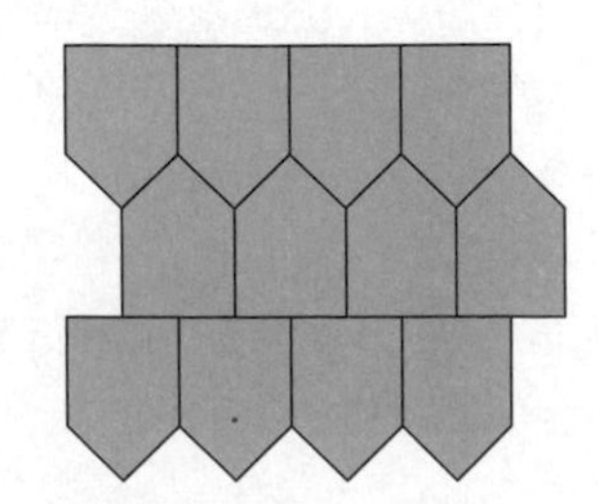

Pentagon (house shape)

Shapes that tessellate with other shapes

Obviously you can draw any shapes that leave no gaps but we will just look at some regular shapes.

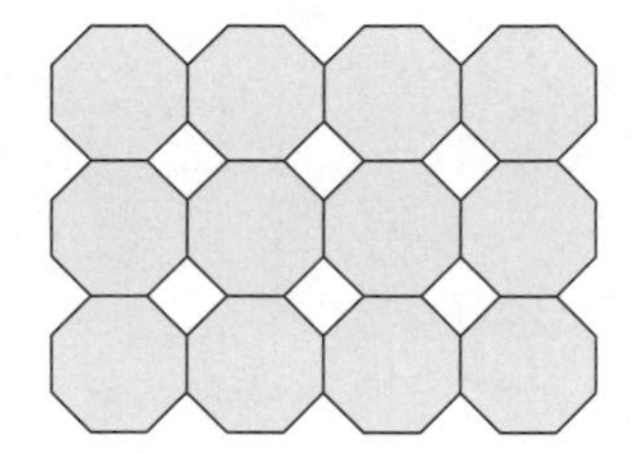

Regular octagons with squares

Different rhombi

Remember: There are only two rules to think about when dealing with tessellations.

1. A tessellation is a repeating pattern of shapes that covers the plane and leaves no gaps.
2. At any point where the vertices of shapes meet, the total of the internal angles of the shapes will be 360°.

EXAMPLE 1

Tetrominoes are shapes made of four squares.
There are five different tetromino shapes.

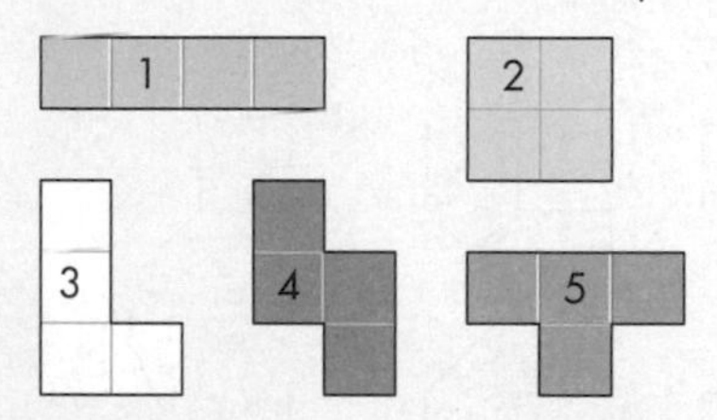

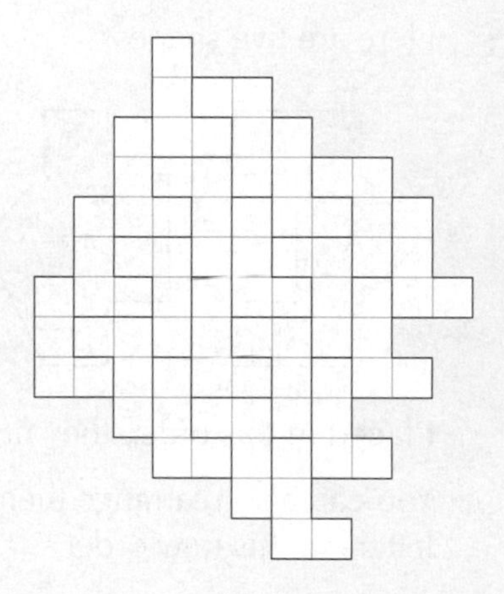

Each of these shapes will tessellate; for example, on the right is a tessellation of shape 3.

On a grid, show how shape 4 will tessellate.
Draw at least six shapes to make your pattern clear.

SOLUTION

This is just one solution.
There are many others.

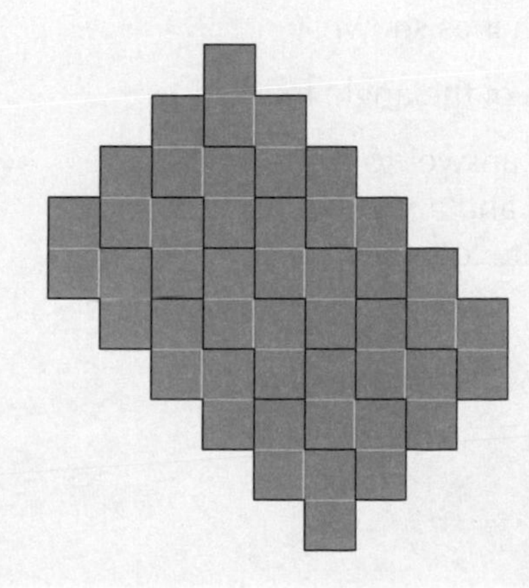

EXAMPLE 2

a Explain why a regular octagon will not tessellate.

b A tessellation of regular dodecagons and equilateral triangles is shown opposite.

By referring to the angles at point P, explain why the shapes tessellate.

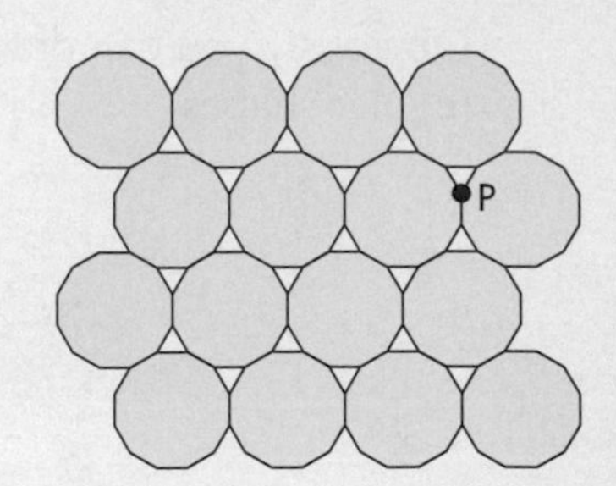

SOLUTION

a The interior angle of a regular octagon is 135°. This does not divide exactly into 360°.

b The interior angle of a dodecagon is 150°.

The interior angle of an equilateral triangle is 60°.

There are two dodecagons and one equilateral triangle at P, so:

$150° + 150° + 60° = 360°$

EXERCISE 10A

E

PS 1 Here are five shapes.

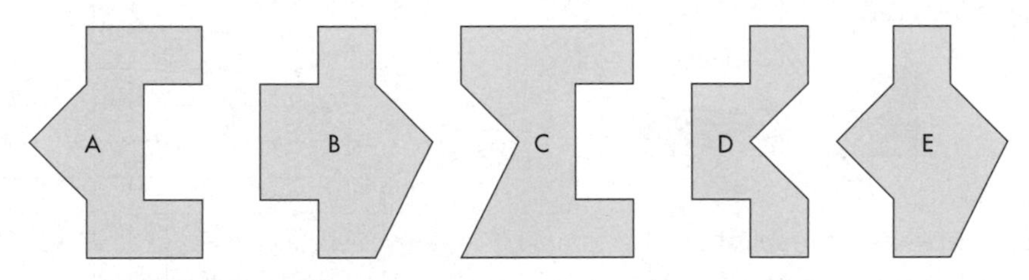

Placed in this order, they fit together with no gaps.

You can also rearrange them so that they fit together in a different order. Write down the letters in this new order.

D

2 A regular hexagon and a regular octagon have sides the same length.

They are placed together as shown.

a Work out the size of the angle marked x.

b Explain why your answer to **a** shows that a regular octagon and a regular hexagon will never tessellate together.

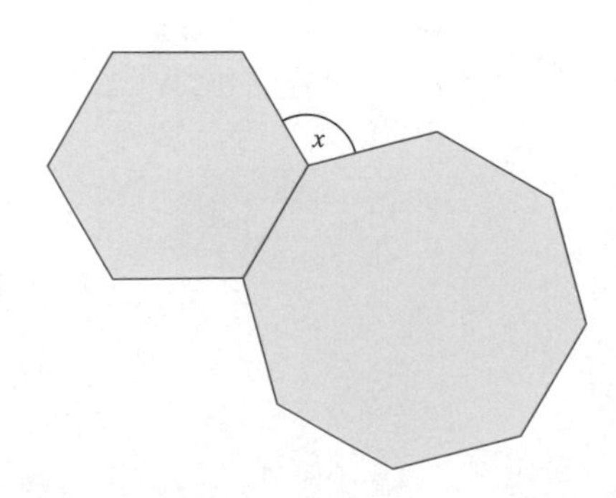

3 The diagram shows a tessellation.

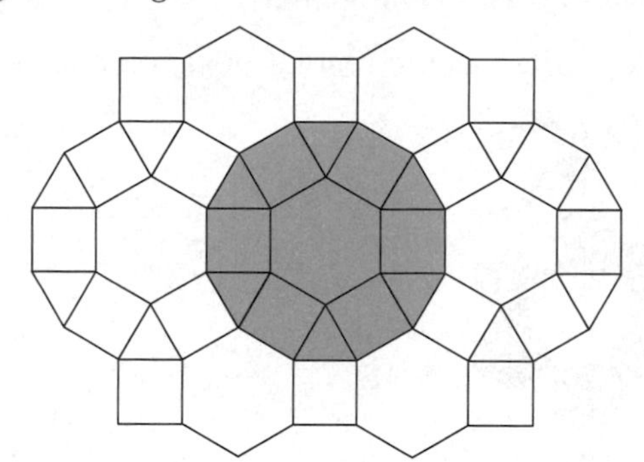

The shaded area contains all the regular polygons that make up the tessellation.

a One of the polygons is a regular hexagon.

Name the other two regular polygons that make up the tessellation.

b Use a vertex of the tessellation to explain why the interior angle of a regular hexagon is 120°.

4 **a** Show, by means of a diagram, why isosceles triangles tessellate.

b Show, by means of a diagram, why squares tessellate.

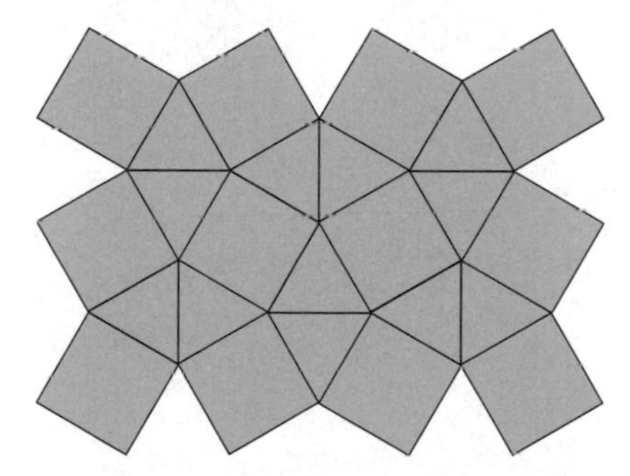

c The diagram above shows a tessellation of equilateral triangles and squares. At any vertex, three equilateral triangles and two squares meet. By reference to the angles around the vertex, explain why this is a tessellation.

5

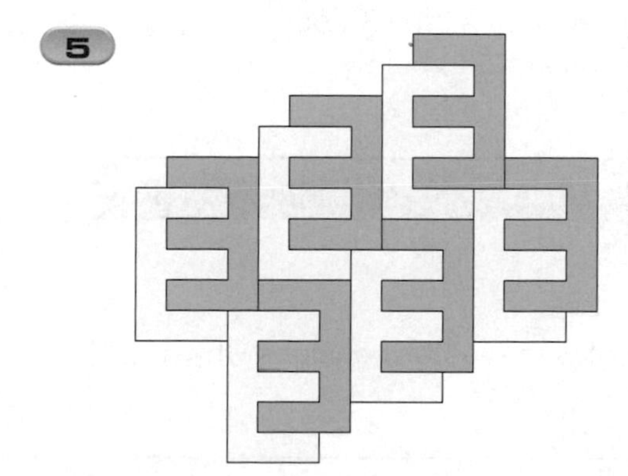

The diagram shows a tessellation of an E-shape. Some of the shapes have been shaded to emphasise the pattern.

On a grid, show how an F-shape (below) will tessellate. Draw at least six shapes to show the pattern clearly. One F has been drawn.

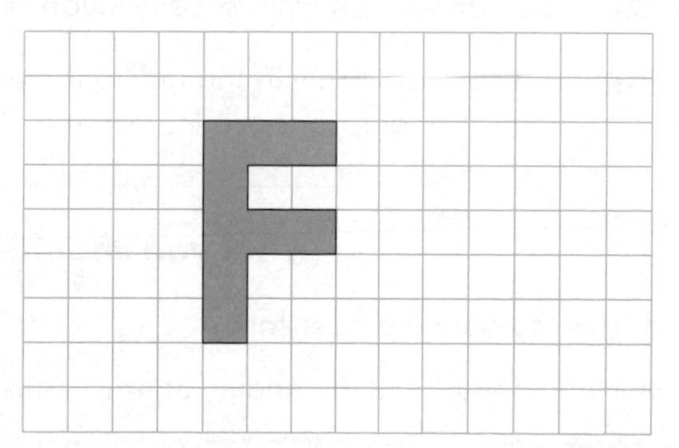

C

6 **a** Explain why regular pentagons will not tessellate together.

b This tessellation is made from a six-pointed star, a regular hexagon and pentagons.

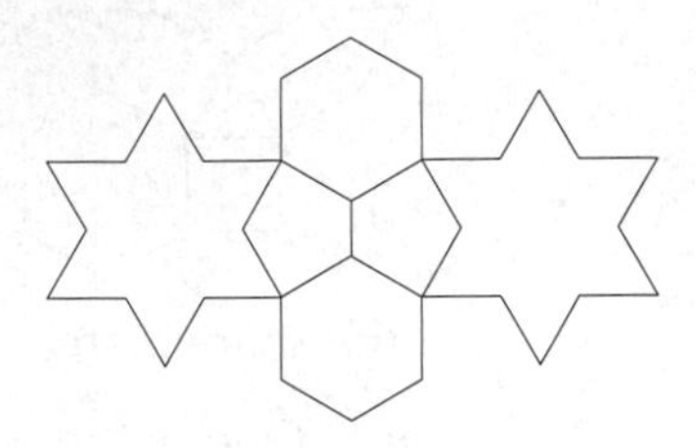

The diagram (above, right) shows part of the tessellation. Explain clearly how you know that the pentagon is not regular.

PS 7 The tessellation shown (right) is formed by three rhombuses, two of which are congruent.

Two angles of x and $3x$ are marked on the diagram. Calculate the value of x.

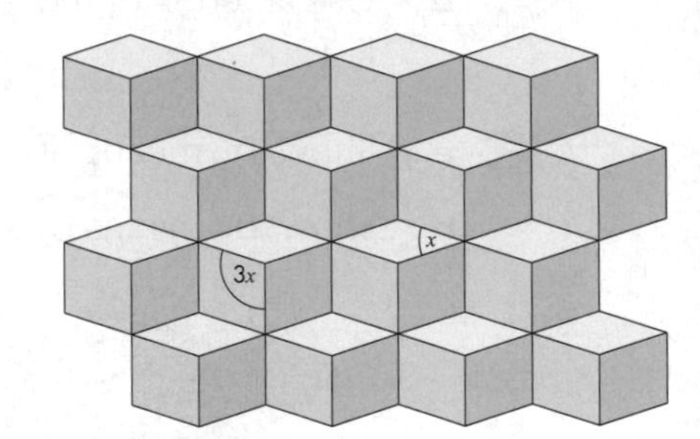

AU 8 Shapes are made by joining points on a 3 by 3 dotty grid, for example:

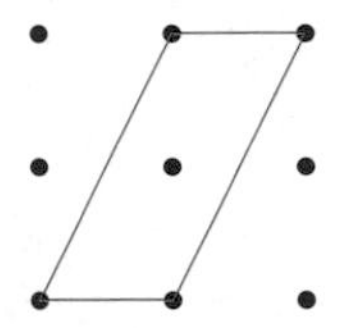

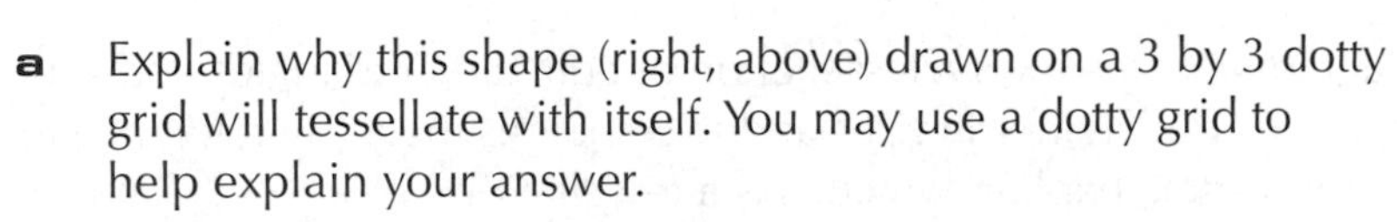

a Explain why this shape (right, above) drawn on a 3 by 3 dotty grid will tessellate with itself. You may use a dotty grid to help explain your answer.

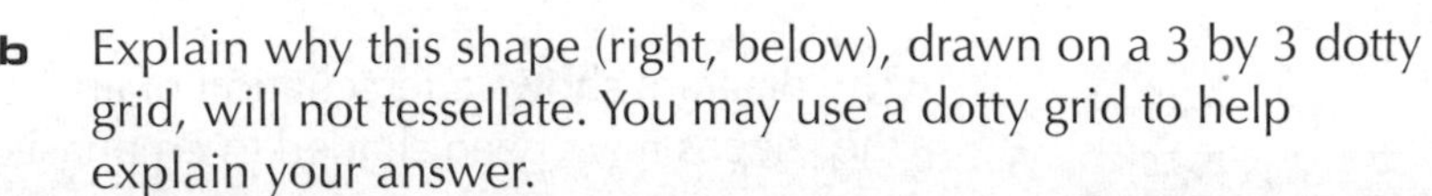

b Explain why this shape (right, below), drawn on a 3 by 3 dotty grid, will not tessellate. You may use a dotty grid to help explain your answer.

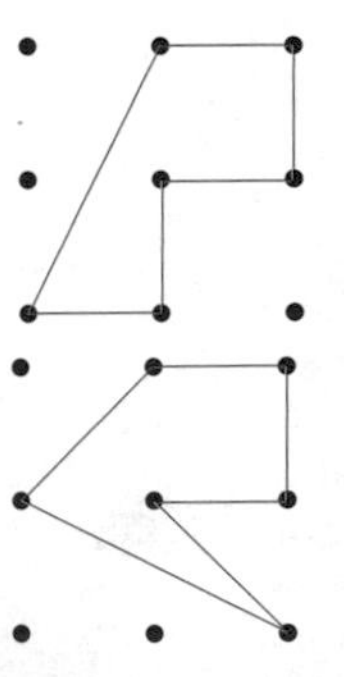

GRADE BOOSTER

D You can explain what a tessellation is

C You can explain by reference to the angles in regular polygons why a pattern of polygons is a tessellation

What you should know now

- How to define a tessellation
- How to work out the angles at any vertex of a tessellation

PS **1** Five shapes fit together in a straight line in the order A, B, C, D, E.

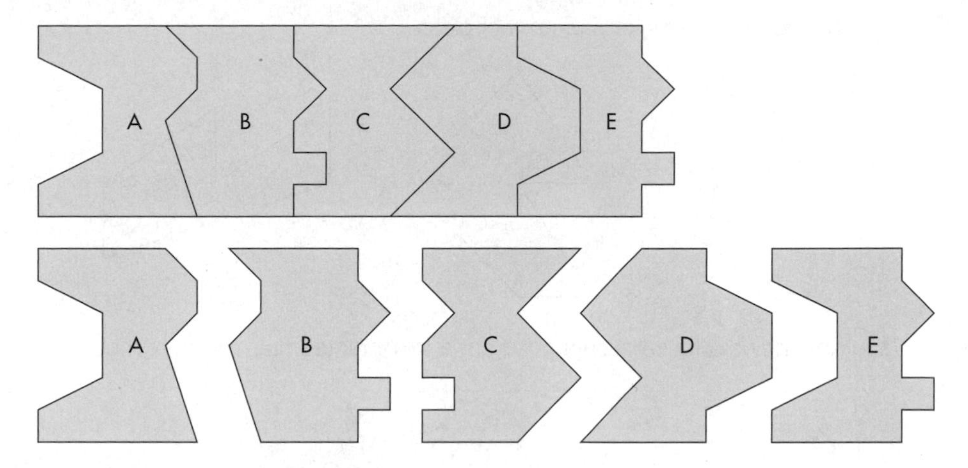

Find a different order in which the five shapes will fit together in a straight line.

2 The diagram shows a tessellation of a regular dodecagon and two other regular polygons.

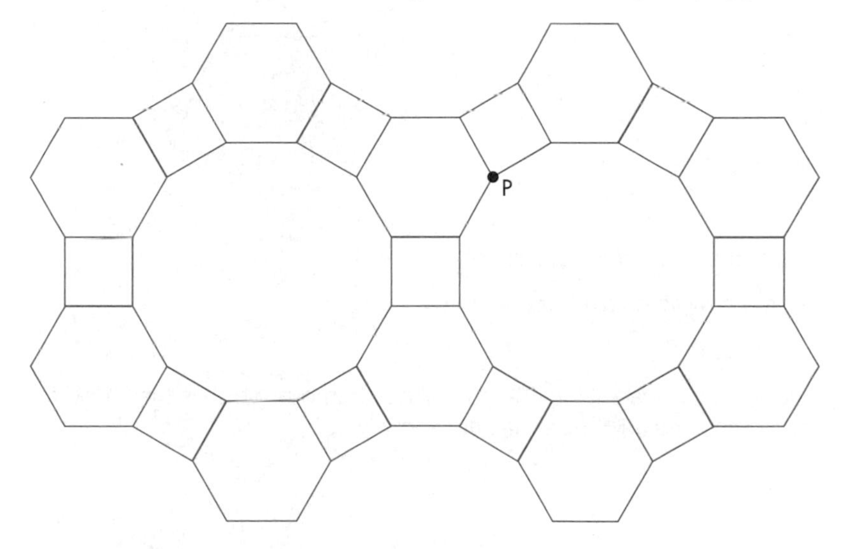

a What other two regular polygons are used in this tessellation?

b Use the angles at the point P to work out the internal angle of a dodecagon.

3 Show, by means of a diagram, that a kite will tessellate.

Draw at least six kites to make your answer clear.

PS 4 Show, by means of a diagram, why these two shapes will tessellate.

Draw enough shapes to make your answer clear.

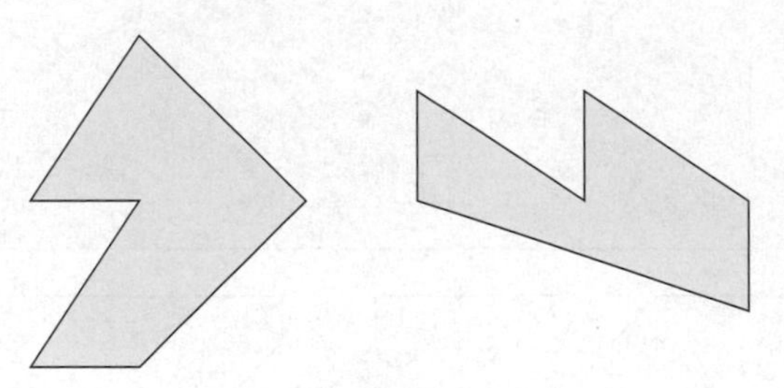

AU 5 The diagram shows a tessellation made from a four-pointed star and a rhombus.

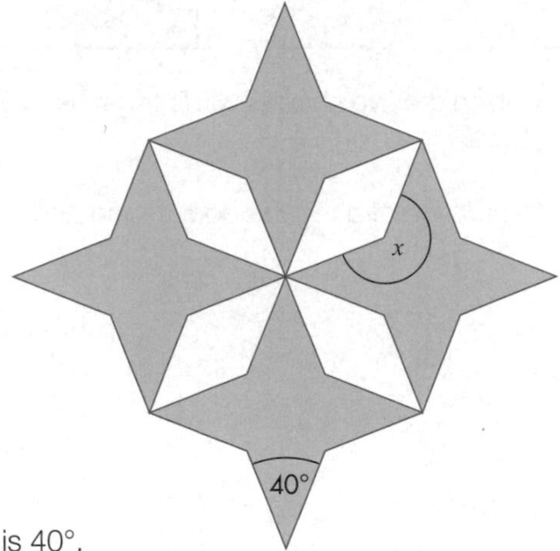

The angle at the point of the star is 40°.

Work out the size of the angle marked x.

AU 6 A kite and an arrowhead fit together to make a rhombus which is then used to form a tessellation. Some of the angles of the kite and arrowhead are shown.

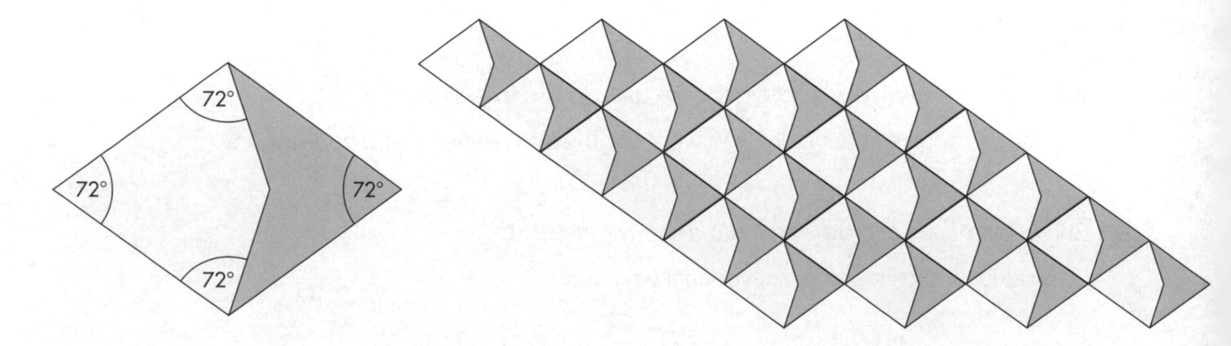

Two kites and an arrowhead can be put together to form a quadrilateral which is then used to form a tessellation.

By reference to the angles at any vertex explain why this quadrilateral will tessellate.

Worked Examination Questions

D **1 a** Explain what is meant by a tessellation.

b The diagram shows a tessellation.

i Six congruent shapes surround point X. What is the special name of this shape?

ii Explain, with reference to the angles at point Y, why these shapes tessellate.

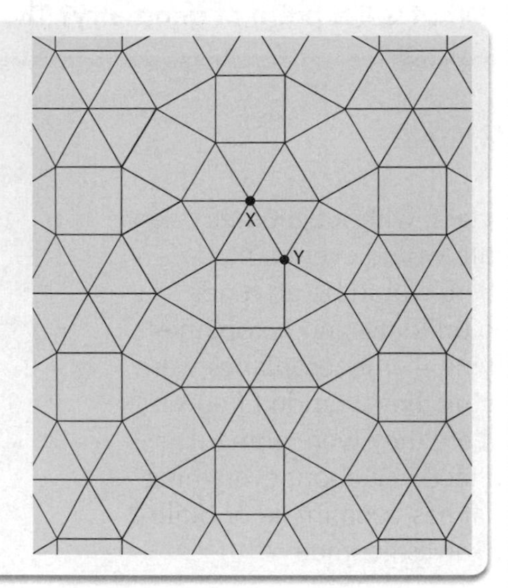

a A tessellation is a regular pattern of shapes that covers the plane without leaving any gaps.

You should learn the definition of a tessellation as this question is likely to be asked frequently. A clear explanation will gain **1 mark**.

b i Equilateral triangle

'Triangle' on its own would not earn the mark. You must say it is an equilateral triangle, to gain **1 mark**.

ii The shapes are two squares and three equilateral triangles so the angles are $90° + 90° + 60° + 60° + 60° = 360°$ so the shapes fit without any gaps.

This explanation gains another **2 marks**

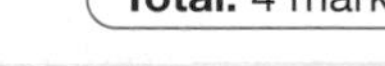
Total: 4 marks

PS C **2** The diagram shows part of a tessellation of an isosceles trapezium.

a Write down the size of the angle x shown in the diagram.

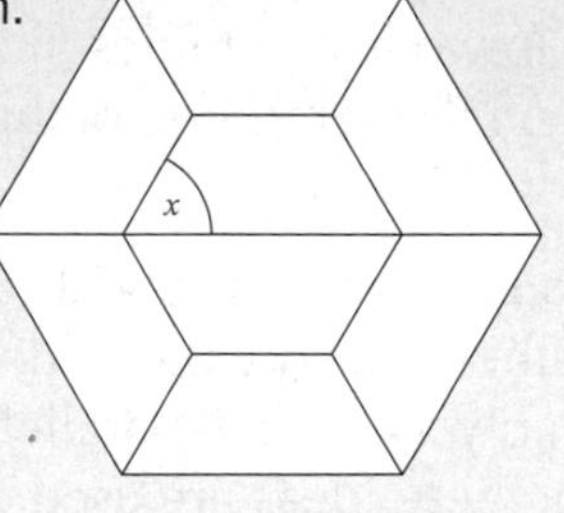

b The two parallel sides of the trapezium are 5 cm and 10 cm.

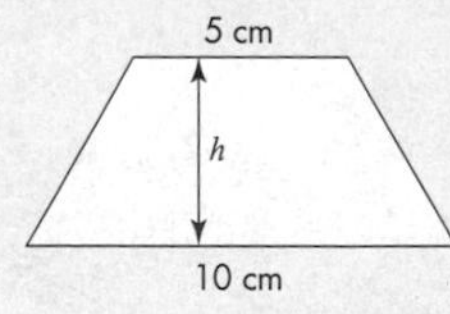

The total area of the part of the tessellation shown is 260 cm^2, to the nearest whole number. Work out the height h of the trapezium.

a 60°

You should realise that the sides of the trapezium make a regular hexagon so the angle is half of 120°. You will gain **1 mark** for this answer.

b Area of one trapezium

$= 260 \div 8$

$= 32.5 \text{ cm}^2$

You will gain **1 method mark** for working out the area of one trapezium.

$\frac{1}{2}h(10 + 5) = 32.5$

You will gain **1 method mark** for putting the area and the sides into the formula for the area of a trapezium.

$h = 4.333...$

You will gain **1 mark** for the answer.

Total: 4 marks

Why this chapter matters

What is the point of geometry? The answer is – where two lines meet!

In fact, without geometry very little would ever be made. There would be no roads, no buildings, no aeroplanes flying – or if aeroplanes were flying they wouldn't know where they were going. In fact, just about everything that has been made or built involves geometry.

Nobody knows when geometry was first used but it was first recorded in about 3000BC to meet the need for formalised methods in construction. Most of this work was practical and concerned lengths, areas and volumes. In the third century BC, Euclid wrote the first book on geometry. Much of the geometry we do in school today is still based on Euclid's elements. Here are the five most important of Euclid's statements.

- Any two points can determine a straight line.
- Any finite straight line can be extended in either direction.
- A circle can be determined from any centre with any radius.
- All right angles are equal.
- If two straight lines in a plane are crossed by a transversal, and the sum of the interior angles on the same side of the transversal is less than two right angles, then the two lines, if extended will intersect.

Doubtless you know all of these results, as they are very basic ideas. The last one simply states that parallel lines never meet, but these ideas are used as the basis for geometrical proofs and theorems that have, over the centuries, led to some very important results which have enabled humankind to explore space and understand more about the universe.

Chapter

Geometry

The grades given in this HIGHER chapter are target grades.

This chapter will show you ...

- **A** how to prove the congruency of triangles
- **A** how to prove and use the intercept theorem
- **A** how to prove and use the midpoint theorem
- **A** how to prove and use the intersecting chords theorem

Visual overview

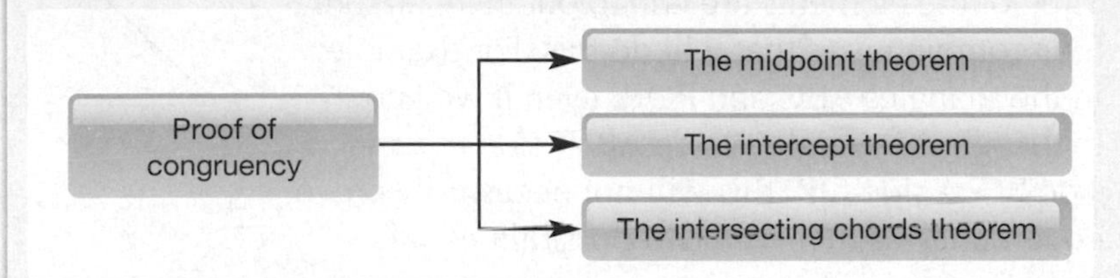

What you should already know

- How to divide a quantity into a given ratio **(KS3 level 7, GCSE grade C)**
- The properties of radii and tangents of circles **(KS3 level 8, GCSE grade B)**
- The circle theorems **(KS3 level 8, GCSE grade B)**
- The alternate segment theorem **(GCSE grade A)**

Quick check

1 Divide 12 in the ratio 1 : 4.

2 PA and PB are two tangents to a circle, centre O.

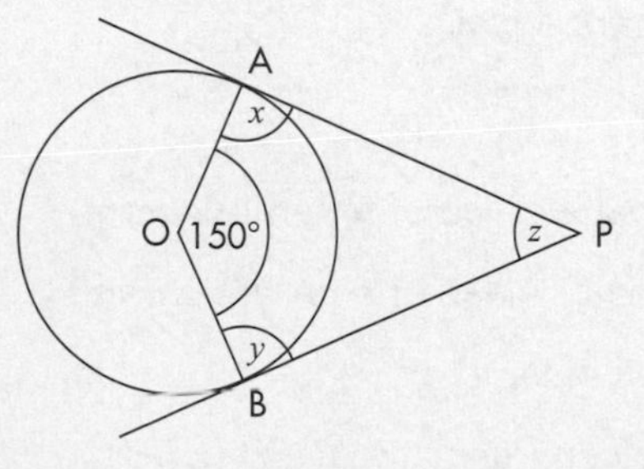

Angle AOB = 150°.

Work out and write down the value of:

a x **b** y **c** z

11.1 Proof of congruency

This section will show you how to:
- prove that two triangles are congruent

Key words
congruent
proof
prove

Congruent triangles are triangles that are *geometrically equivalent*. That means that all three corresponding sides are equal and all three corresponding angles are equal. These are the six *elements* of the triangle. To solve a triangle, using trigonometry or the sine and cosine rules, we need a minimum of three elements. That means any combination of three of the six elements gives enough information to find the other three elements, with one exception – when we only know three angles.

To **prove** congruency we also need to show that three of the elements are equal, but there are only four combinations that will do this. For example, in the triangles ABC and PQR, even if we know that angle A = angle P, side AB = side PQ and side BC = side QR, this will not necessarily prove congruency as there are two possibilities.

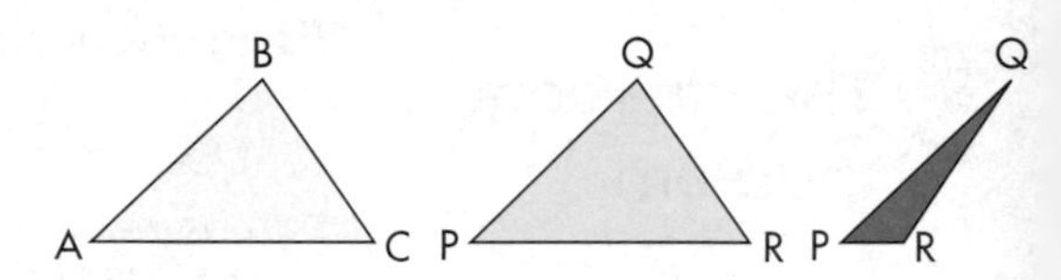

There are two possible ways that the three elements could be arranged in triangle PQR. This is known as angle, side, side, abbreviated to ASS. It is easy to remember that this is not a condition for congruency.

The four conditions are described below.

1 Side, side, side (SSS)

Two triangles are congruent if we can prove that all three pairs of corresponding sides are equal.

EXAMPLE 1

PQRS is a parallelogram. Prove that triangles PQS and QRS are equal.

SOLUTION

PQ = SR (opposite sides of a parallelogram)

PS = QR (opposite sides of a parallelogram)

SQ is a common side.

Hence ΔPQS is congruent to ΔQRS (SSS).

Note: Always give a reason for each statement you make, even if the information is given in the question or is obvious. **Proofs** have to be rigorous and mathematically precise. Notice also that you can use the symbols ∠ for angle and Δ for triangle.

2 Side, angle, side (SAS)

Two triangles are congruent if we can prove that two corresponding sides and the included corresponding angle are equal. Note that the angle must be the one between the two sides.

EXAMPLE 2

PQRS is a kite. Prove that triangles PQS and QRS are equal.

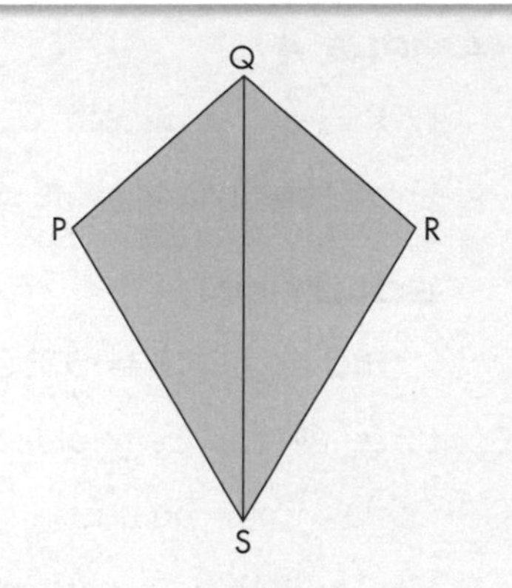

SOLUTION

PQ = QR (short sides of a kite)

PS = RS (long sides of a kite)

$\angle QPS = \angle QRS$ (opposite angles of a kite)

Hence ΔPQS is congruent to ΔQRS (SAS).

Note: We could have proved this congruency by SSS. Very often there are alternative ways of showing congruency for the standard quadrilaterals.

3 Angle, side, angle or Angle, angle, side (ASA or AAS)

Two triangles are congruent if all three pairs of corresponding angles are equal and one pair of **corresponding** sides are equal. Because of the nature of triangles (the sum of the angles of a triangle is 180°), we only have to prove AA (angle, angle) to prove AAA (angle, angle, angle); thus, we can rewrite this condition as AAS (angle, angle, side) or ASA (angle, side, angle).

EXAMPLE 3

In the diagram, AB and CD are parallel and AB = CD.

The lines AC and BD intersect at E.

Prove that triangles ABE and CDE are congruent.

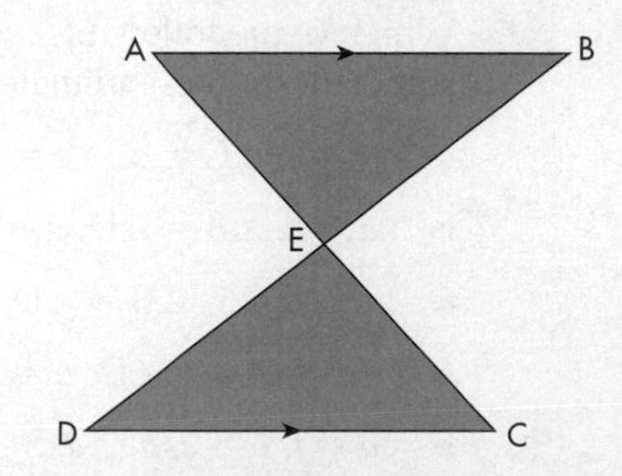

SOLUTION

$\angle EAB = \angle ECD$ (alternate angles)

$\angle EBA = \angle EDC$ (alternate angles)

AB = CD (given)

Hence ΔABE is congruent to ΔDEC (ASA).

Note: An alternative proof would be:

$\angle EAB = \angle ECD$ (alternate angles)

$\angle AEB = \angle DEC$ (opposite angles)

AB = CD (given)

Hence ΔABE is congruent to ΔDEC (AAS).

4 Right angle, hypotenuse, side (RHS)

This particular condition is specific to right-angled triangles. If the hypotenuses and a pair of corresponding short sides are equal, the two right-angled triangles are congruent.

EXAMPLE 4

PQRS is a rectangle.

Prove that triangles PQS and QRS are equal.

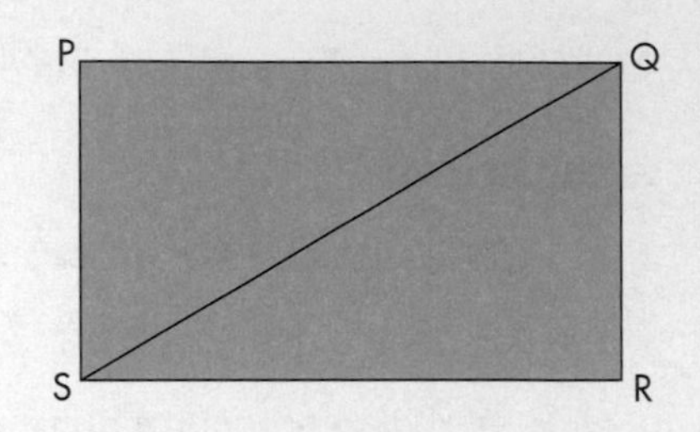

SOLUTION

$\angle SPQ = \angle QRS = 90°$ (angles in a rectangle)

$PS = QR$ (opposite sides of a rectangle)

SQ is a common side.

Hence ΔPQS is congruent to ΔQRS (RHS).

EXERCISE 11A

AU 1 **a** Are all equilateral triangles congruent? Explain your answer.

b Are all right-angled triangles congruent? Explain your answer.

c Are all right-angled triangles with a hypotenuse of 10 cm congruent? Explain your answer.

d Are all right-angled triangles with a hypotenuse of 10 cm and one angle of 60° congruent? Explain your answer.

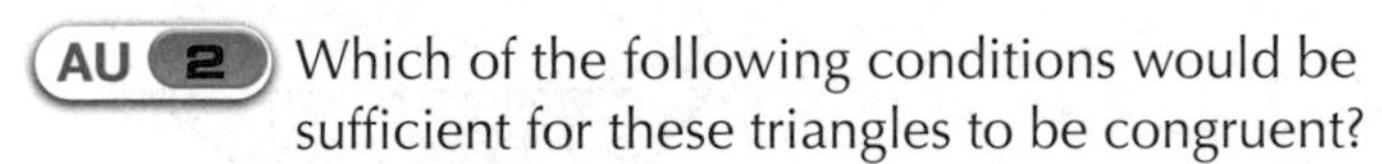

AU 2 Which of the following conditions would be sufficient for these triangles to be congruent?

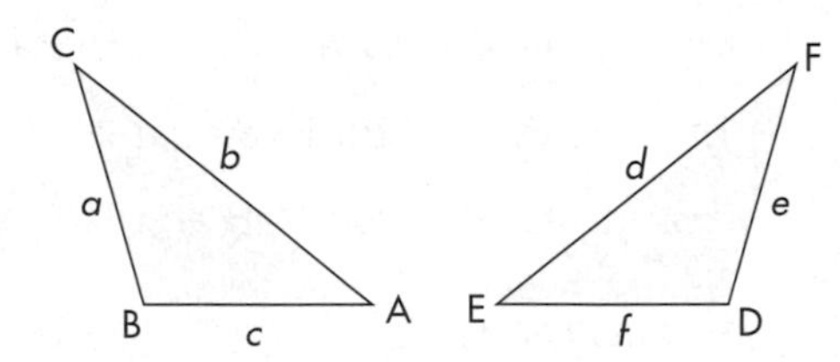

a $a = e$, $\angle C = \angle F$, $c = f$

b $a = e$, $\angle B = \angle D$, $\angle A = \angle E$

c $\angle C = \angle F$, $\angle B = \angle D$, $\angle A = \angle E$

d $c = f$, $\angle B = \angle D$, $a = e$

e $a = f$, $c = e$, $b = d$

3 ABCD is an isosceles trapezium. Prove that triangles ACD and BDC are congruent.

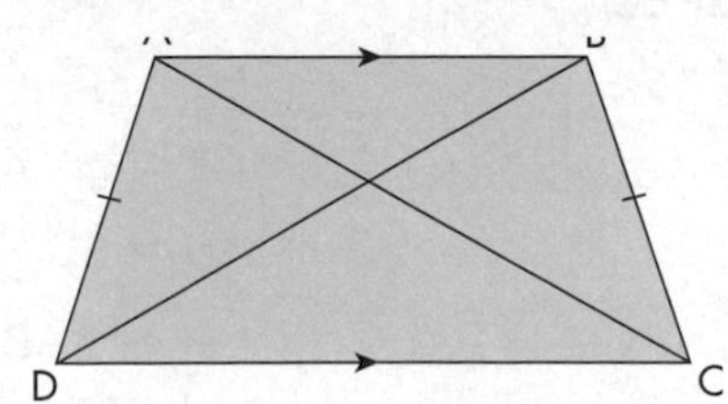

4 PA and PB are tangents to a circle, centre O. Prove that triangles PBO and PAO are congruent.

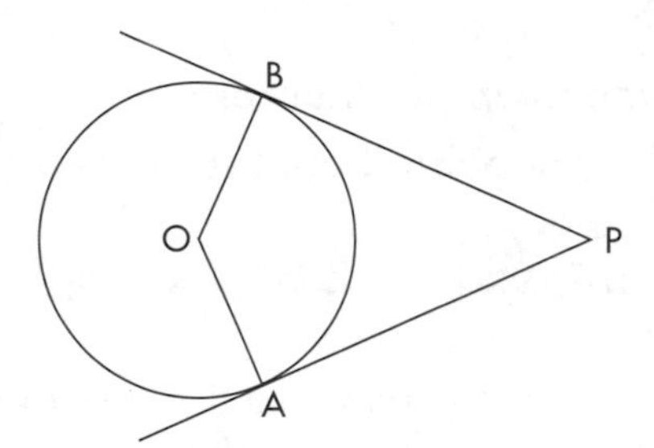

5 ABCDE is a regular pentagon. Prove that triangles ABC and AED are congruent.

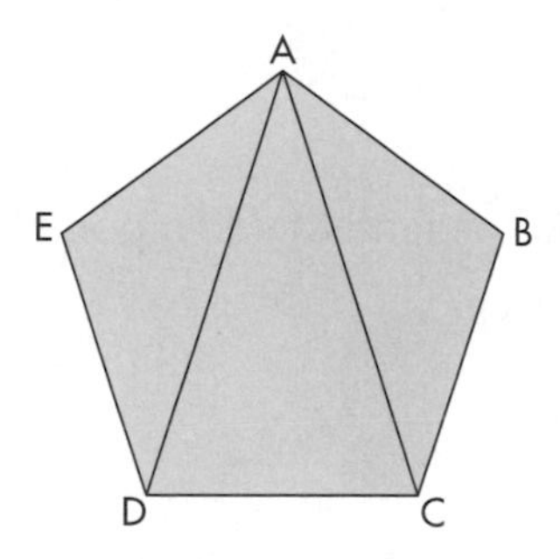

6 PQRS and PBCD are two identical squares. The squares overlap and intersect at points P and N. N is the midpoint of both SR and CB.

a Prove that triangles PBN and PSN are congruent.

b Prove that triangles PSD and PQB are congruent.

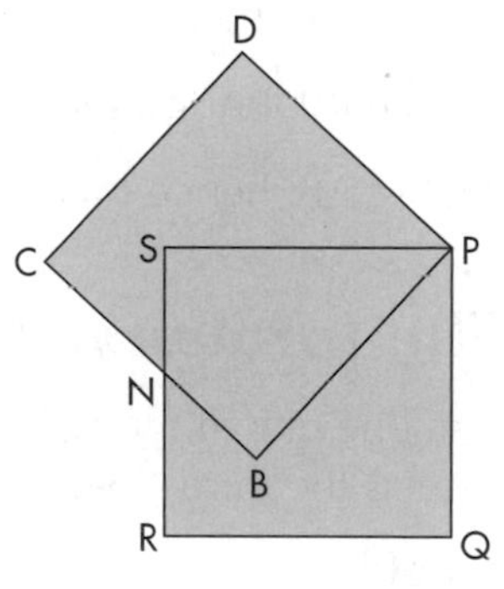

PS 7 A, B and C are points on a circle. ∠ABC = ∠ACB. PB and PC are tangents from a common point P. Prove that triangles APB and APC are congruent.

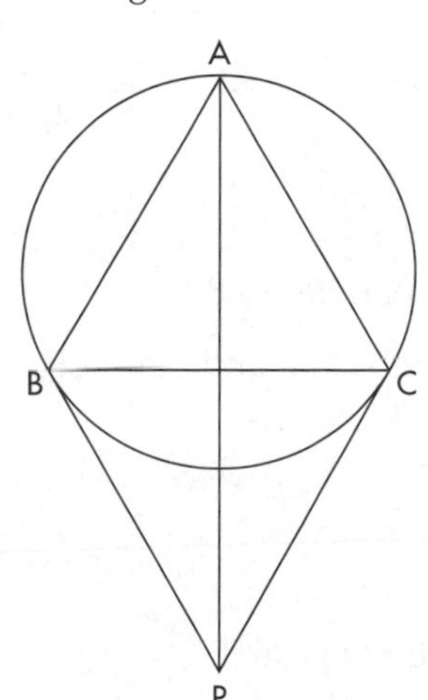

A

A*

11.2 The midpoint theorem

This section will show you:

- the proof of the midpoint theorem
- how to use the midpoint theorem to solve problems

Key words
corollary
midpoint theorem
proof

The **midpoint theorem** states that the line joining the midpoints of any two sides of a triangle is parallel to the third side of the triangle and equal to half its length.

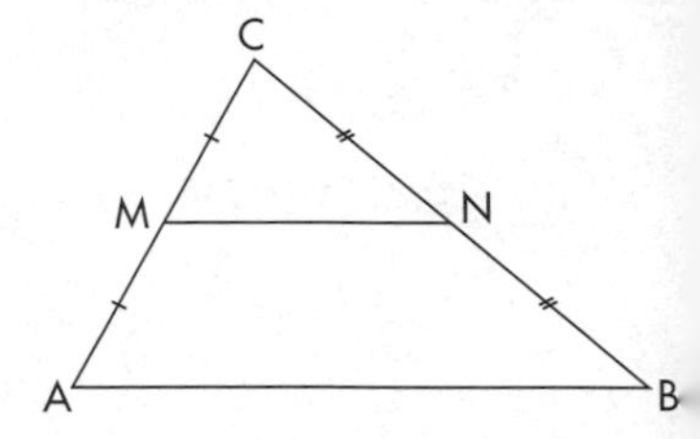

Proof

Draw a line from B parallel to AC. Extend MN to a point D that is on the line from B.

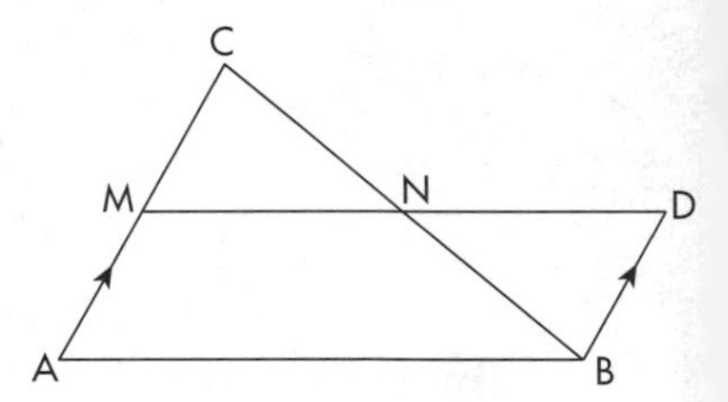

∠CNM = ∠BND (opposite angles)

∠MCN = ∠NBD (alternate angles)

CN = NB (Given)

Hence triangles MNC and NDB are congruent (ASA).

Therefore MN = ND and CM = BD = AM

Hence MDBA is a parallelogram so MN is parallel to AB.

MD = AB, hence MN = $\frac{1}{2}$AB

The corollary of the midpoint theorem

The line drawn through the midpoint of one side of a triangle and parallel to another side bisects the third side.

Proof

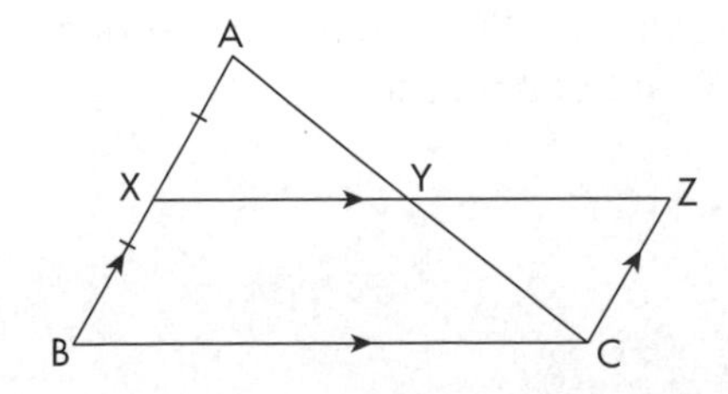

A **corollary** is a statement that follows directly from an earlier statement. Many theorems in geometry have corollaries.

Produce XY to Z, such that CZ is parallel to BX.

BCZX is a parallelogram (two sides equal and parallel)

BX = CZ = AX

∠YCZ = ∠YAX (alternate angles)

∠CYZ = ∠AYX (opposite angles)

Hence triangles CYZ and AYX are congruent.

Therefore CY = AY

Note: You will **not** be expected to reproduce these proofs in an examination.

EXAMPLE 5

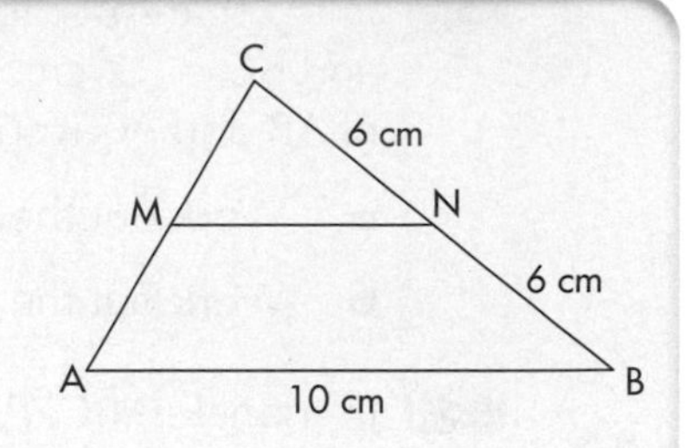

M is the midpoint of the side AC of the triangle ABC.

The lengths of the sides AC and BC are in the ratio 3 : 4

a Work out the length of CM.

b Work out the length of MN.

SOLUTION

a CM : CN = 3 : 4

$CM = 6 \div 4 \times 3 = 4.5$ cm

b As M and N are midpoints of the sides then MN is half the length of AB.

MN = 5 cm

EXAMPLE 6

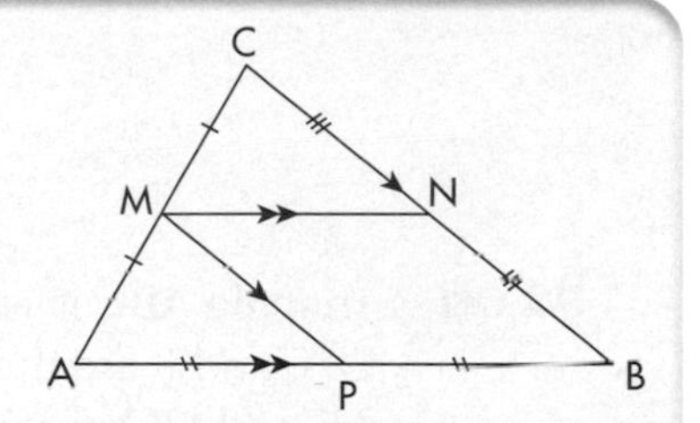

In triangle ABC, M, N and P are the midpoints of the sides AC, CB and AB respectively.

Prove that the area of the triangle CMN is one quarter of the area of the triangle ABC.

SOLUTION

$\angle PAM = \angle NMC$ (corresponding angles)

Area $\Delta ABC = \frac{1}{2} \times AB \times AC \times \sin \angle PAM$

Area $\Delta NMC = \frac{1}{2} \times MN \times AC \times \sin \angle NMC$

$= \frac{1}{2} \times \frac{1}{2}AB \times \frac{1}{2}AC \times \sin \angle PAM$

$= \frac{1}{4} \times \frac{1}{2} \times AB \times AC \times \sin \angle PAM = \frac{1}{4} \times$ area ΔABC

EXERCISE 11B

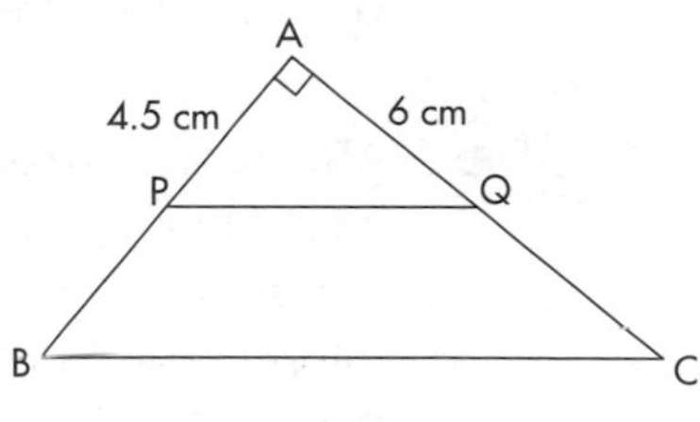

1 In triangle ABC, P and Q are the midpoints of the sides AB and AC respectively. Angle A = 90°.

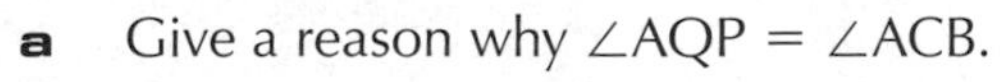

a Give a reason why $\angle AQP = \angle ACB$.

b AP = 4.5 cm and AQ = 6 cm. Find the length of BC.

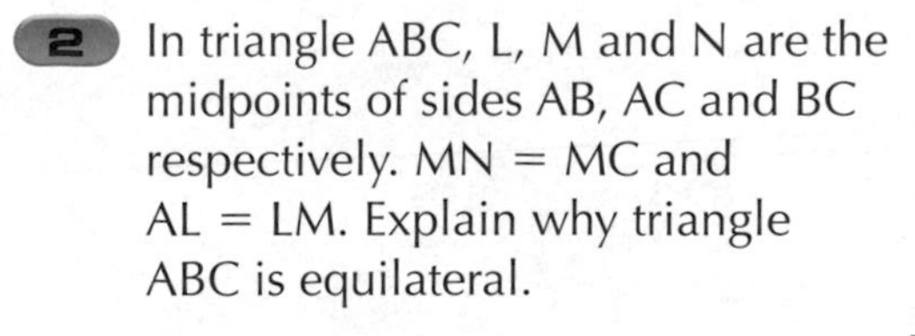

2 In triangle ABC, L, M and N are the midpoints of sides AB, AC and BC respectively. MN = MC and AL = LM. Explain why triangle ABC is equilateral.

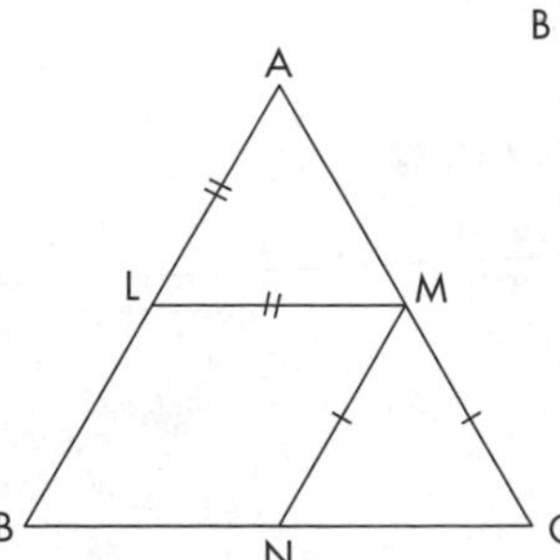

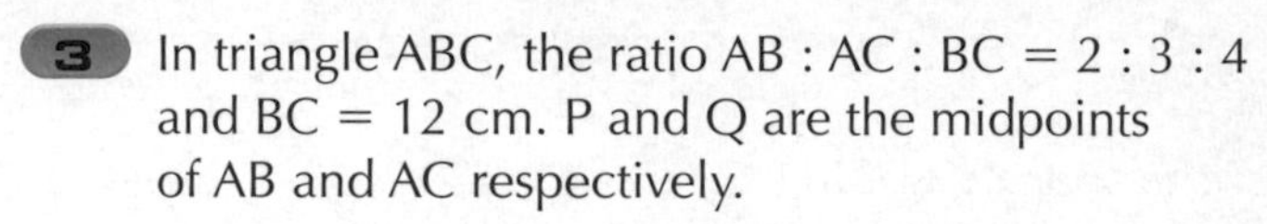

3 In triangle ABC, the ratio AB : AC : BC = 2 : 3 : 4 and BC = 12 cm. P and Q are the midpoints of AB and AC respectively.

a Work out the length of AC.

b Work out the length of AP.

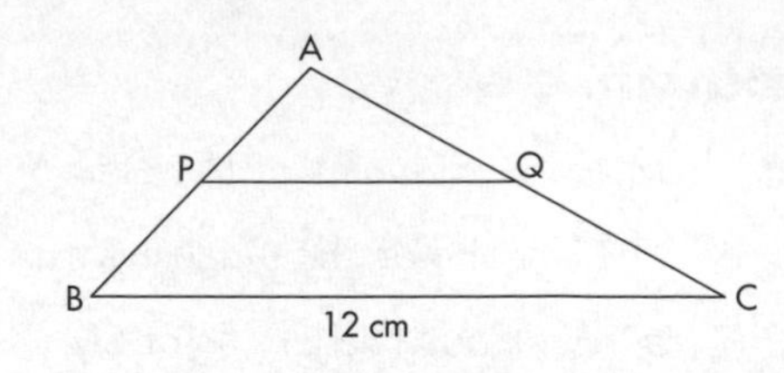

4 In triangle ABC, P, Q and R the midpoints of sides AB, AC and BC respectively. Explain why PQRB is a parallelogram.

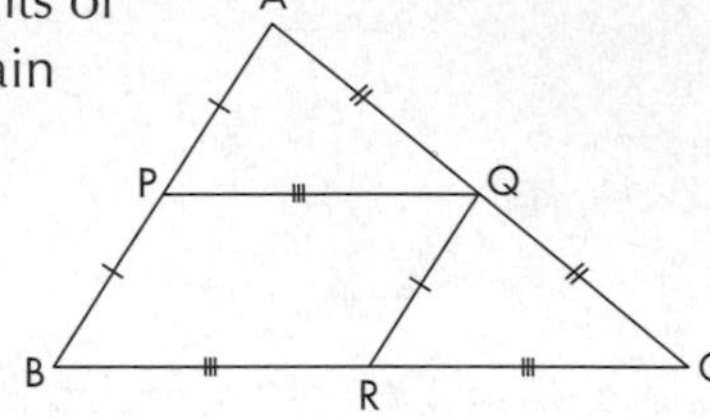

PS 5 ABC is a right-angled triangle. P is the midpoint of AC. Q is the midpoint of AB. Prove that PB = PA = $\frac{1}{2}$AC.

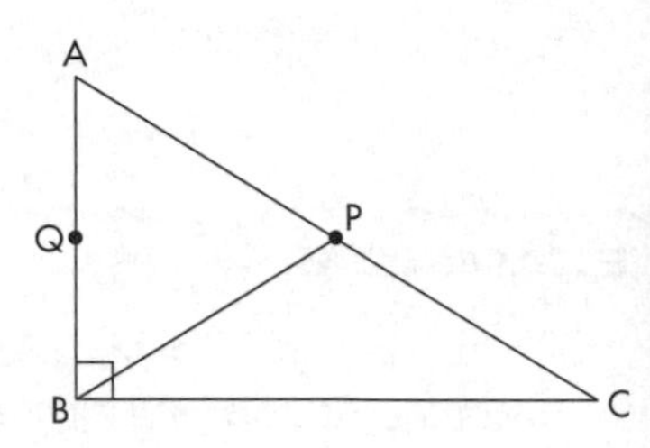

PS 6 Triangle ABC is isosceles with AB = AC. D, E and F are the midpoints of BC, AB and AC respectively. Prove that AD is perpendicular to EF and is bisected by it.

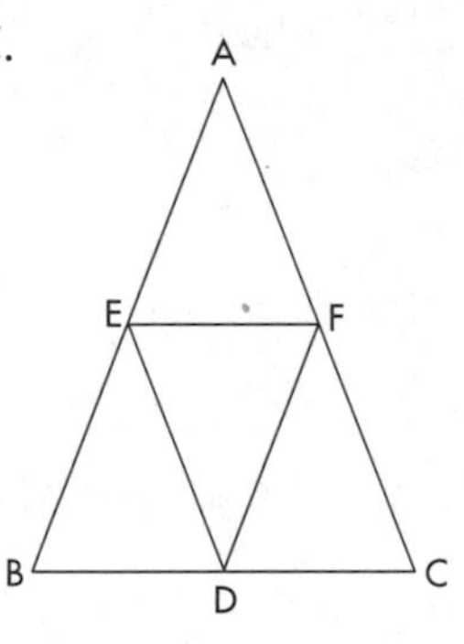

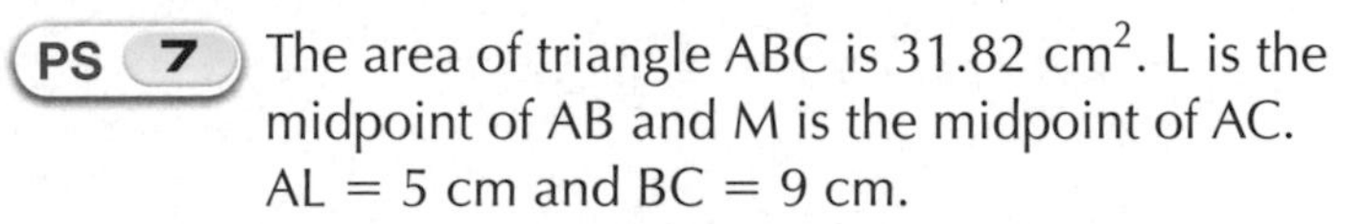

PS 7 The area of triangle ABC is 31.82 cm^2. L is the midpoint of AB and M is the midpoint of AC. AL = 5 cm and BC = 9 cm.

a Work out the size of the angle at B.

b Calculate the length of AC.

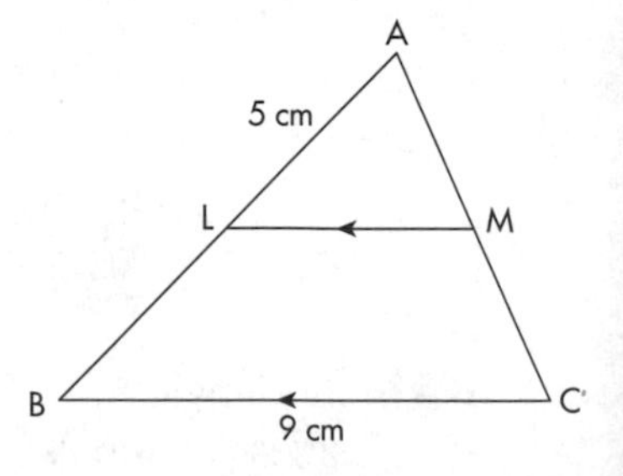

AU 8 ABC is an isosceles triangle with AB = AC. L is the midpoint of AB. M is the midpoint of CB. AP is parallel to CB and CQ is parallel to AB. Prove that triangles APL and CMQ are congruent.

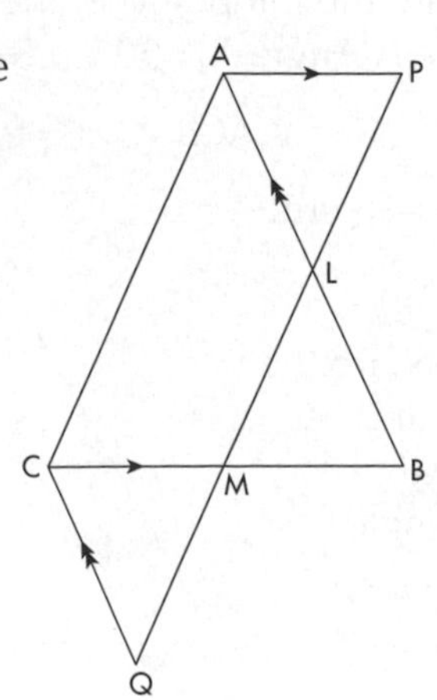

The intercept theorem

This section will show you:

- the proof of the intercept theorem
- how to use the intercept theorem to solve problems

Key words
intercept theorem
proof
transversal

The **intercept theorem** states that if three or more parallel straight lines make intercepts on one **transversal**, then the intercepts on any other transversal are such that the ratios of lengths on the transversals are equal.

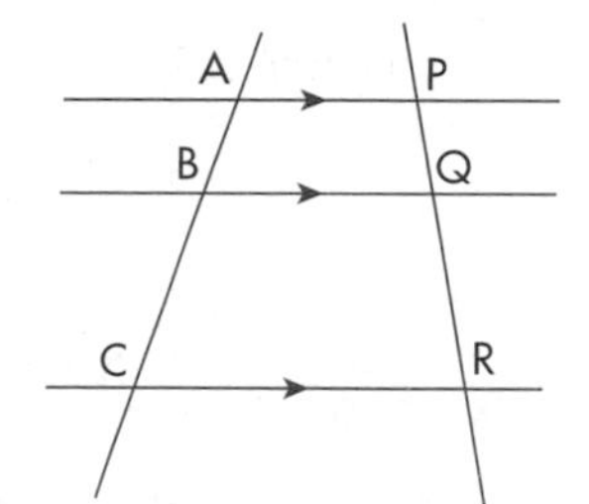

Proof

Draw a transversal parallel to AC passing through P, and label the points where this transversal crosses the parallel lines X and Y.

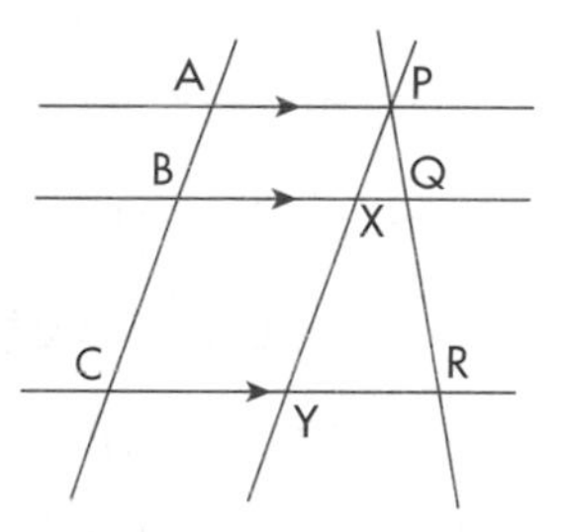

Clearly, triangle PXQ is similar to triangle PYR so

PX : XY = PQ : QR

Hence the ratio AB : BC = PQ : QR **QED**

The corollary of the intercept theorem (sometimes known as the ratio theorem)

A line MN drawn parallel to the side AB of the triangle ABC divides the sides AC and BC such that AM : MC = BN : NC

Proof

CMN and CAB are similar triangles so

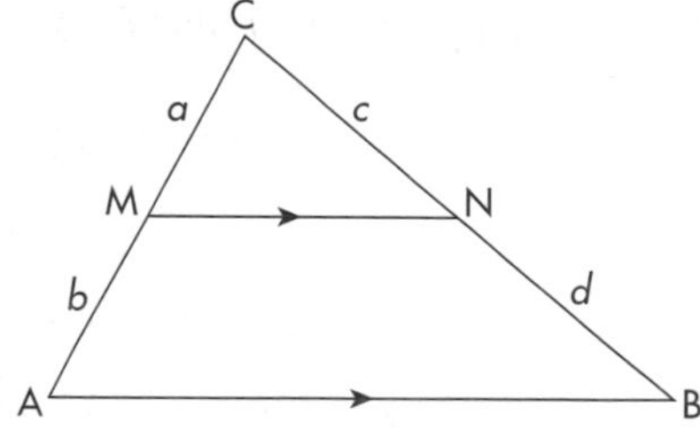

$$\frac{AC}{MC} = \frac{BC}{NC}$$

Hence $\dfrac{a+b}{a} = \dfrac{c+d}{c}$

$$1 + \frac{b}{a} = 1 + \frac{d}{c}$$

$$\frac{b}{d} = \frac{d}{c}$$ **QED**

Note: the length of the line MN = $\dfrac{a}{a+b} \times AB$

Link to the midpoint theorem

If X and Y are the midpoints of the sides PS and QR of the trapezium PQRS then XY = $\frac{1}{2}$(PQ + SR) or $c = \frac{1}{2}(a + b)$

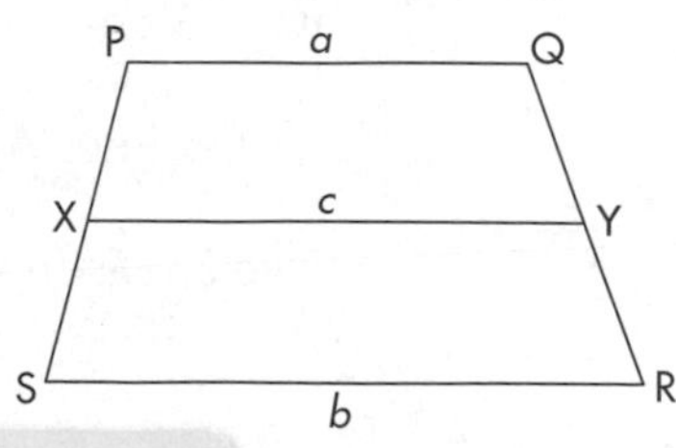

Note: You will **not** be expected to reproduce these proofs in an examination.

EXAMPLE 7

AB, CD and EF are parallel lines.

AC = 4 cm, CE = 6 cm and BD = 3 cm.

Calculate the length DF.

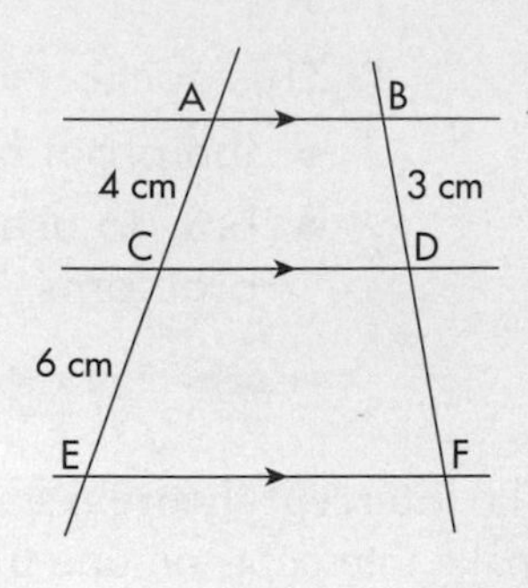

SOLUTION

AC : BD = 4 : 3

Hence CE : DF = 4 : 3

DF = 6 ÷ 4 × 3 = 4.5 cm

EXAMPLE 8

ABCD is a trapezium.

XY is a line parallel to the sides AB and CD.

AB = 8 cm, DC = 10 cm, AX = 6 cm, XD = 2 cm.

Work out the length of XY.

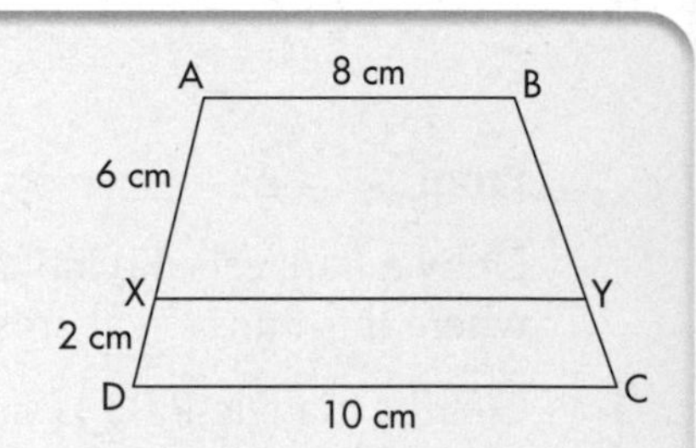

SOLUTION

The sides AX and XD of the trapeziums are in the ratio 6 : 2 = 3 : 1

Hence the length of XY = $\frac{1}{4} \times 8 + \frac{3}{4} \times 10 = 9.5$ cm

EXERCISE 11C

1 AB, CD and EF are parallel lines.

CE = 9 cm and BD = 6 cm.

The ratio AC : CE = 4 : 5

Calculate the length of:

a DF **b** AC.

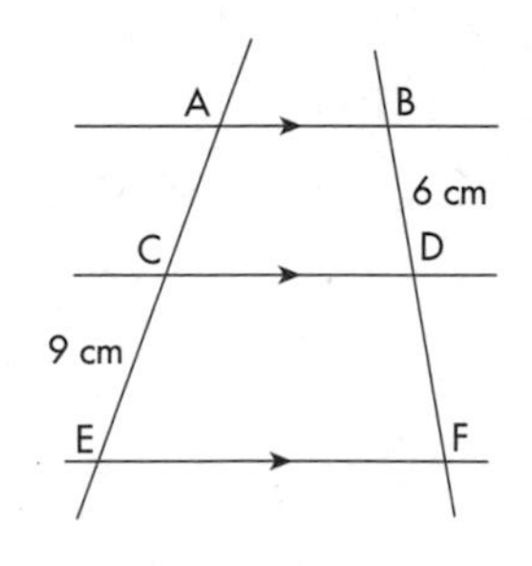

AU 2 AB, CD and EF are three lines that intersect XY and PQ.
AC = 5 cm, CE = 9 cm, BD = 4 cm and DF = 7 cm.

Are the lines AB, CD and EF parallel?

Give a reason for your answer.

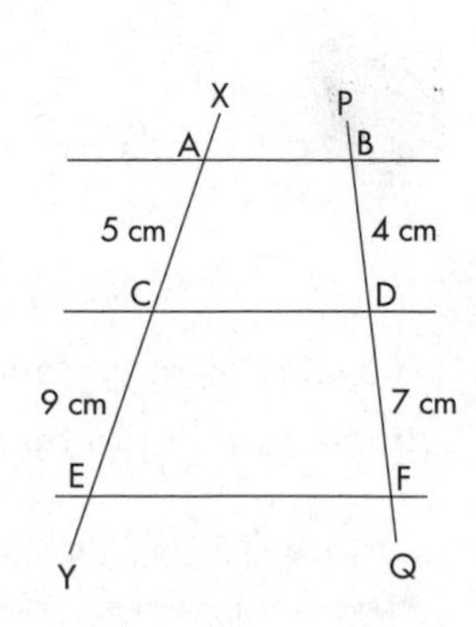

3 The lines ABC, DEF and GHJ are parallel. AD = 12 cm, DG = 7.5 cm, CF = 9 cm and EH = 6 cm.

Work out the length of:

a BE **b** FJ.

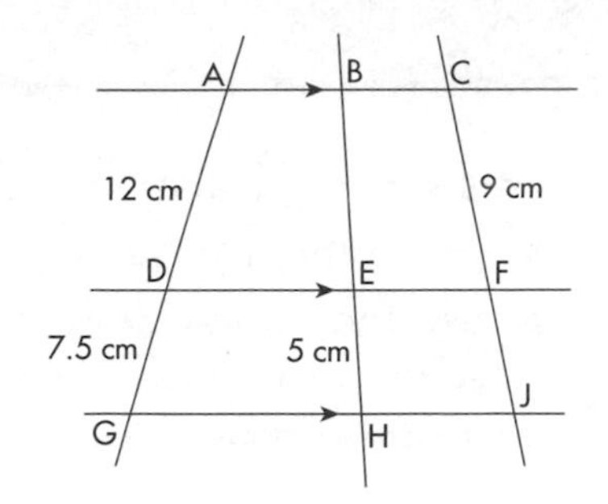

PS 4 In trapezium ABCD, AB = 5 cm and CD = 12 cm.

GH is parallel to CD.

BE is parallel to AD.

GH and BE intersect at P.

AG : GD = 3 : 1

Work out the length of PH.

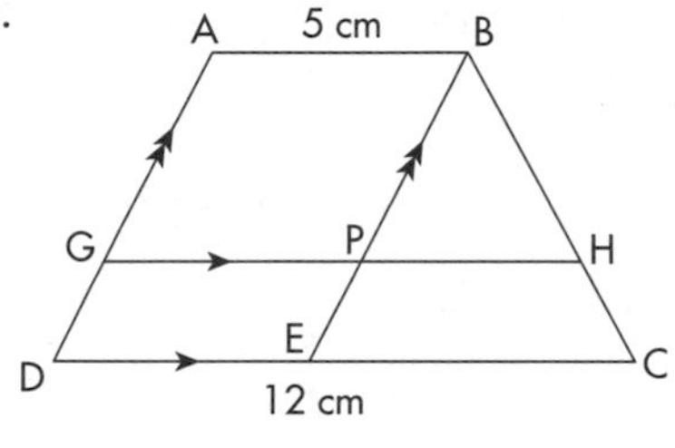

5 In trapezium ABCD, XY is a line parallel to the sides AB and CD.

AB = 12 cm, CD = 15 cm, AX = 8 cm and XD = 2 cm.

Work out the length of XY.

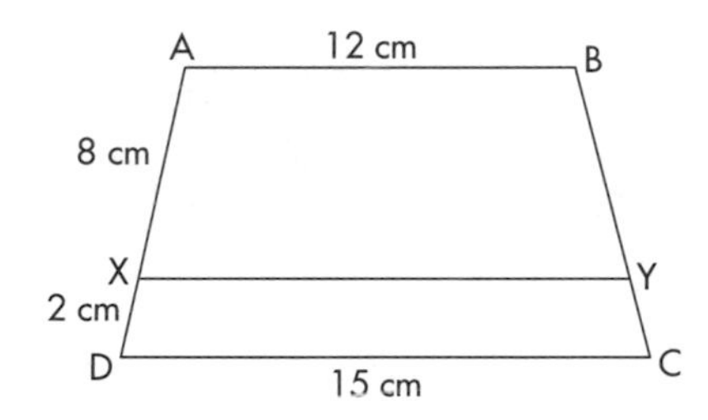

6 In triangle ABC, P is a point on AB such that AP : PB = 4 : 5.

PR is parallel to BC and RQ is parallel to AB. BC = 18 cm.

Work out the length of BQ.

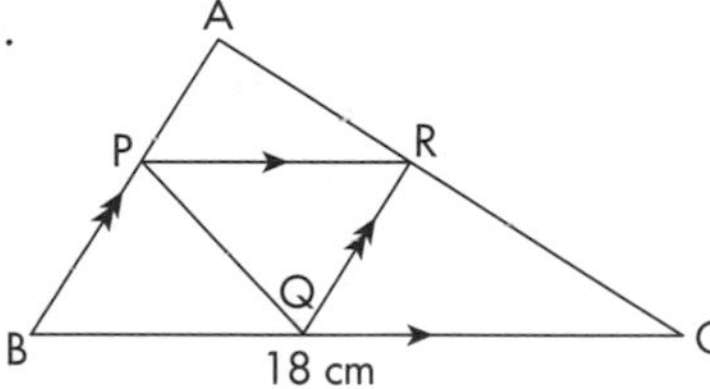

7 In trapezium ABCD, AB = 6 cm and CD = 10 cm. XY and LM are parallel to CD.

BX : XL : LC = 1 : 3 : 1

Work out the length of:

a XY **b** LM.

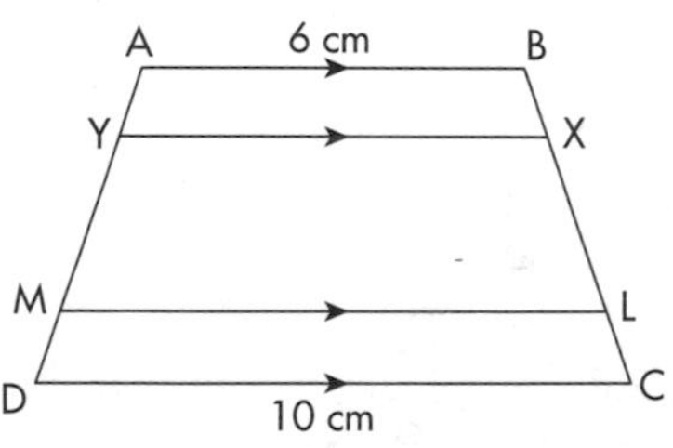

PS 8 In parallelogram ABCD, LM is parallel to AD and XY is parallel to AB.

LM and XY intersect at P.

AL : LB = 1 : 3

AX : XD = 2 : 3

Find the ratio of the areas:

a ALPX : LBYP **b** LBYP : PYCM **c** ALPX : PYCM.

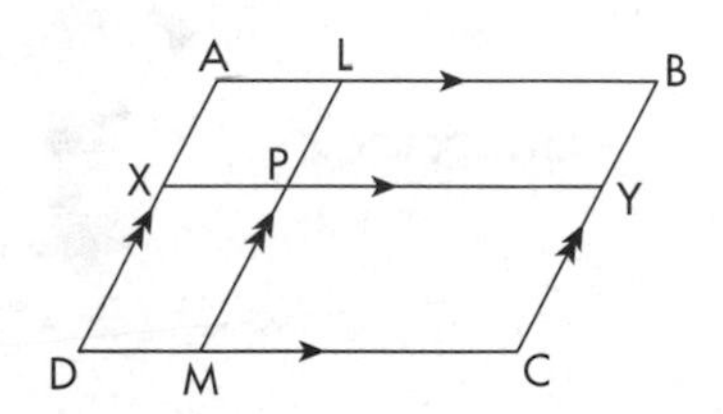

11.4 The intersecting chords theorem

This section will show you how to:

- prove the intersecting chords theorem
- use the intersecting chords theorem to solve problems

Key words
intersecting chords
proof

The **intersecting chords** theorem states that if, in a circle, two chords, AD and BC, intersect at P, then AP × PD = BP × PC.

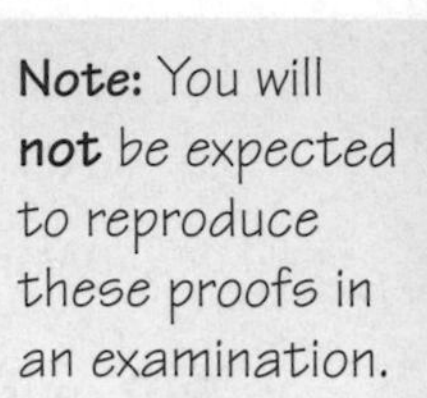

Proof

∠BAP = ∠DCP (angles in same segment)

∠ABP = ∠CDP (angles in same segment)

∠BPA = ∠DPC (opposite angles)

Hence ΔAPB is similar to ΔCPD (equiangular).

Hence $\frac{AP}{BP} = \frac{PC}{PD}$ and so AP × PD = BP × PC

Note: You will **not** be expected to reproduce these proofs in an examination.

EXAMPLE 9

In the circle, AB and QP are two chords that intersect at X.

XP = 6 cm, QX = 9 cm and XB = 12 cm

Work out the length of AX.

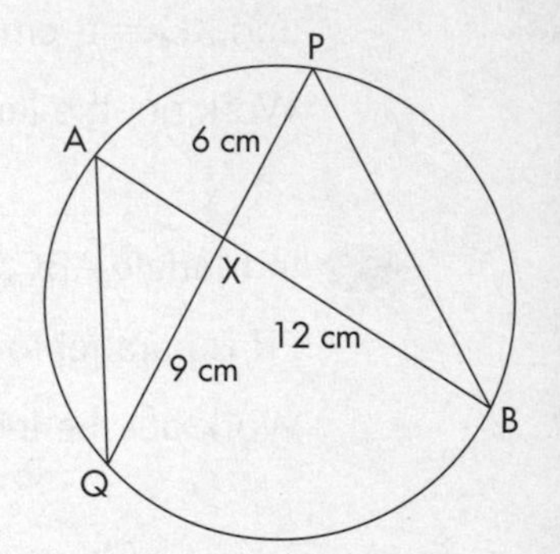

SOLUTION

By the intersecting chords theorem:

PX × XQ = AX × XB

So 6 × 9 = AX × 12

Hence AX = 54 ÷ 12 = 4.5 cm

EXAMPLE 10

AB is the diameter and CD is a chord of a circle.

AB and CD are extended to meet at X.

AX is 30 cm, CD is 11 cm and DX is 9 cm.

Calculate the length of AB, the diameter of the circle.

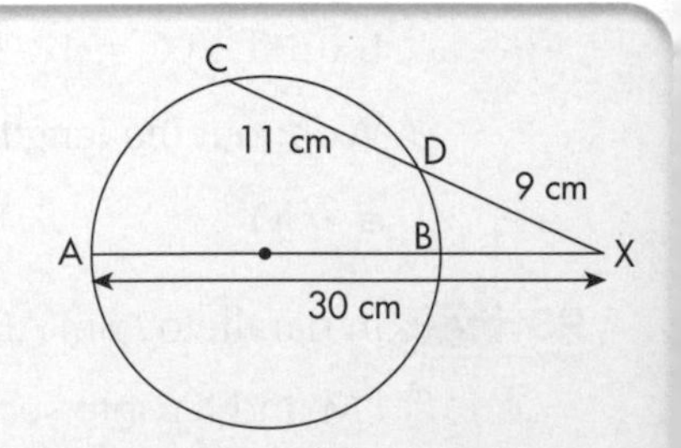

SOLUTION

By the intersecting chords theorem:

CX × DX = AX × BX

So 20 × 9 = 30 × BX

Hence BX = 180 ÷ 30 = 6 So length of diameter = AB = 30 − 6 = 24 cm.

A practical use of the intersecting chord theorem

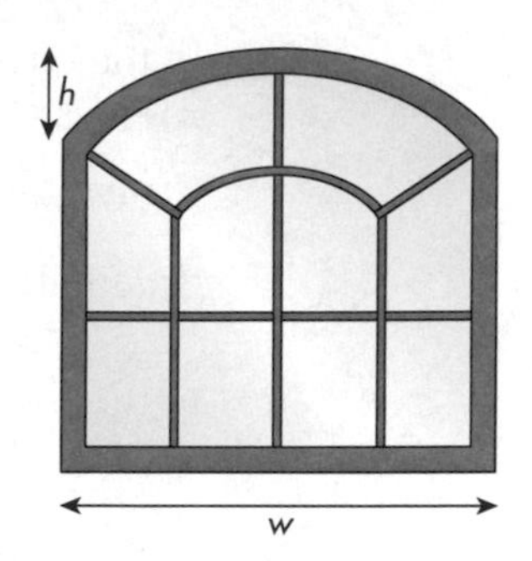

When we know the width, w, and height, h, of a segment of a circle, such as the top of a window, we can use the intersecting chord theorem to work out the radius. This would be useful for a joiner making a new window frame, for example.

The rule for the radius is $r = \frac{h}{2} + \frac{w^2}{8h}$

Proof

Assume that the segment is part of a circle, as shown.

By the intersecting chords theorem:

$$h \times x = \frac{w}{2} \times \frac{w}{2}$$

$$\text{So } x = \frac{w^2}{4h}$$

The diameter is $d = h + x$

The radius is $r = \frac{h}{2} + \frac{x}{2} = \frac{h}{2} + \frac{w^2}{8h}$

Note: You are **not** expected to know this formula.

EXAMPLE 11

An arc of a circle has a width of 12 cm and a height of 3 cm.

Work out the radius of the circle that contains this arc.

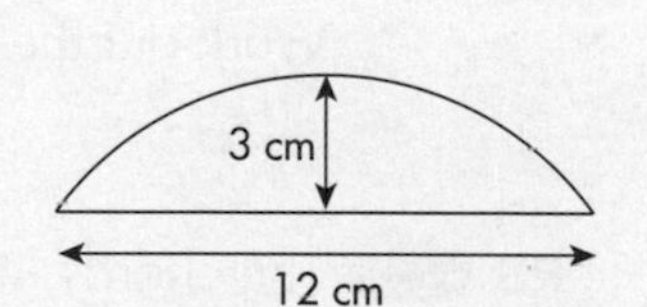

SOLUTION

Using the rule above:

$$r = \frac{3}{2} + \frac{12^2}{8 \times 3} = 1.5 + 6 = 7.5 \text{ cm}$$

EXERCISE 11D

1 In the circle, AB and CD are two chords that intersect at X.

CX = 3 cm, DX = 9 cm, XB = 4 cm

Work out the length of AX.

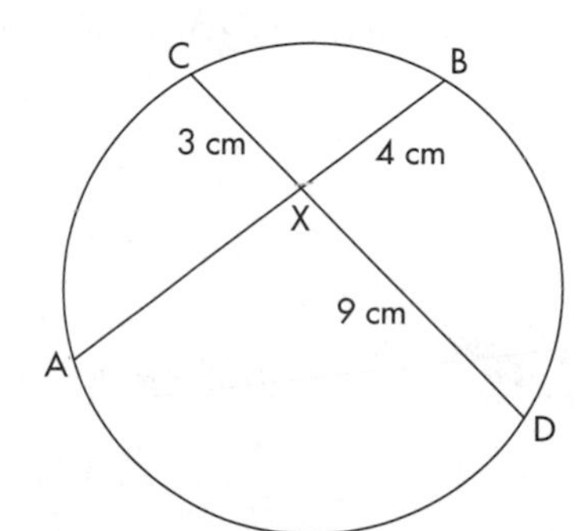

B

2 In the circle, AB is a chord and CD is a diameter.
AB and CD intersect at P.

AP = 4 cm, PB = 9 cm, PD = 3 cm

Calculate the radius of the circle.

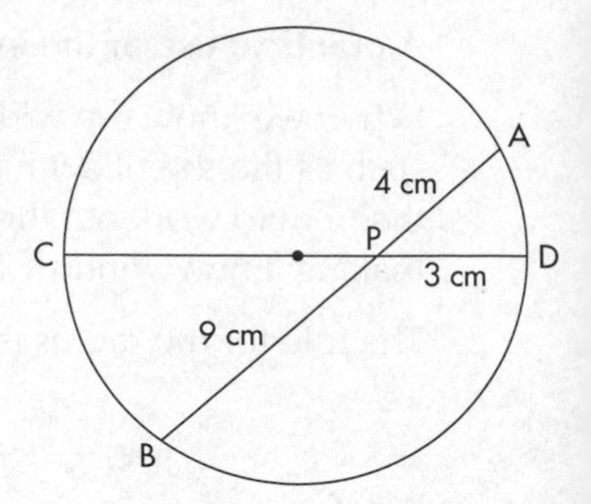

A

3 AB, CD and EF are chords of a circle.

AB and EF intersect at P.

CD and EF intersect at Q.

AP = 6 cm, PB = 3 cm, EP = 2 cm, CQ = 5 cm, QF = 3 cm

PQ = a and QD = b

Work out the value of:

a a

b b.

4 The chords AB and CD intersect at a point X outside the circle.

AB = 5 cm, BX = 9 cm, DX = 10 cm

Work out the length CD.

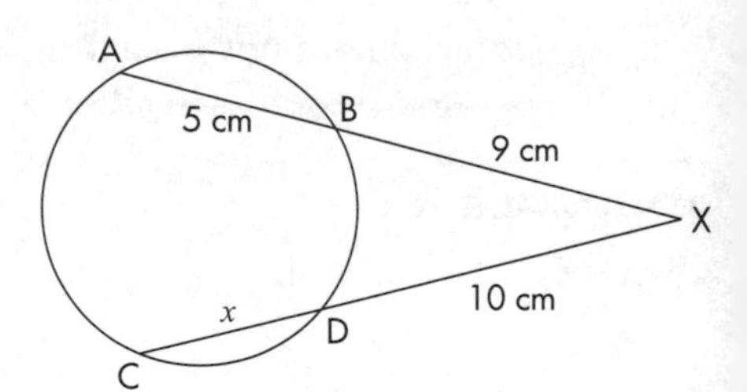

PS 5 The chords AB and CD intersect at a point X outside the circle. PQ is a diameter of the circle. PQX is a straight line.

AB = 6 cm, BX = 9 cm,
DX = 10 cm, QX = 7.5 cm

a Work out the length of DC.

b Work out the radius of the circle.

6 A segment of a circle is bounded by an arc. The height of the arc is 4 cm and the width of the arc is 20 cm. Calculate the radius of the circle that contains this arc.

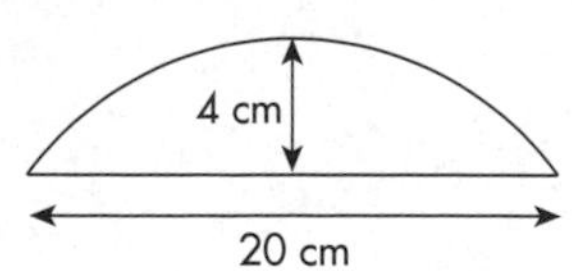

AU 7 Could these segments be part of the same circle?

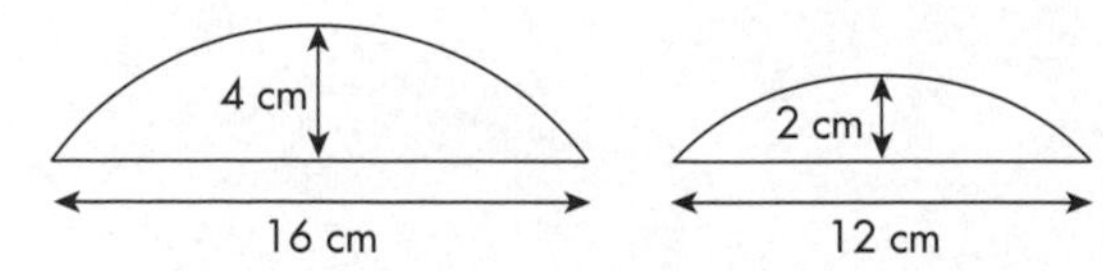

A*

8 The chords AB and CD intersect at a point X outside the circle.

PQ is a diameter of the circle, of length 19 cm. PQX is a straight line.

AB = 14 cm, BX = 6 cm, CD = 7 cm

Work out the length of:

a DX

b QX.

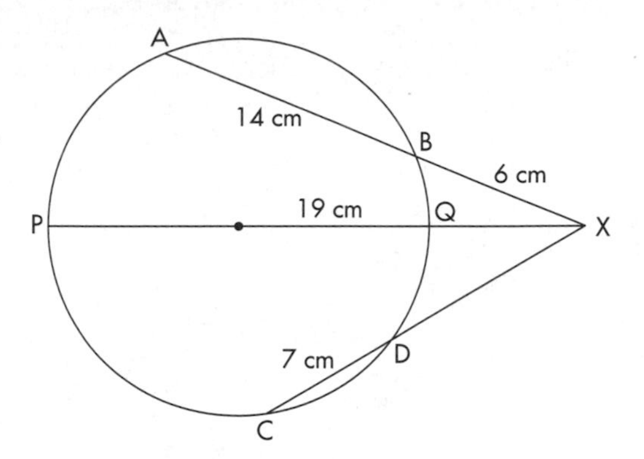

GRADE BOOSTER

- **B** You understand and can use the midpoint theorem and its corollary
- **B** You understand and can use the intercept theorem and its corollary
- **B** You understand and can use the intersecting chords theorem
- **A** You can use the midpoint and intercept theorems in more complex problems
- **A** You can use the intersecting chords theorem when the chords intersect outside the circle
- **A** You understand the conditions for congruency and can use these to prove the congruence of two triangles
- **A*** You understand and can use the angle property of the intersecting chords theorem
- **A*** You can prove the congruence of two triangles in more complex problems

What you should know now

- The midpoint theorem and its related facts
- The intercept theorem and its related facts
- The intersecting chords theorem and its related facts
- How to prove two triangles are congruent

EXAMINATION QUESTIONS

1 In this circle, two chords AB and CD intersect at E.

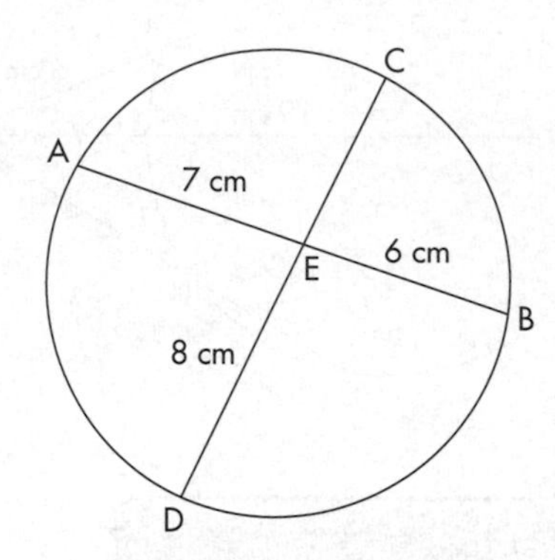

AE = 7 cm, EB = 6 cm, DE = 8 cm

Calculate the length EC.

2 ABC is a triangle with AC : AB : CB = 3 : 5 : 7.

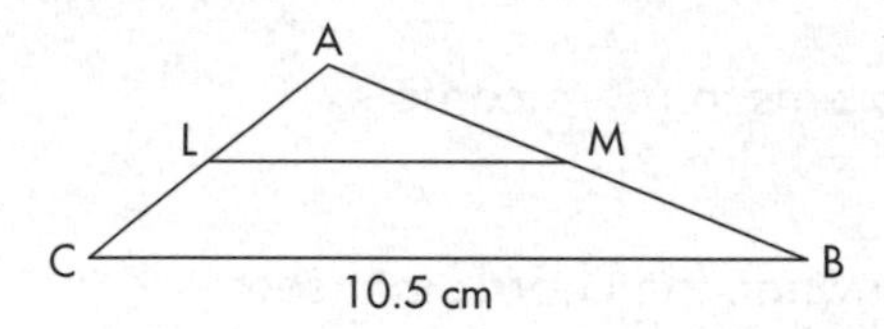

L and M are the midpoints of the sides AC and AB respectively.

CB = 10.5 cm

Work out the length of:

a AC

b AM.

3 ABCD is a trapezium with DC = 9 cm, AB = 13.5 cm, and EF is parallel to AB.

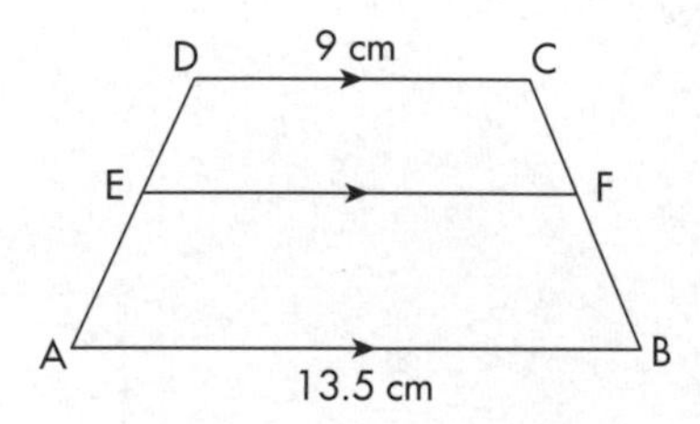

The ratio DE : EA = 4 : 5.

Work out the length of EF.

4 In triangle ABC, CB = 8 cm and AB = 9 cm.

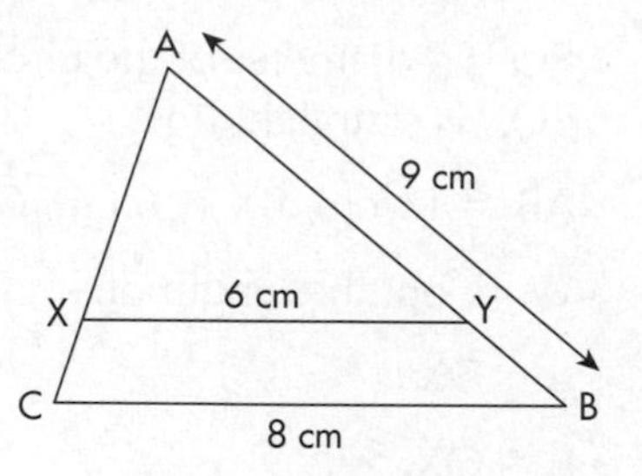

XY is parallel to CB and is 6 cm long.

Calculate the length of AY.

5 Show that the quadrilateral formed by joining the midpoints of the sides of a rectangle is a rhombus.

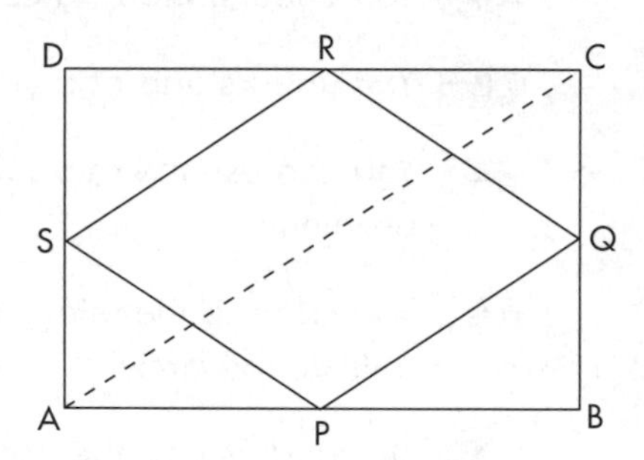

6 The diagonals of kite ABCD intercept at P.

M and N are the midpoints of the sides CD and CB.

Prove that triangles MNC and MNP are congruent.

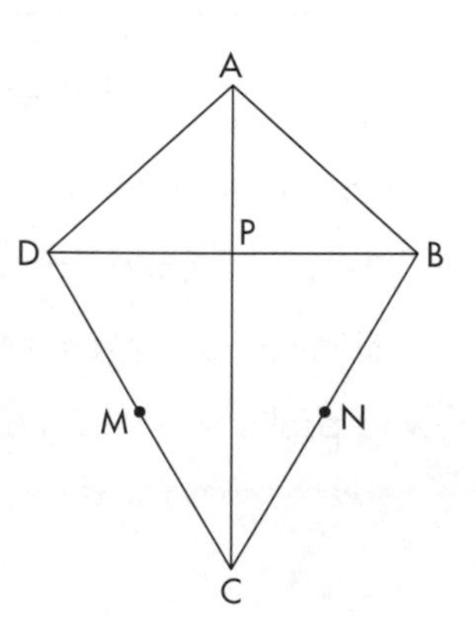

Worked Examination Questions

A **1** The diagonals of quadrilateral ABCD are perpendicular. Prove that the quadrilateral PQRS formed by joining the midpoints of its sides, as shown, is a rectangle.

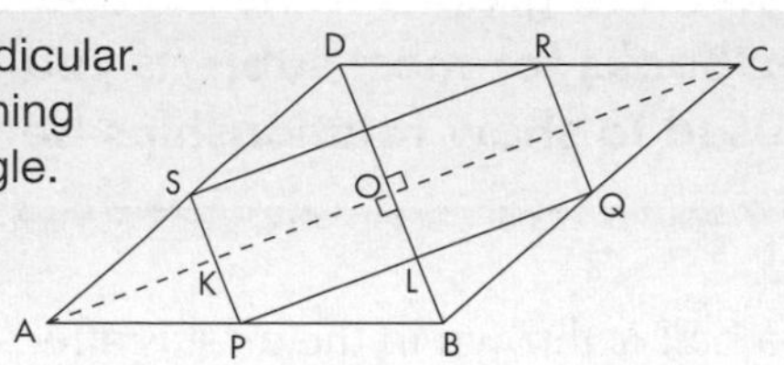

Taking triangle BDC:

RQ is parallel to and half the length of DB. (midpoint theorem)

Make it clear which triangle you are using and give a reason for any facts you write down. This earns **1 mark**.

Similarly, using triangle ABD:

SP = half of DB = RQ

Similarly, using triangles ADC and ACB:

SR = PQ

Saying 'similarly' means we are using exactly the same argument on a different triangle so there is no need to go through the steps again. Hence we have proved that the opposite sides are the same length. This does not prove it is a rectangle yet, but it earns **1 mark**.

As RQ is parallel to DB and RS is parallel to AC then RQ is perpendicular to RS.

Similarly, all other adjoining sides are perpendicular to each other.

Showing that two sides are perpendicular completes the proof. This earns **1 mark**.

Hence RQPS is a rectangle.

This final statement must be made and would earn **1 QWC mark** if all preceding working was accurate and reasons were given for all the statements.

Total: 4 marks

2 A regular octagon is split into six triangles, as shown. Triangles 3 and 6 are congruent.

a Identify two other pairs of congruent triangles.

b Prove that triangles 3 and 6 are congruent.

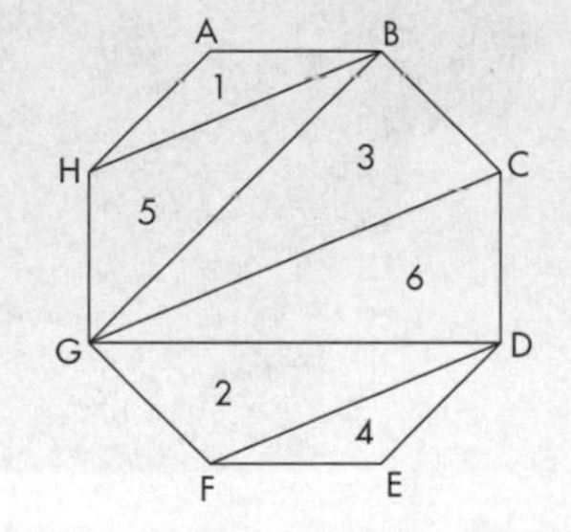

a 1 and 4, 5 and 2

This is just a lead in. Picking out the pairs earns **1 mark**.

b ∠GDE = 180° – ∠DEF = 180° – 135° = 45° (interior angle in a regular octagon and angles in a trapezium)

Use the interior angles and the properties of isosceles trapeziums to work out angles at D. This earns **1 mark**.

∠CDG = 135° – 45° = 90° (interior angle of a regular octagon)

Always state common sides even if it seems obvious. This earns **1 mark**.

Similarly, ∠CBG = 135° – 45° = 90°

CG is a common hypotenuse

Always give a reason in a proof. This earns **1 mark**.

BC = CD (sides of a regular octagon)

Hence triangles 3 and 6 are congruent (RHS).

There are other ways to show the congruency of triangles 3 and 6 but don't forget to state the one you have used. This earns **1 mark**.

Total: 5 marks

Why this chapter matters

You will find graphs in newspapers, on the internet and in textbooks for most subjects you learn in school. Graphs are used to show relationships between variables.

When a ball is thrown in the air it is affected by **gravity**, **wind** and **wind resistance**. The height of the ball, through a **time interval**, can be modelled by a **quadratic graph** such as the one shown here, although the actual height at any time may be slightly different from that shown.

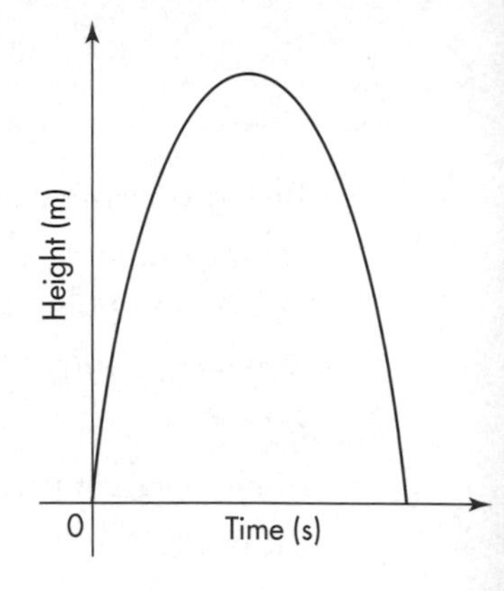

Most real-life graphs are not smooth curves, though. We use mathematical functions to **model** real-life situations.

Graphs give a visual representation of how variables change. They can be used to compare data and give information in a way that simple lists of data cannot.

On 20 April 2010 one of BP's oil rigs in the gulf of Mexico blew up, causing oil to spill into the water. This was carried by the sea and went on to ruin the local fishing industry and pollute the beaches. It wasn't until July that the attempts to cap the leak started to appear successful.

This graph shows the changes in the share price of BP as a result of the oil spill. The effects of the disaster can clearly be seen by the trend of the graph.

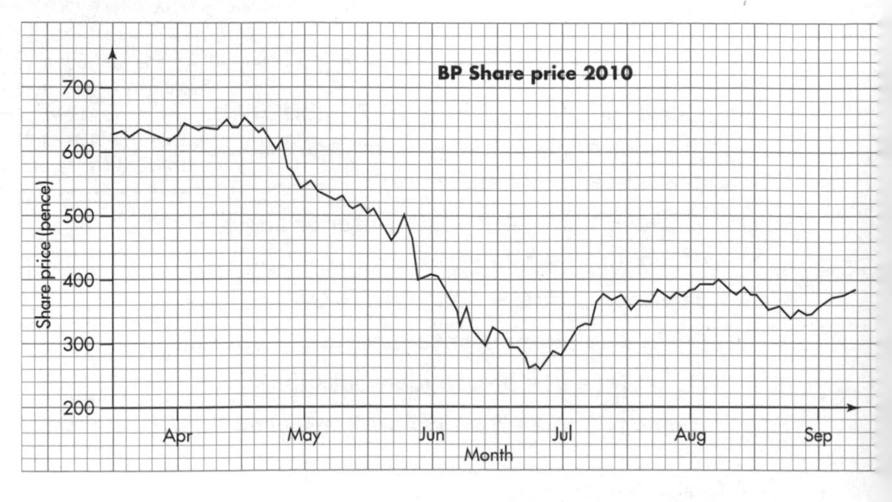

Chapter

Algebra: Graphs

The grades given in this FOUNDATION & HIGHER chapter are target grades.

1. Speed–time graphs
2. Estimation of the area under a curve
3. Gradients of straight-line graphs in practical contexts
4. Interpreting the gradient at a point on a curve as the rate of change
5. Graphs that illustrate direct and inverse proportion

This chapter will show you …

- **C** how to calculate the area under a graph consisting of straight lines and to interpret the meaning of the area under a speed–time graph
- **C** how to interpret the gradients of the straight lines on a speed–time graph
- **B** how to work out speed from a distance–time graph
- **B** how to work out acceleration from a speed–time graph
- **B** how to work out a formula from a graph that represents a real-life context
- **A** how to estimate and interpret the area under a curve (using areas of rectangles, triangles and trapeziums)
- **A** how to work out and interpret a gradient at a point on a curve
- **A*** how to draw and use graphs to illustrate direct and inverse proportion

Visual overview

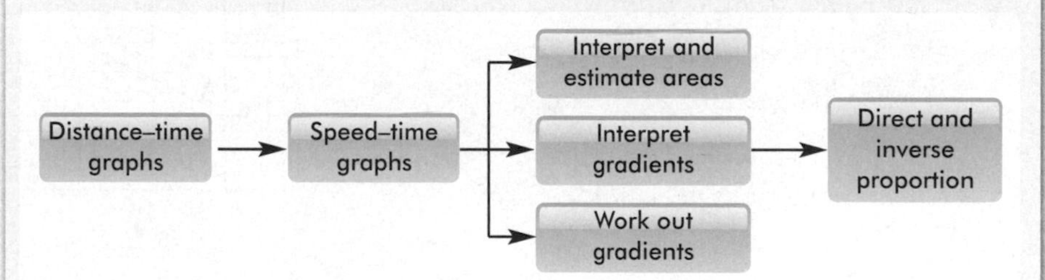

What you should already know

- How speed, distance and time are related (KS3 level 6, GCSE grade D)
- How to interpret a distance–time graph (KS3 level 6, GCSE grade D)
- How to draw straight-line graphs and quadratic graphs (KS3 level 6, Grade D–C)
- How to find the gradient of a line (KS3 level 8, Grade C)

Quick check

1 Work out the average speed of a car that travels 70 miles in 2 hours.

2 a What does a horizontal line mean on a distance–time graph?

b How can you tell from a distance–time graph which part of the journey is fastest?

12.1 Speed–time graphs

This section will show you how to:
- read information from a speed–time graph
- find the distance travelled from a speed–time graph

Key words
average speed
constant speed
negative gradient
positive gradient
speed–time graph
zero gradient

You read a **speed–time graph** in a similar way to a distance–time graph. A **positive gradient** means the speed is increasing. A **zero gradient**, when the line is horizontal, indicates a steady or **constant speed**. A **negative gradient** means the speed is decreasing. Any part of the graph represents the **average speed** maintained over that section.

EXAMPLE 1

Describe the journey represented by this graph.

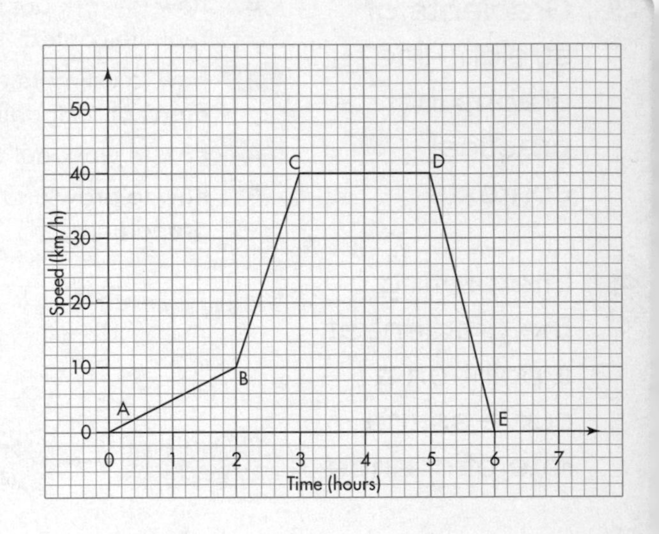

SOLUTION

A to B: This takes 2 hours and the speed increases from 0 km/h to 10 km/h.

B to C: This takes 1 hour and the speed increases from 10 km/h to 40 km/h.

C to D: This takes 2 hours and the speed in constant at 40 km/h.

D to E: This takes 1 hour and the speed decreases from 40 km/h to 0 km/h.

You can also use a speed–time graph to work out the distance covered, by finding the area under the graph. The next example shows why this is the case. The area is found by multiplying speed × time and this is equal to the distance.

EXAMPLE 2

The speed–time graph represents a car journey.

What can you say about:

a the speed

b the time taken

c the distance travelled?

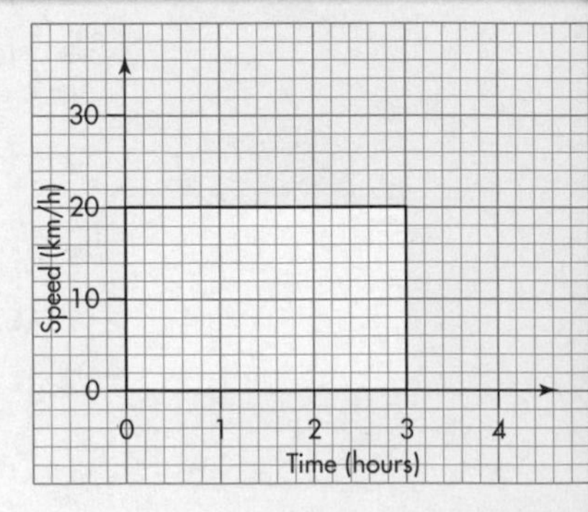

continued

SOLUTION

a As the line is horizontal, the speed is constant (steady) at 20 km/h.

b The time taken is 3 hours.

c Use the formula:

distance travelled = average speed × time taken

Then distance travelled is 20 × 3 = 60 km.

You can see that this is the same as working out the area of the rectangle shown on the graph (3 × 20 = 60 km/h).

EXAMPLE 3

The speed–time graph shows a train journey between two stations.

Work out:

a the distance travelled

b the average speed for the whole journey.

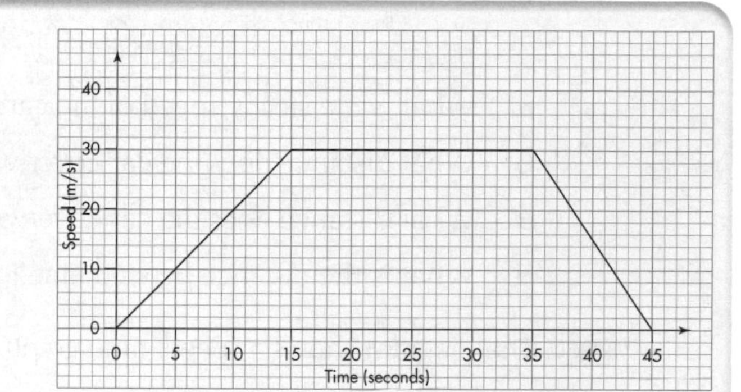

SOLUTION

a The distance travelled can be found from the area under the graph.

Method 1 (Area of a trapezium)

Hint: Remember the formula for the area of a trapezium:
$A = \frac{1}{2}(a + b)h$

Area of the trapezium $= \frac{1}{2}(45 + 20) \times 30 = 975$ metres

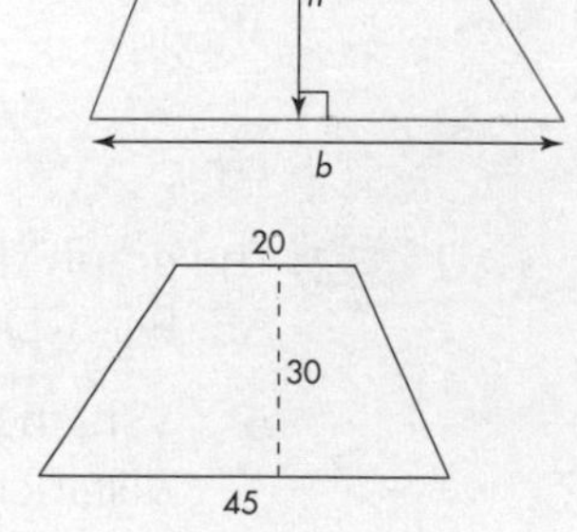

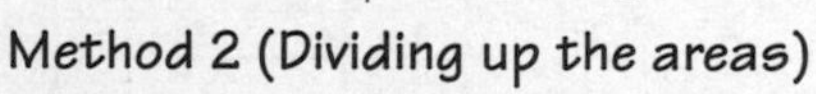

Method 2 (Dividing up the areas)

Hint: Remember the formula for the area of a triangle:
area $= \frac{1}{2} \times$ base $\times$ perpendicular height

Split the trapezium into triangles and rectangles to find the total area.

Area of first triangle
$= \frac{1}{2} \times 15 \times 30$
$= 225$ metres

Area of rectangle
$= 20 \times 30 = 600$ metres

Area of second triangle
$= \frac{1}{2} \times 10 \times 30 = 150$ metres

Total distance travelled = 225 + 600 + 150 = 975 metres

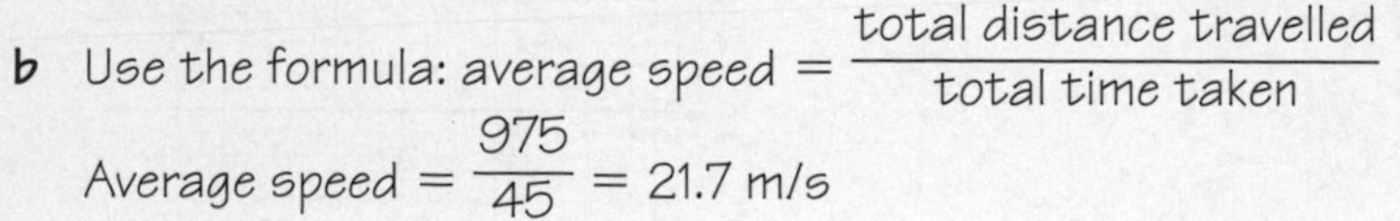

b Use the formula: average speed $= \frac{\text{total distance travelled}}{\text{total time taken}}$

Average speed $= \frac{975}{45} = 21.7$ m/s

C

EXERCISE 12A 

1 The diagram represents a car journey between two junctions.

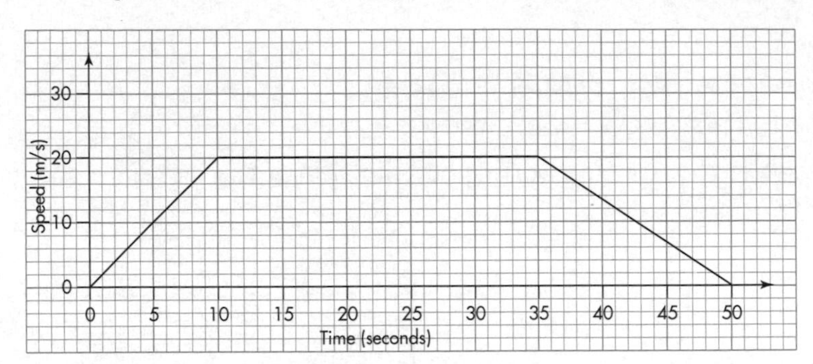

a What is the steady speed of the car?

b What distance does the car cover while speeding up?

c What distance does the car cover while slowing down?

d What is the distance between the junctions?

2 The diagram shows a speed–time graph.

a Work out the total distance travelled.

b Work out the average speed for the whole journey.

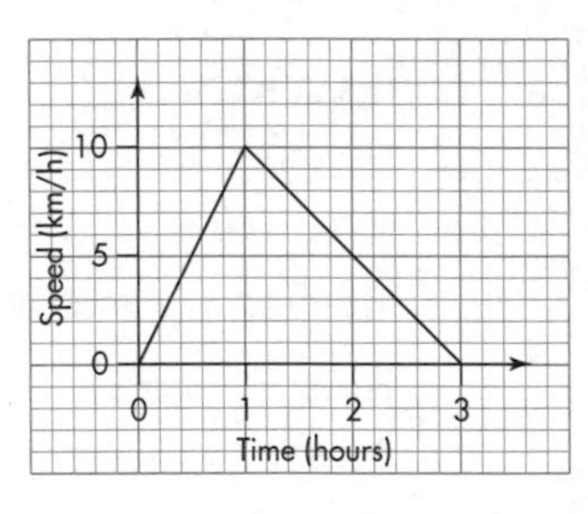

AU 3 The graph shows four parts of a two-hour journey, AB, BC, CD and DE.

a Which part of the journey covers the greatest distance? Give a reason for your answer.

b Work out the distance covered travelling from C to E.

c Work out the total distance covered.

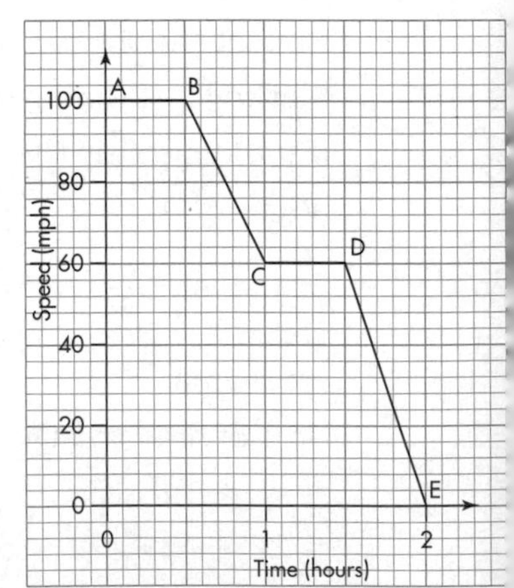

PS 4 The graph shows 40 seconds of a car journey.

The maximum speed is v m/s.

The distance covered is 300 metres.

Work out the value of v.

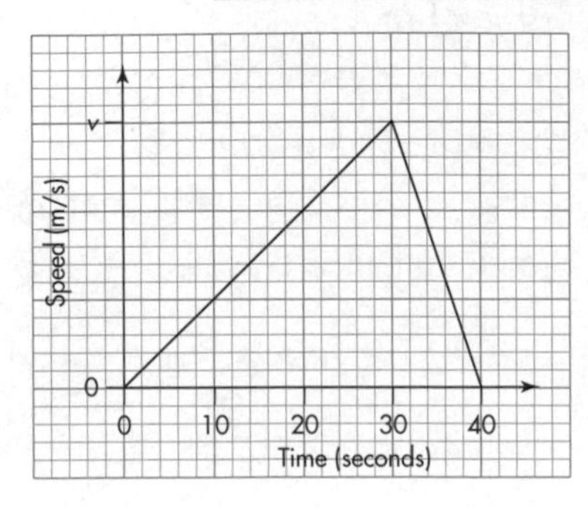

C

5 A cyclist travels from rest to 10 m/s in 15 seconds, increasing speed at a steady rate.

She then travels at a steady speed for 30 seconds.

She then slows down to rest in a further 20 seconds, decreasing speed at a steady rate.

a Draw a graph of speed (m/s) against time (seconds) for the journey.

b Work out the total distance travelled.

AU 6 The diagram shows the journeys of two trains, A and B.

In each case write down whether for these journeys the statement:

Must be true   Could be true or false   Must be false

a The trains are travelling in opposite directions.

b The trains both cover the same distance.

c Train A is speeding up and train B is slowing down.

d Train A is travelling up a slope.

e Train A overtakes train B.

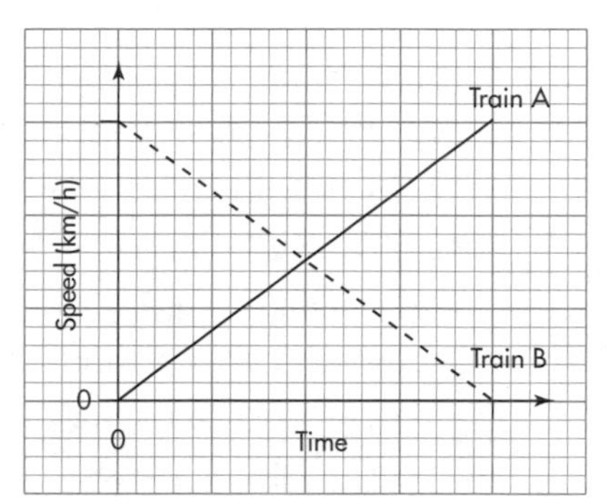

12.2 Estimation of the area under a curve

This section will show you how to:

- use areas of rectangles, triangles and trapeziums to estimate the area under a curve
- interpret the meaning of the area under a curve

Key words
acceleration
area under a curve
deceleration
estimate
under-estimation

A speed–time graph made up of straight lines is generally not as realistic as one that is curved. This is because the curve can show gradual changes in **acceleration** or **deceleration**.

Calculating the **area under a curve** accurately, though, to work out the total distance travelled on a speed–time graph, would be too complicated at this stage. However, it is possible to **estimate** the area by dividing the area approximately into simpler shapes such as rectangles, triangles and trapeziums.

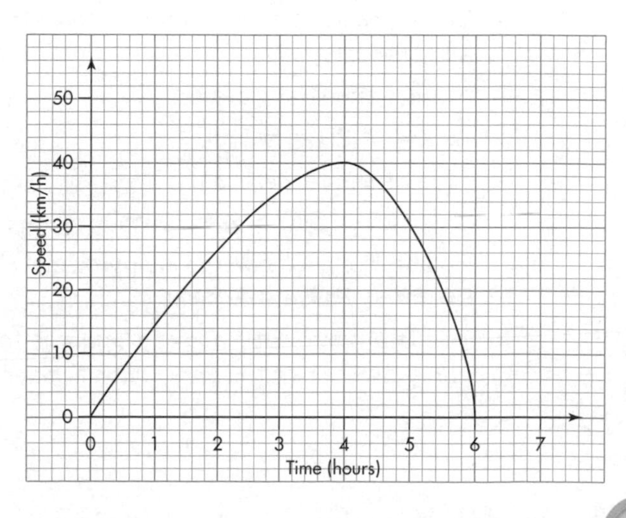

EXAMPLE 4

Estimate the total distance travelled for the journey shown in the graph on the previous page.

SOLUTION

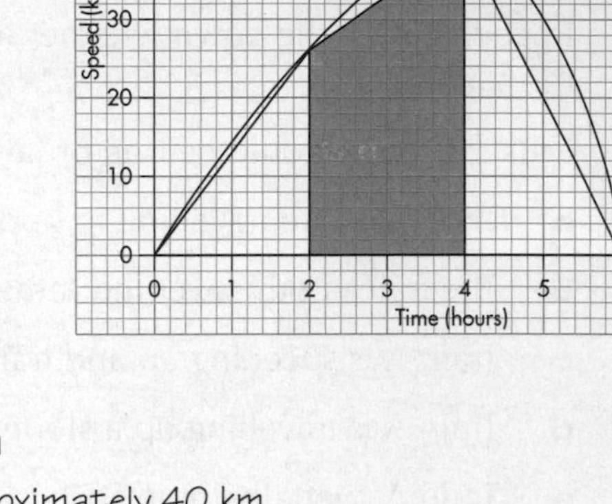

We can estimate the total distance travelled by working out the areas of the triangles and the trapezium, as shown.

Area of first triangle

$= \frac{1}{2} \times 2 \times 26 = 26$ km

So distance travelled in the first 2 hours is approximately 26 km.

Area of the trapezium

$= \frac{1}{2} \times (26 + 40) \times 2 = 66$ km

So distance travelled in the next 2 hours is approximately 66 km.

Area of second triangle $= \frac{1}{2} \times 2 \times 40 = 40$ km

So distance travelled in the final 2 hours is approximately 40 km.

This gives a total distance travelled of $26 + 66 + 40 = 132$ km.

Note: as all the shapes are inside the curved area this will be a slight **under-estimation** of the true distance covered.

EXAMPLE 5

The speed–time graph represents a journey.

a Estimate the distance travelled.

b State whether your estimate is an under-estimate or an over-estimate.

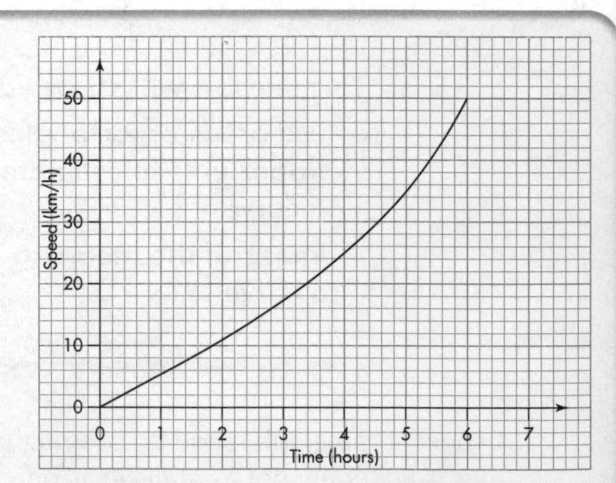

SOLUTION

a Divide the area approximately into a triangle and trapezium.

Area of the triangle is

$\frac{1}{2} \times 4 \times 25 = 50$ km

Area of the trapezium is

$\frac{1}{2} \times (25 + 50) \times 2 = 75$ km

So the distance covered is approximately 50 km + 75 km = 125 km.

b This is an over-estimate as the areas found are greater than the area under the curve.

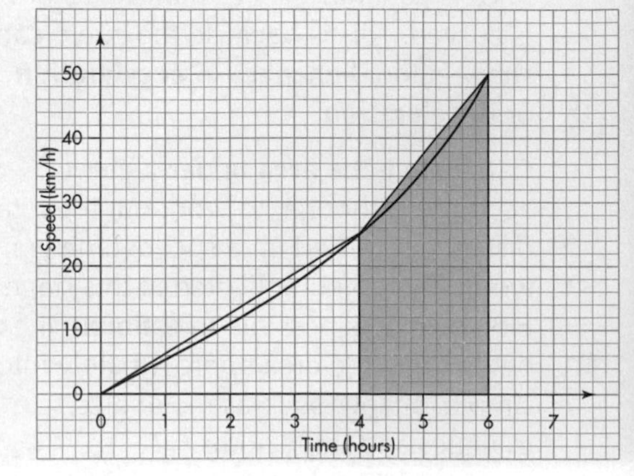

EXERCISE 12B 

For each speed–time graph in questions 1 to 7, estimate the distance travelled and state whether your estimate is an underestimate or an overestimate.

1

2

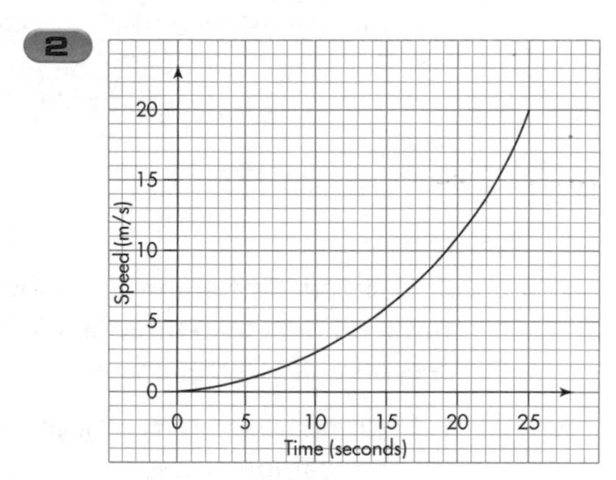

3

4

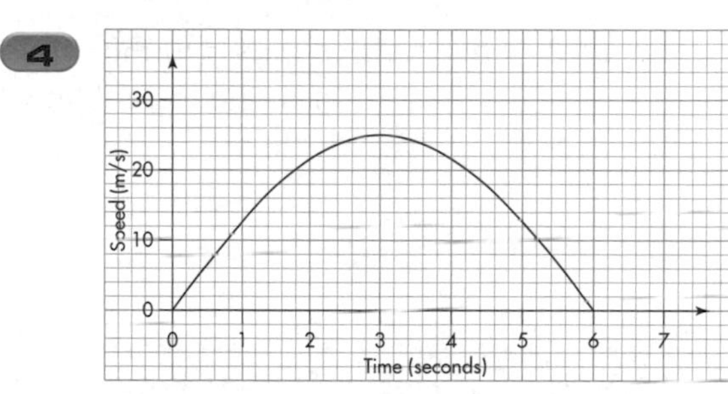

5

6

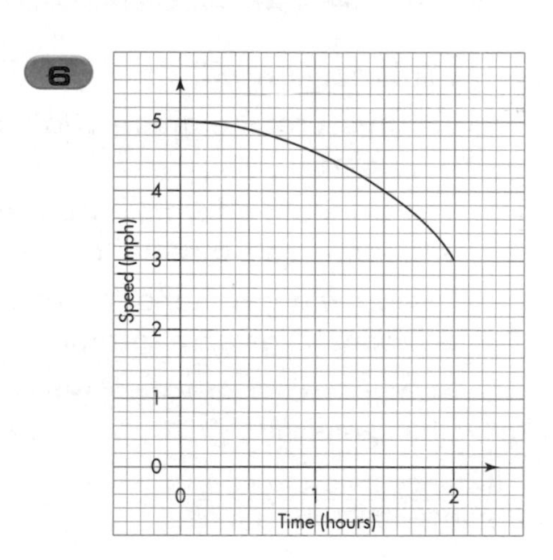

7

8 **a** The speed–time graph shows a car journey over one minute.

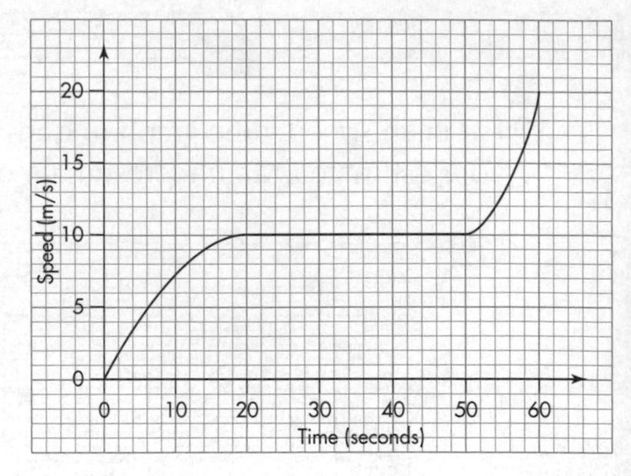

AU **i** Describe the journey.

ii Estimate the distance travelled in the first 20 seconds.

b This speed–time graph also shows, as a dashed line, the journey of a lorry, which sets off at the same time as the car.

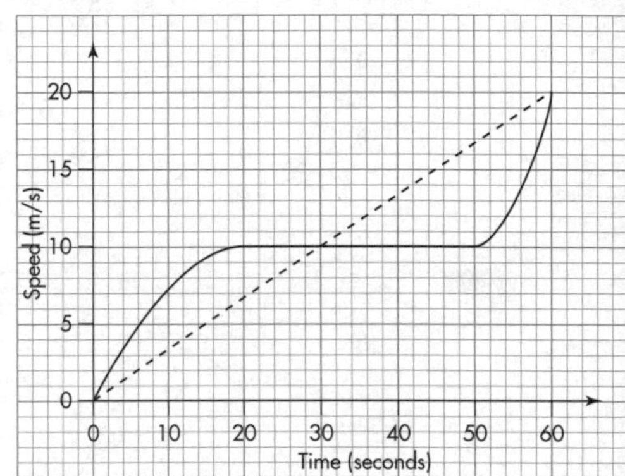

AU **i** Compare the lorry journey with the car journey.

PS **ii** Which travels further in one minute, the car or the lorry? Show how you decide.

12.3 Gradients of straight-line graphs in practical contexts

This section will show you how to:

- interpret the graph of a line for a real-life context
- work out a formula from a graph that represents a real-life context
- draw a graph that represents a real-life context given a formula

Key words

acceleration
deceleration
gradient
units
y-intercept

You know that: **gradient**

$$= \frac{\text{distance measured up}}{\text{distance measured along}}$$

Remember: When a line slopes up from left to right it has a positive gradient, when it slopes down from left to right it has a negative gradient.

The point where a graph crosses the *y*-axis is called the ***y*-intercept**.

EXAMPLE 6

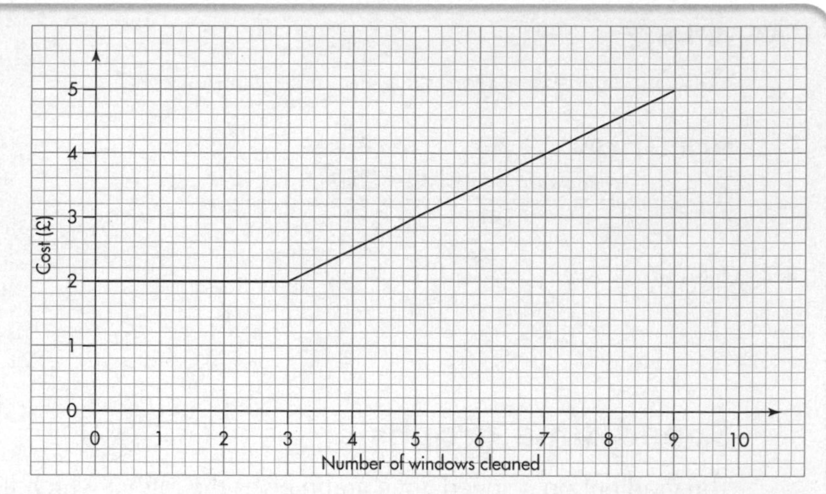

The graph shows how a window cleaner works out his costs.

He has a fixed charge for cleaning up to three windows.

He then charges for each extra window he cleans.

a What is the fixed charge?

b How much is the extra charge per window?

SOLUTION

a The fixed charge is shown by the y-intercept. This is £2.

b The extra charge per window is worked out using the gradient.

From the graph, when four extra windows are cleaned the cost increases by £2.

$$\text{gradient} = \frac{\text{distance measured up}}{\text{distance measured along}} = \frac{£2}{4} = 50\text{p}$$

So the extra cost is 50p per window.

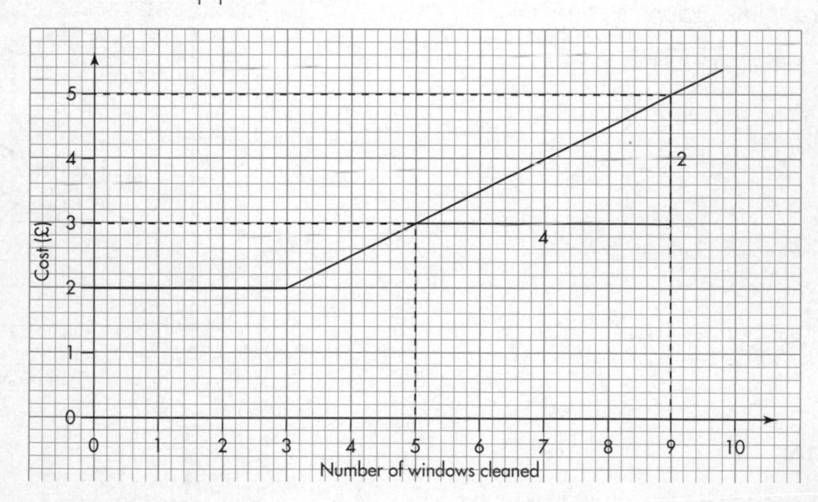

Distance–time graphs

You can find the speed from the gradient of a distance–time graph, since:

$$\text{speed} = \frac{\text{distance travelled}}{\text{time taken}}$$

If the gradient is positive the direction is forwards. If the gradient is negative the direction is backwards. The **units** for speed are, for example, metres per second (m/s), kilometres per hour (km/h) or miles per hour (mph).

EXAMPLE 7

Work out the speed from this distance–time graph.

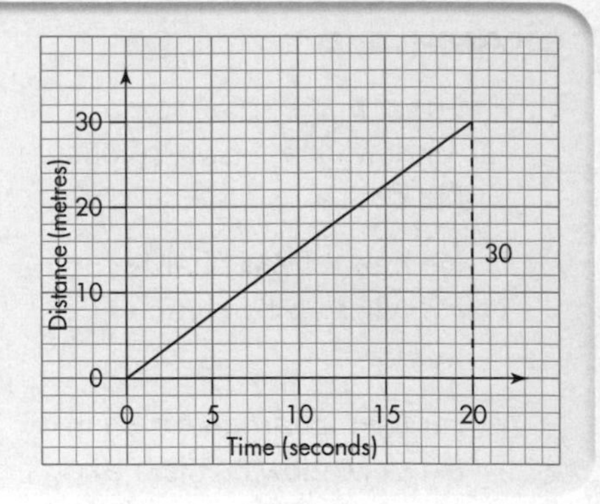

SOLUTION

$$\text{Speed} = \frac{\text{distance travelled}}{\text{time taken}} = \frac{30}{20} = 1.5 \text{ m/s}$$

Speed–time graphs

The gradient on a speed–time graph gives the rate at which the speed is increasing or decreasing in a given time. If the gradient is positive it is an **acceleration**. If the gradient is negative it is a **deceleration**.

$$\text{Acceleration or deceleration} = \frac{\text{difference in speed}}{\text{difference in time}}$$

The units for acceleration and deceleration are, for example, metres per second per second (m/s^2) or kilometres per hour per hour (km/h^2).

EXAMPLE 8

In this speed–time graph:

a find the initial speed

b work out the acceleration.

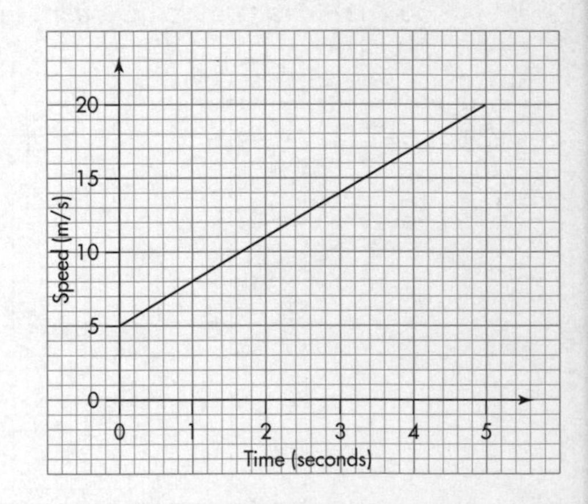

SOLUTION

a Initial speed is the speed at the start of the journey and is found from the y-intercept.

Initial speed = 5 m/s

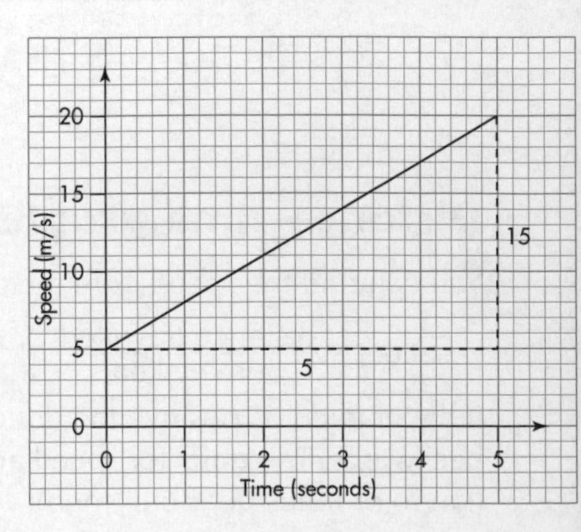

b $$\text{Acceleration} = \frac{\text{difference in speed}}{\text{difference in time}}$$

$$= \frac{15}{5} \text{ m/s}^2$$

$$= 3 \text{ m/s}^2$$

EXERCISE 12C

AU 1 The graph shows the charges made by a taxi company.

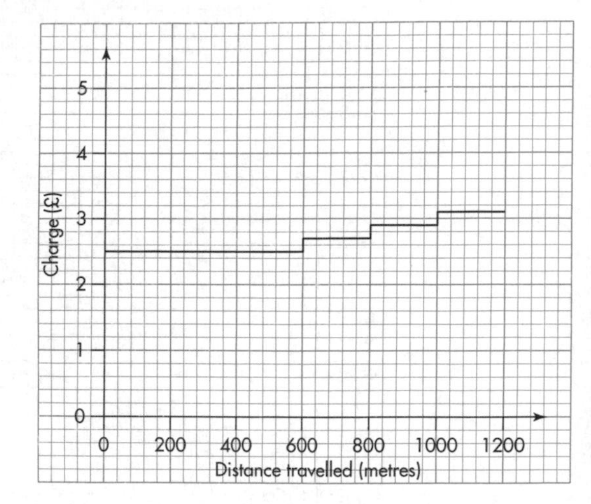

a Describe the charges in words by completing this sentence.

It costs … for the first … metres and then … for every extra … metres.

b Explain why the graph goes up in steps.

2 A plumber has a £35 call out charge and then charges £15 per half-hour.

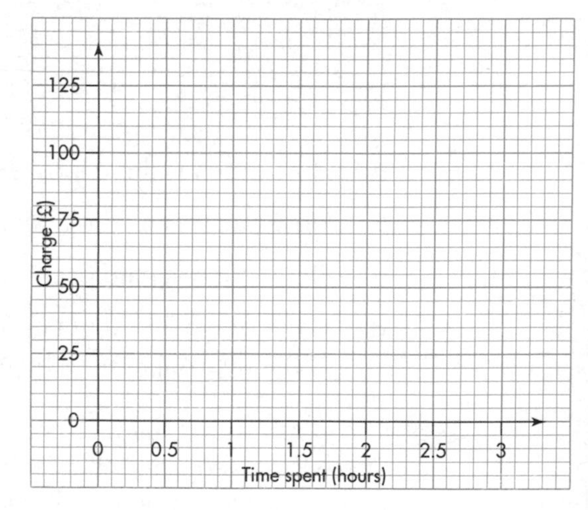

a Copy the grid and represent this information as a graph.

b Work out the cost of a job lasting 2 hours.

FM 3 The graph shows a mobile phone tariff, made up of a fixed charge (including free calls) and a charge for extra minutes of calls.

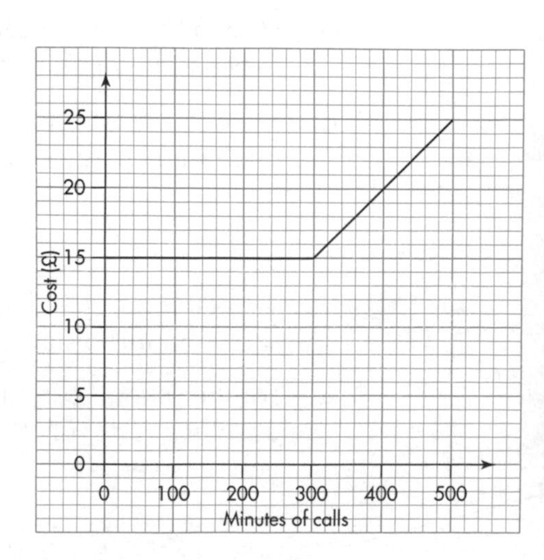

a How much is the fixed charge?

b How many free minutes of calls are there?

c Work out the charge per minute of the extra calls.

D

C

4 The graph shows the deposit and the monthly payments for a sofa.

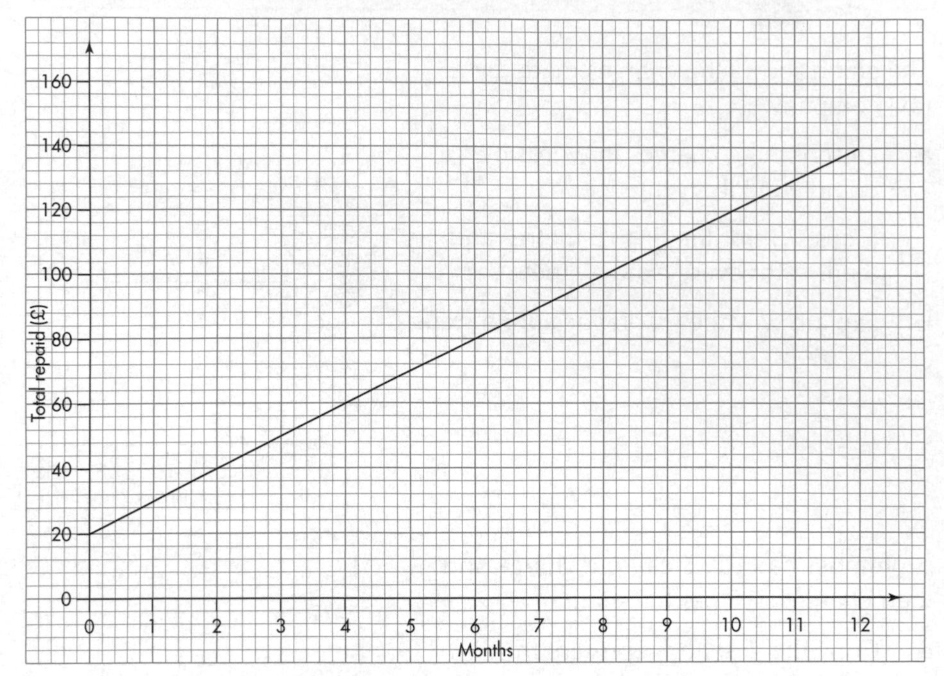

a How much is the deposit?

b How much is repaid altogether over the 12 months?

c How much is paid each month?

d Write down a formula connecting the total repaid, T, the deposit, d, and the monthly payment, m.

5 The graph shows the hire charge for a meeting room. There is a fixed basic charge, then it increases, based on the number of people attending.

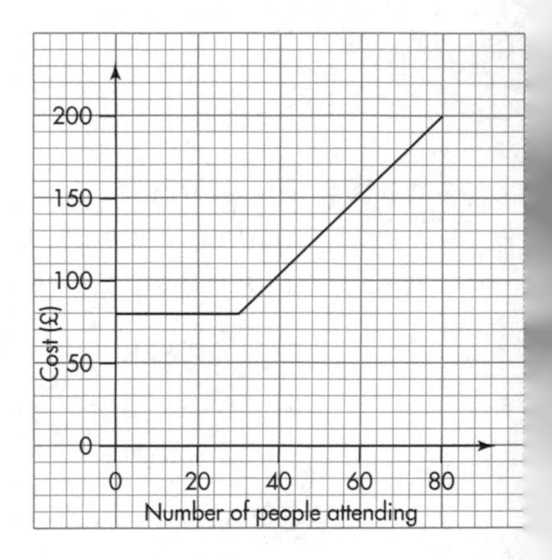

a How much does it cost if 20 people attend?

b How many people can attend before the charge exceeds the minimum (fixed) charge?

c How much extra is it per person, when the minimum charge is exceeded?

AU 6 **a** Work out the speed for the first part of the journey shown in this distance–time graph.

b For the second part of the journey, was the vehicle travelling faster or more slowly? Give a reason for your answer.

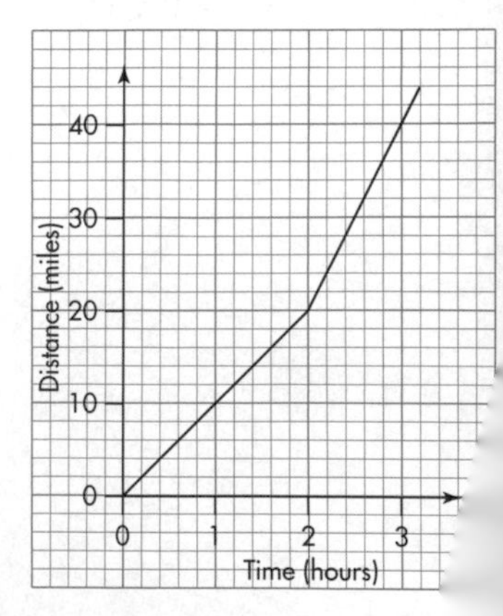

In questions **7** and **8**, work out the acceleration or deceleration for each section of the graph.

7

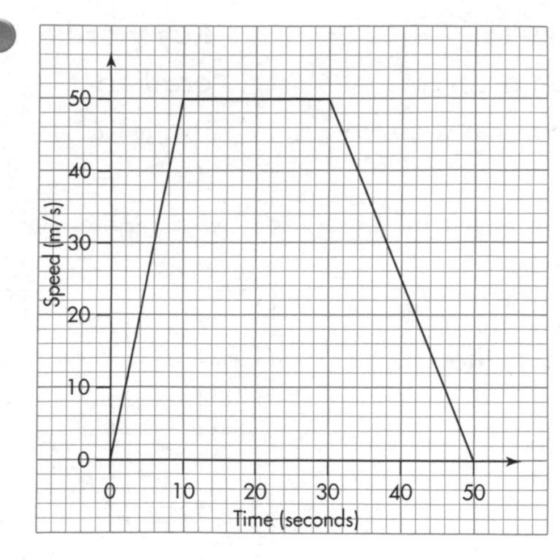

8

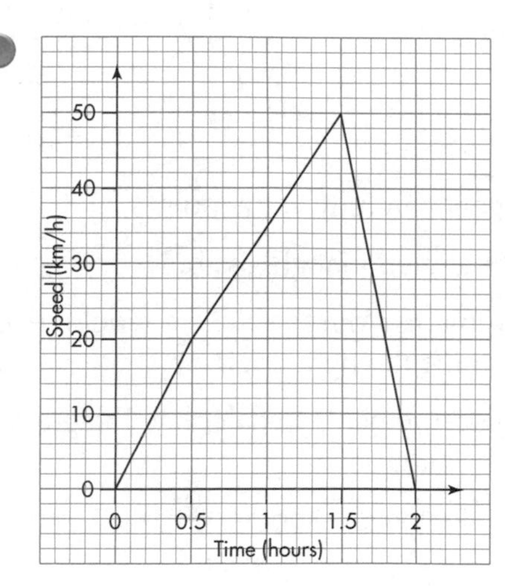

PS 9 The graph shows the speed of a car between two junctions.

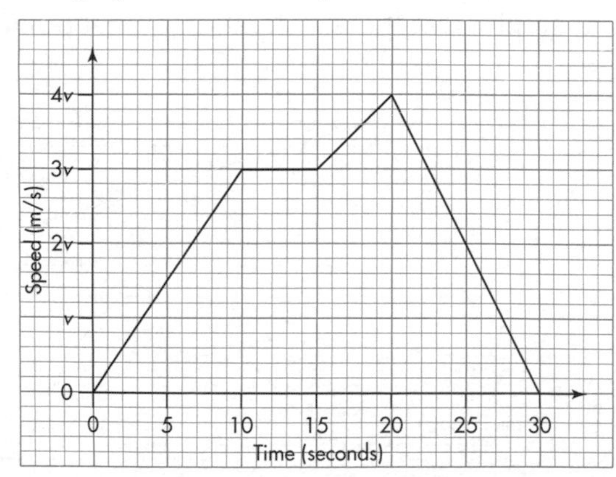

a Work out the acceleration in the first 10 seconds, in terms of v.

b The distance travelled in the first 10 seconds is 75 metres. Work out the total distance travelled between the junctions.

12.4 Interpreting the gradient at a point on a curve as the rate of change

This section will show you how to:
- draw a tangent at a point on a curve
- work out a gradient at a point on a curve
- interpret the gradient

Key words
gradient
tangent

A **tangent** is a line that touches a curve at a point.

The **gradient** of a curve at a point is the same as the gradient of the tangent at that point.

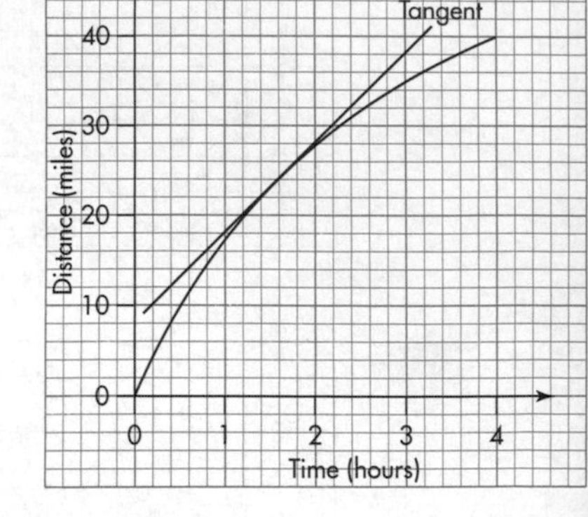

Distance–time graphs

The gradient at any point on a distance–time graph gives the speed at that point. Follow these steps to calculate the speed.

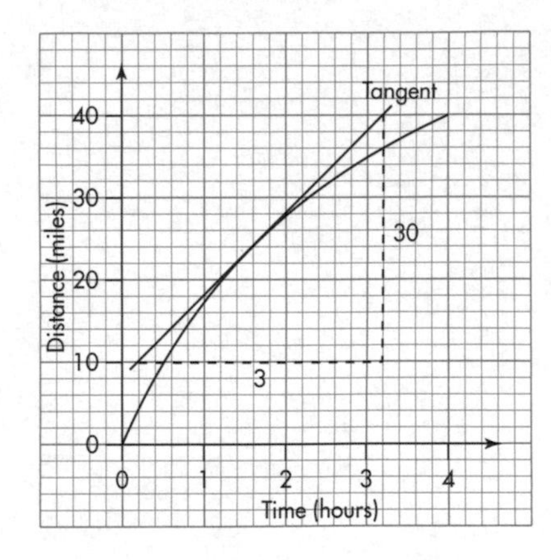

Draw a tangent carefully at the point.

Make a right-angled triangle as shown.

Use measurements from the axes to label the length of the sides forming the right angle.

Use these measurements to calculate the gradient of the tangent.

In this example, the gradient is $\frac{30}{3}$ = 10 mph.

Speed–time graphs

The gradient at any point on a speed–time graph gives the acceleration at that point. Follow the same steps as described above.

EXAMPLE 9

Estimate the acceleration at 1 hour, in the journey shown in this graph.

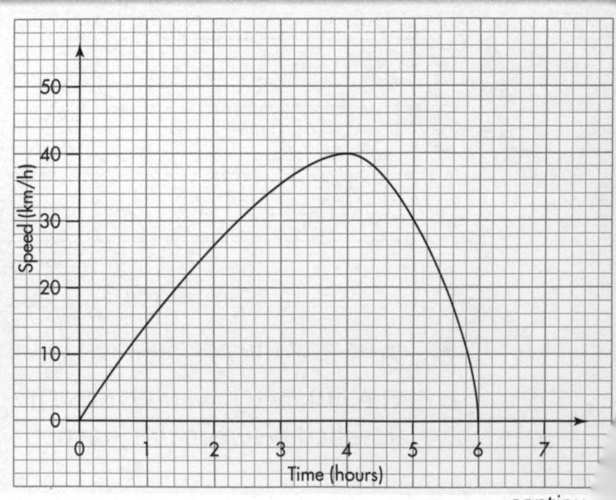

continu

SOLUTION

Draw a tangent carefully at the point where time = 1 hour.

Add a right-angled triangle.

Use measurements from the axes to label the length of the sides forming the right angle.

Use these measurements to calculate the gradient of the tangent.

Acceleration is $\frac{36}{3}$ = 12 km/h²

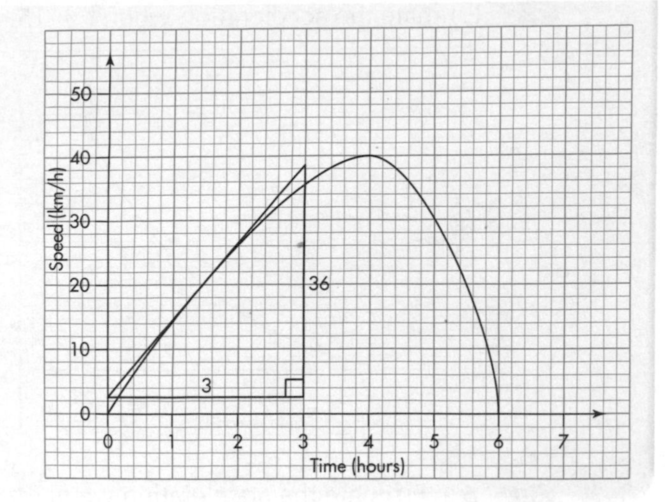

EXERCISE 12D

1 The graph shows the height of a ball as it is thrown into the air.

a Draw a tangent at the point where t = 1.

b Use your tangent to estimate the speed of the ball after one second.

c Write down the speed of the ball after 2 seconds.

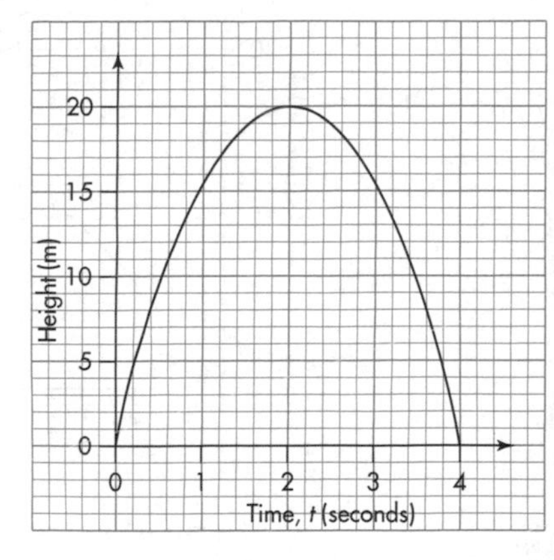

AU 2 The graph shows a distance–time graph.

a Estimate the speed when:

i t = 1 **ii** t = 4

b At what time is the speed zero?

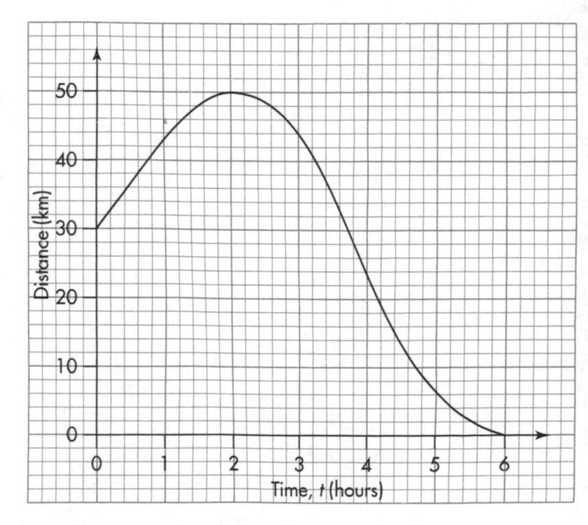

3 Estimate the acceleration when $t = 15$.

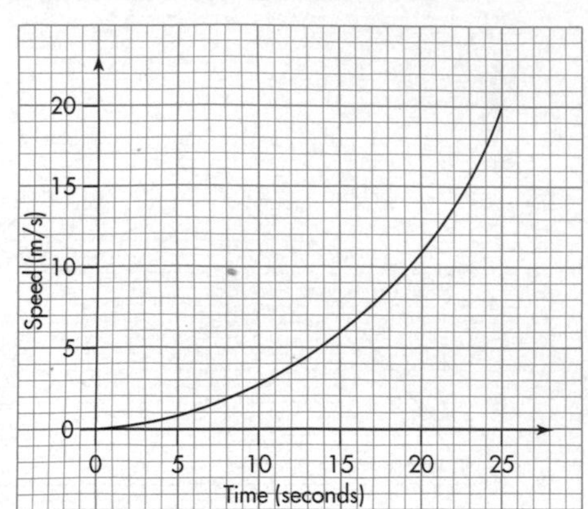

4 Estimate the acceleration when $t = 10$.

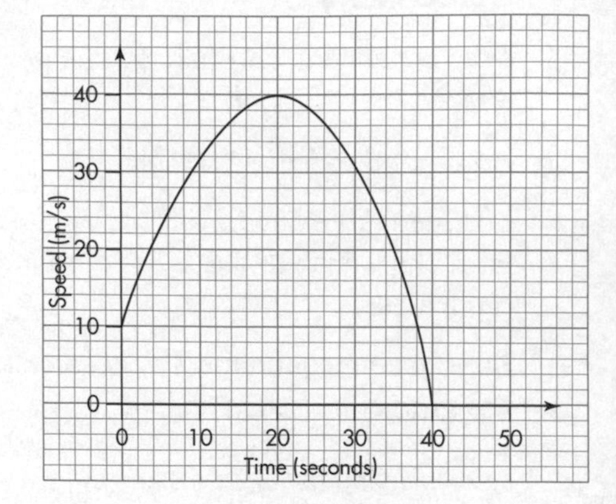

5 **a** Estimate the acceleration when $t = 2$.

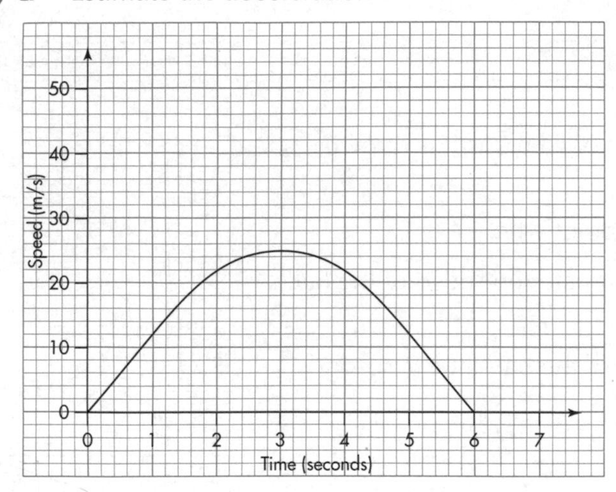

b Write down the deceleration when $t = 4$.

6 Estimate the acceleration when $t = 1$.

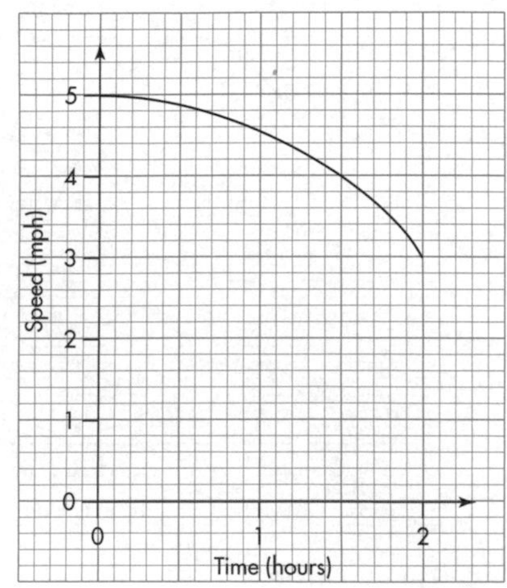

7 Look again at the graph for question **1**. Find two times when the speeds are numerically equal but in opposite directions.

12.5 Graphs that illustrate direct and inverse proportion

This section will show you how to:
- work out the constant of proportionality from a given graph
- match graphs to equations that represent direct proportion
- match graphs to equations that represent indirect proportion (Higher tier)

Key words
constant of proportionality, k
direct proportion
direct variation
indirect proportion (higher)
indirect variation (higher)
reciprocal (higher)

Direct proportion

Direct proportion and **direct variation** mean the same. There is a direct proportion between two variables when one variable is a constant multiple of the other. This multiple is called the **constant of proportionality** and is often represented as $\boldsymbol{k}$.

For example, the statement: 'Distance travelled is directly proportional to the time taken' can be written as:

distance travelled $\propto$ time taken

Note: The symbol $\propto$ means 'is directly proportional to'.

This implies that distance travelled $= k \times$ time taken where k is the constant of proportionality. On a sketch graph this would be a straight line.

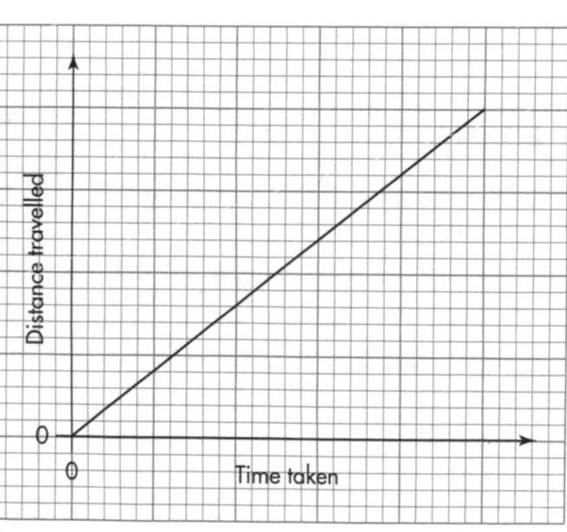

The statement: 'Area of a circle is directly proportional to the square of the radius' can be written as:

area of a circle $\propto$ square of radius

This implies that the area of a circle $= k \times (\text{radius})^2$

$A = kr^2$

On a sketch graph of area against radius, this would be a quadratic curve.

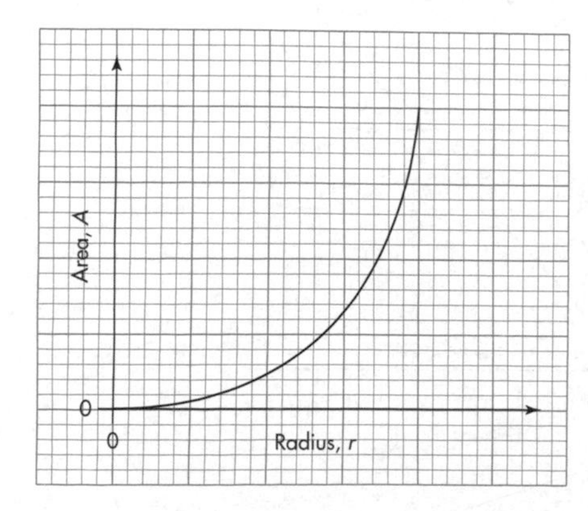

To summarise, at Foundation tier you need to know and recognise these direct proportion graphs.

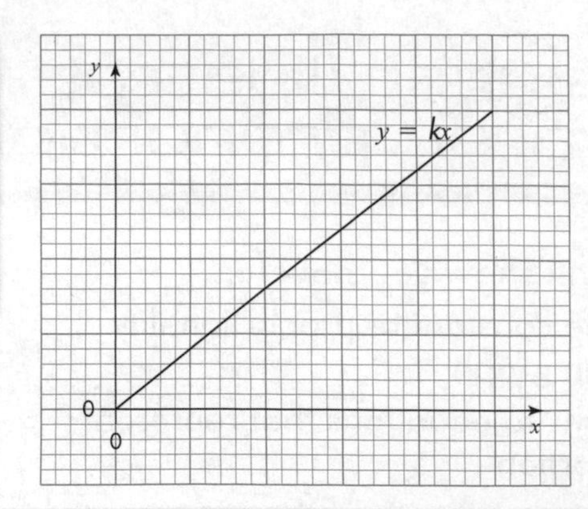

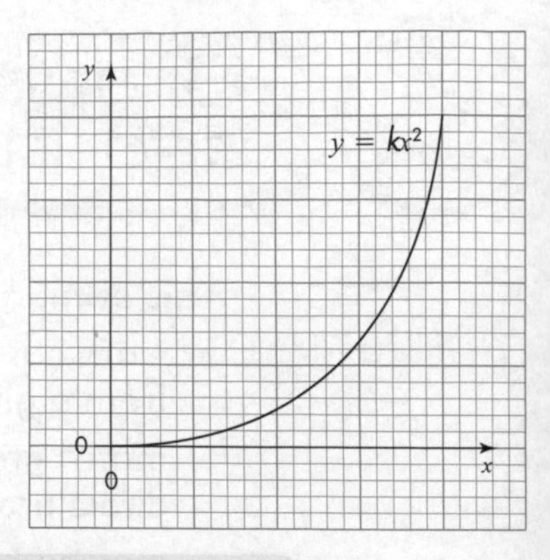

At Higher tier you also need to know and recognise this direct proportion graph.

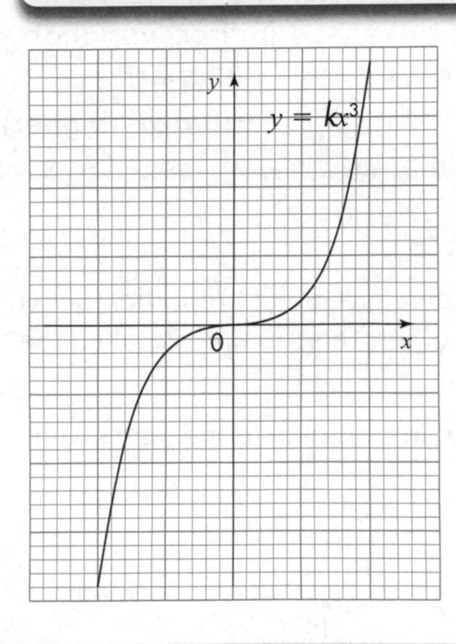

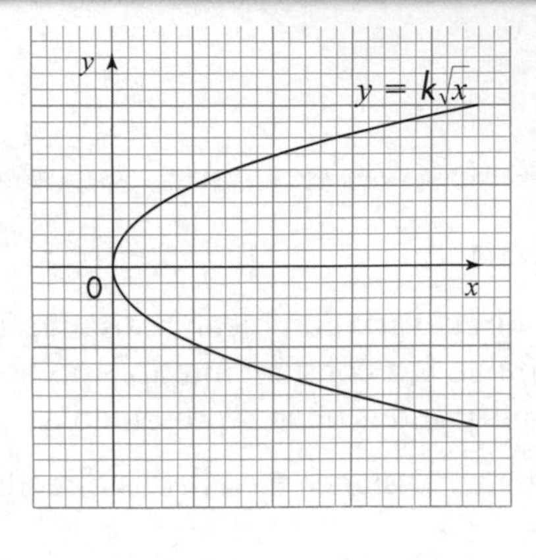

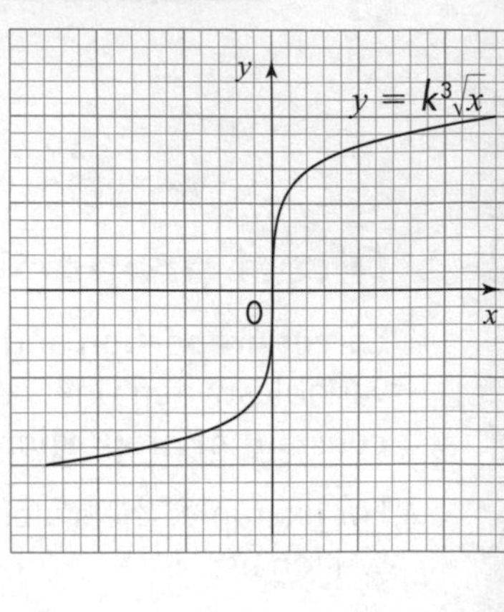

EXAMPLE 10

a Sketch a graph to represent the relation 1 kilogram = 2.2 pounds.

b Write down an equation to represent your graph.

SOLUTION

a

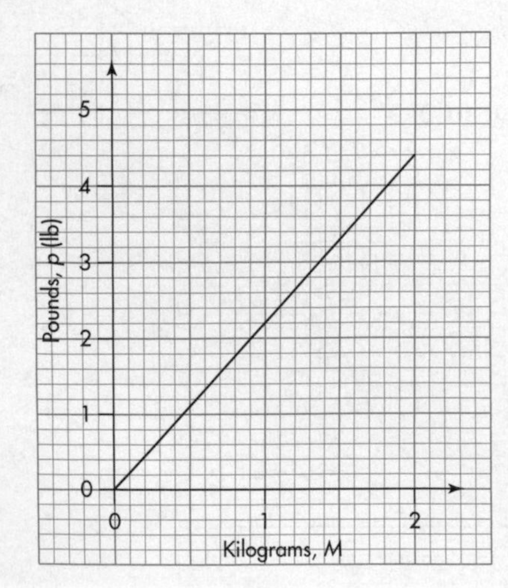

b $p = 2.2M$

EXAMPLE 11

The graph represents the relationship $y = kx^2$.

Use the graph to work out the value of k.

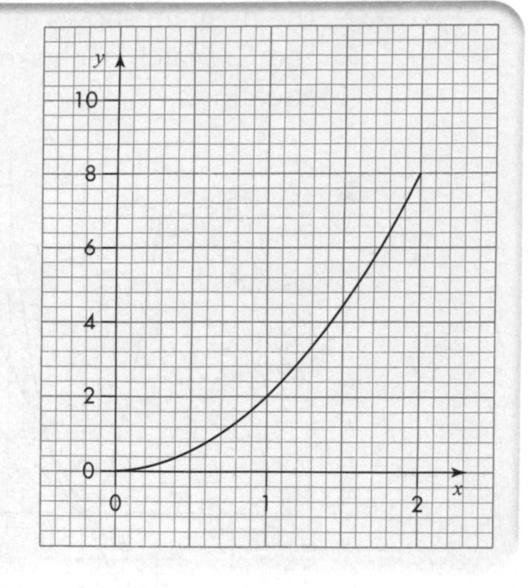

SOLUTION

Choose a point of the graph other than (0, 0), for example, (1, 2).

Substitute $x = 1$ and $y = 2$ into the equation $y = kx^2$.

So $2 = k \times 1^2$ giving $k = 2$.

Indirect proportion

Indirect proportion and **indirect variation** mean the same. There is indirect proportion between two variables when one variable is a constant multiple of the **reciprocal** of the other.

You need to know and recognise these indirect proportion graphs.

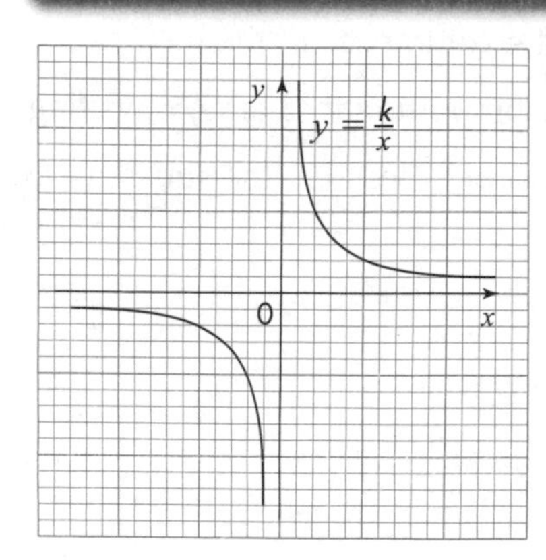

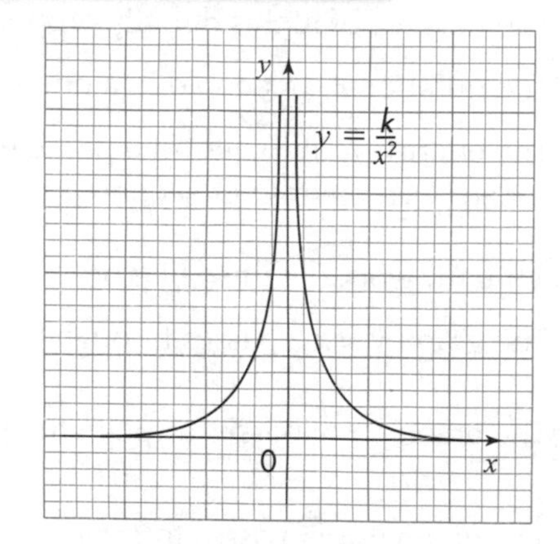

EXERCISE 12E

1 For each equation, copy the grid, then draw and label a sketch graph.

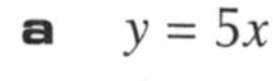

a $y = 5x$

b $y = 2x^2$

c $y = \frac{1}{2}x$

d $y = 3x^2$

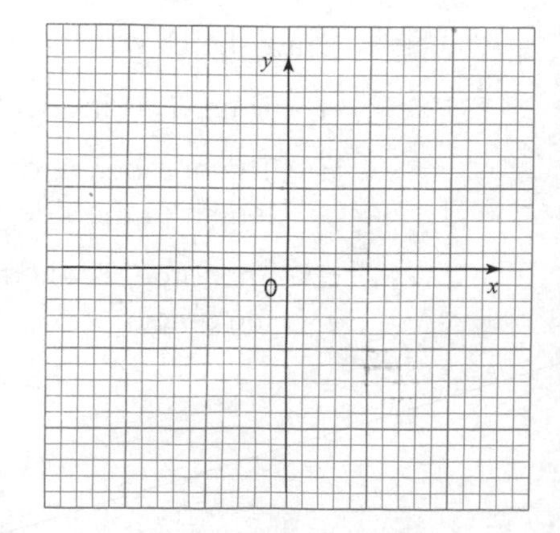

PS **2** The diagram shows three sketch graphs, A, B and C. Match each sketch graph to the appropriate equation.

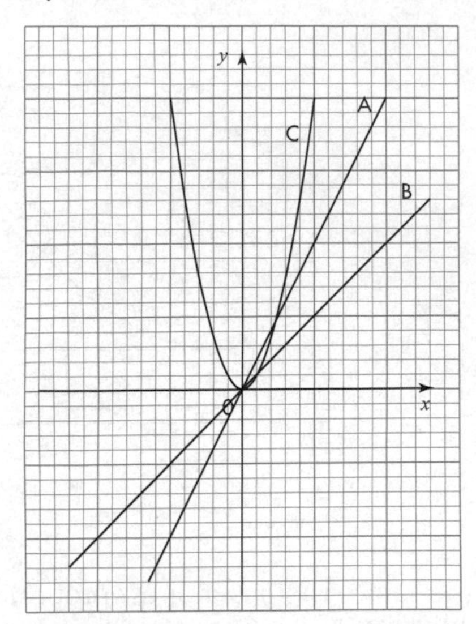

Equation 1: $y = 4x^2$ Equation 2: $y = x$ Equation 3: $y = 2x$

AU **3** Draw a sketch graph to represent each of the following statements.

a The cost, C, of fuel is directly proportional to the volume, V, of the fuel.

b The number, n, of people in a queue is directly proportional to time, t, waiting.

c The energy, J, in joules, of a particle is directly proportional to the square of the speed, s, in metres per second (m/s).

4 In each part, work out the value of k.

a The graph $y = kx^3$ passes through the point (2, 16)

b The graph $y = \frac{k}{x}$ passes through the point (5, 4)

c The graph $y = \frac{k}{x^2}$ passes through the point (2, 9)

d The graph $y = \frac{k}{x^3}$ passes through the point (5, 2)

AU FM **5** Draw sketch graphs to represent each of the following.

a Time, T, is inversely proportional to speed, s.

b The number, n, of square tiles needed is inversely proportional to the square of the length of the side, s, of the square.

c The force, F, exerted by a magnet is inversely proportional to its distance, d, from the object.

6 The kinetic energy, K, of an object is directly proportional to the square of its velocity, v.

a Which of the following graphs represents the relationship between kinetic energy and velocity?

Graph A

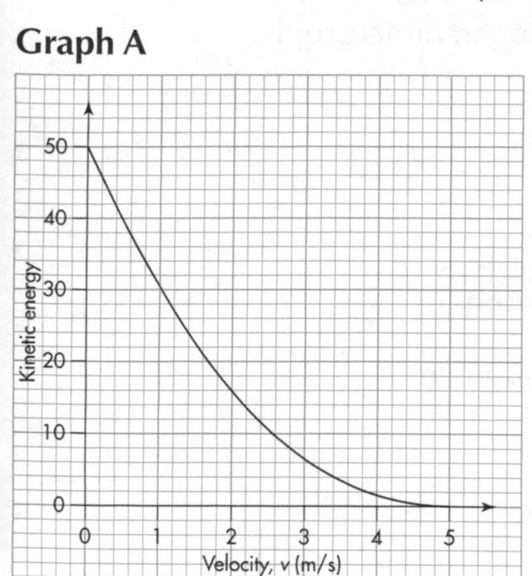

Graph B

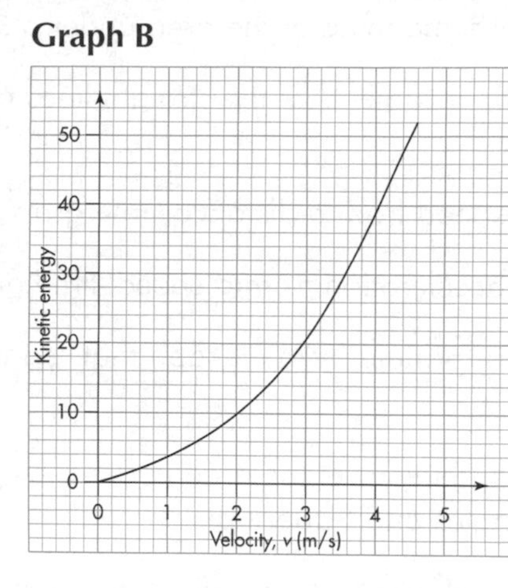

Graph C

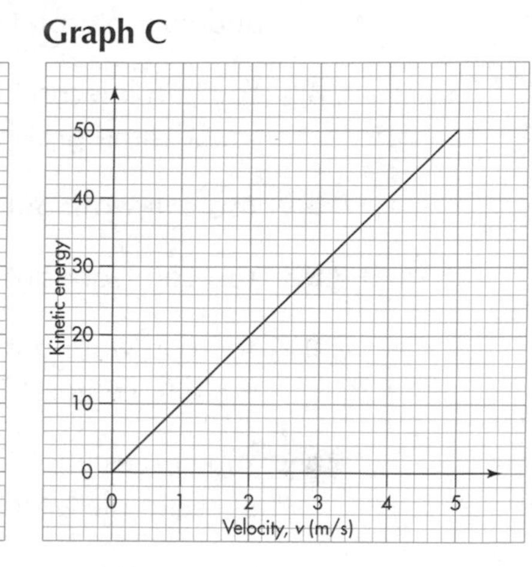

b The formula for the kinetic energy of an object is $k = \frac{1}{2}mv^2$, where m is the mass of the object.

Use the graph you chose in part **a** to work out the mass of the object.

7 Boyle's law states that, for a fixed amount of gas at a fixed temperature, the pressure is inversely proportional to the volume.

a Which of the following graphs represents the relationship between pressure and volume?

Graph A

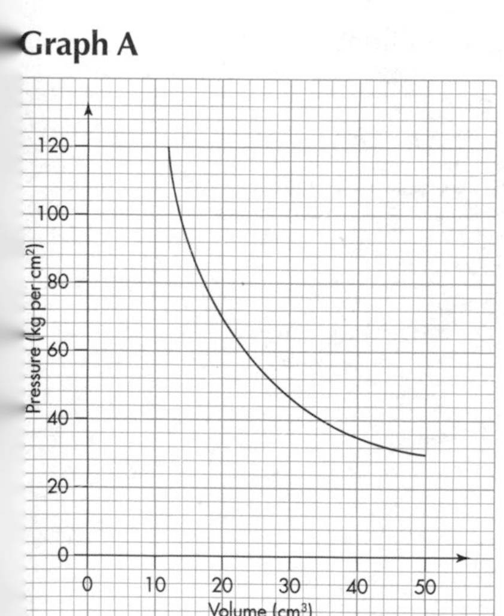

Graph B

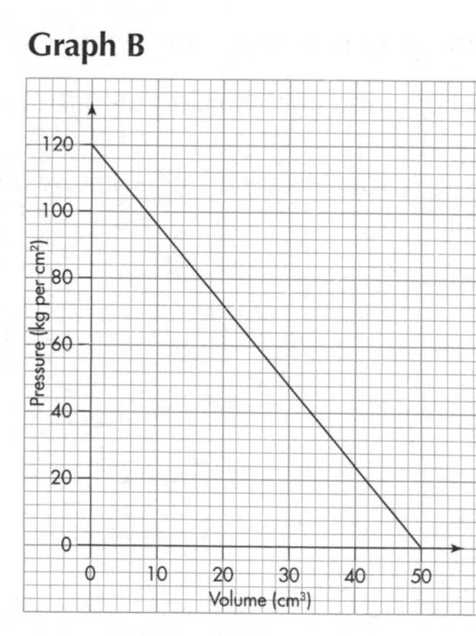

Graph C

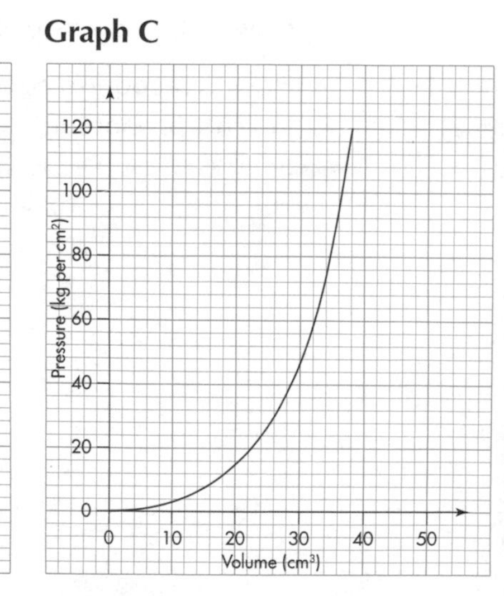

b Boyle's law can be stated algebraically as $PV = kT$, where P is the pressure, V is the volume, T is the temperature and k is a constant.

The data was measured when the temperature was kept at a constant 20°C.

Use the graph you chose in part **a** to work out the approximate value of k.

GRADE BOOSTER

- **C** You can calculate the area under a graph consisting of straight lines and can interpret the meaning of the area under a speed–time graph
- **C** You can interpret the gradients of the straight lines on a speed–time graph
- **B** You can work out speed from a distance–time graph
- **B** You can work out acceleration from a speed–time graph
- **B** You can work out a formula from a graph that represents a real-life context
- **A** You can estimate and interpret the area under a curve (using areas of rectangles, triangles and trapeziums)
- **A** You can work out and interpret a gradient at a point on a curve
- **A*** You can draw and use graphs to illustrate direct and inverse proportion

What you should know now

- How to draw and interpret distance–time and speed–time graphs fully
- How to find and interpret the gradient of a straight line and a curved line
- How to find areas under graphs with straight lines and curved lines
- How to recognise equations and curves relating to direct and indirect proportion

AU PS 1 The graph shows the charges for a bill from company A.

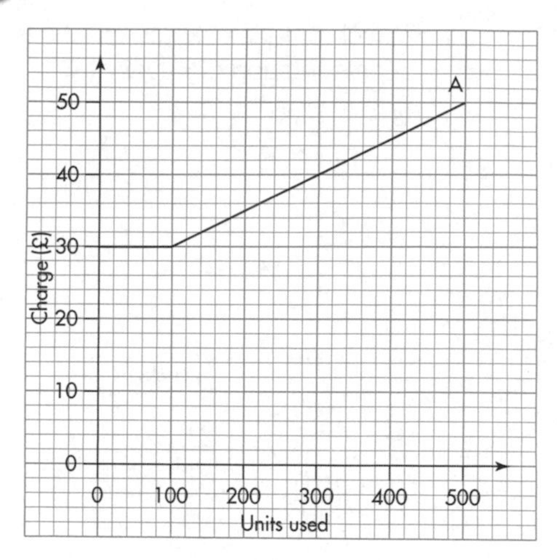

a What is the fixed charge?

b Work out the cost per unit.

c Company B has a fixed charge of £20 for the first 200 units and then charges 10p per unit.

  i Show this information on a copy of the graph.

  ii Which company should a person choose if they use more than 500 units? Give a reason for your answer.

2 The diagram shows a speed–time graph for a journey.

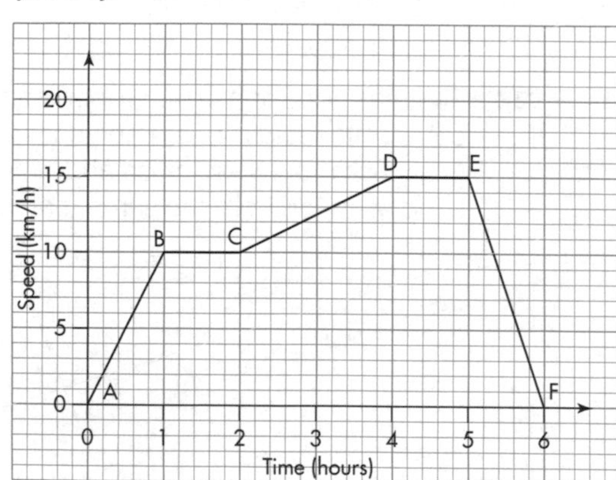

a During which parts of the journey is the speed constant?

b During which parts of the journey is the speed increasing?

c Work out the total distance travelled.

d Work out the average speed for the whole journey.

AU 3 The distance travelled by a car is directly proportional to the time taken.

a Represent this information on a sketch graph.

b What can you say about the speed in this case?

4 A train travels between two stations as shown.

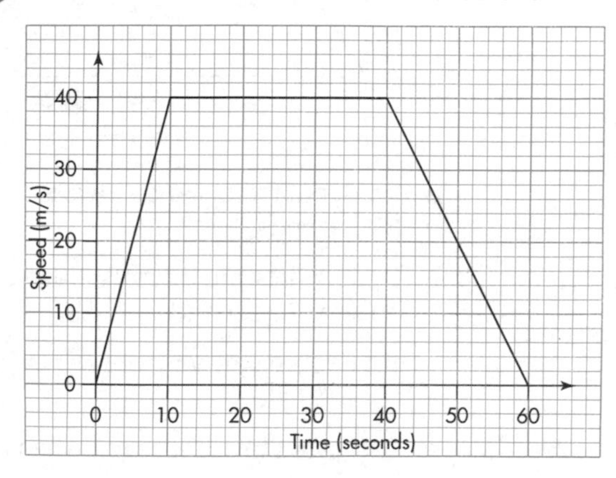

Work out the distance between the stations, giving your answer in kilometres.

5 The diagram shows a speed–time graph.

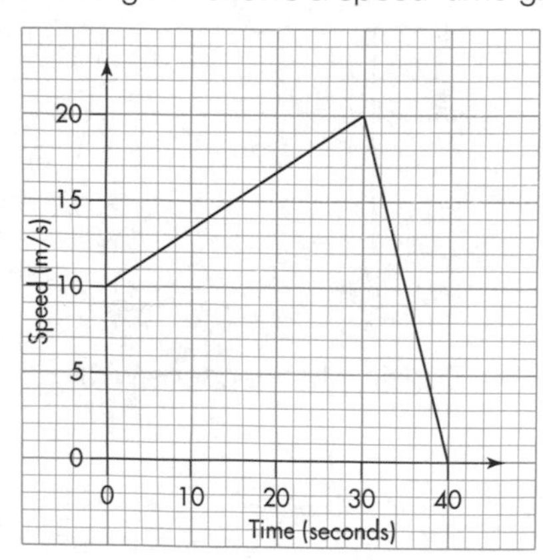

a Work out the total distance travelled.

b Work out the acceleration in the first 30 seconds.

AU **6** The diagram shows a speed–time graph.

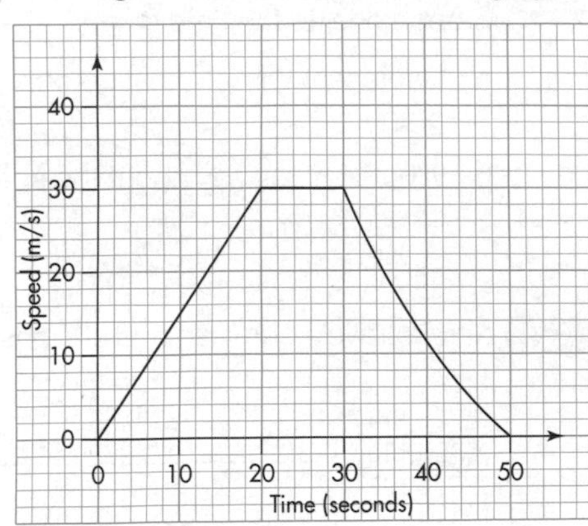

a Work out the distance travelled in the first 30 seconds.

b **i** Estimate the distance travelled in the final 20 seconds.

ii Explain why your answer to part i is an estimate.

AU **7** **a** Estimate the acceleration when $t = 10$.

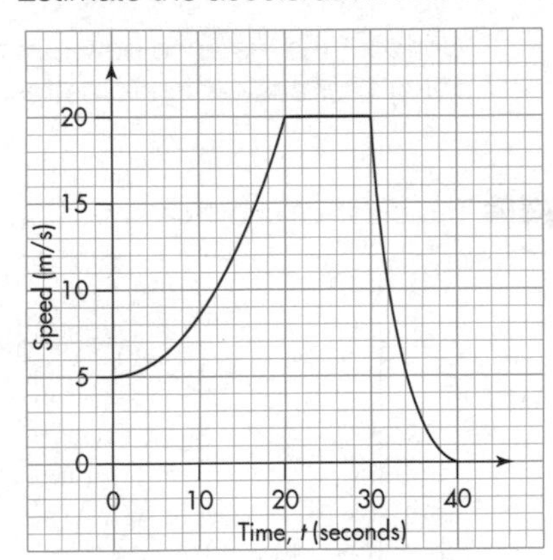

b Explain what happens when $t = 20$.

AU **8**

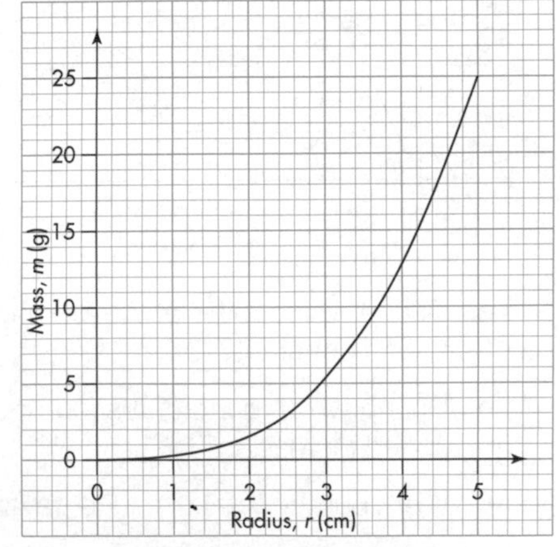

a Estimate the acceleration when $t = 0.5$ hours.

b At what time is the acceleration greatest? Explain your answer.

AU **9** The frequency, f, of sound is inversely proportional to the wavelength, w.

a Which of these sketch graphs is the correct representation of f against w?

A

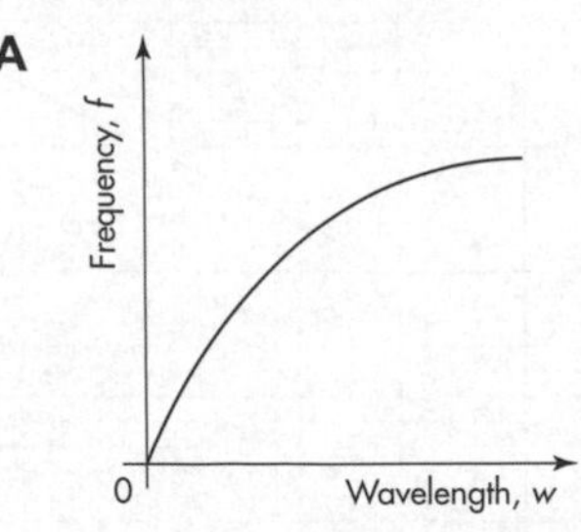

B

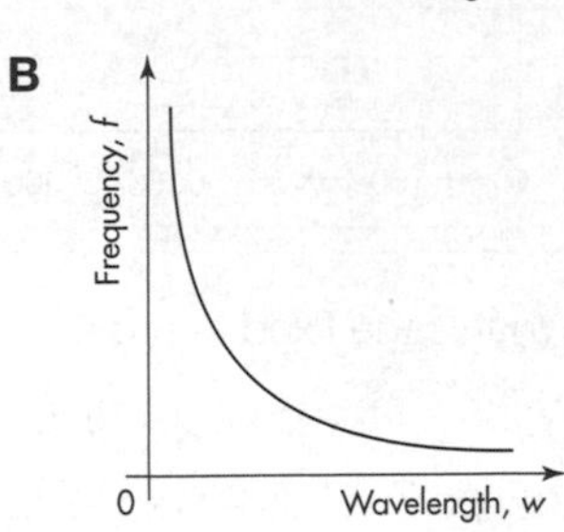

C

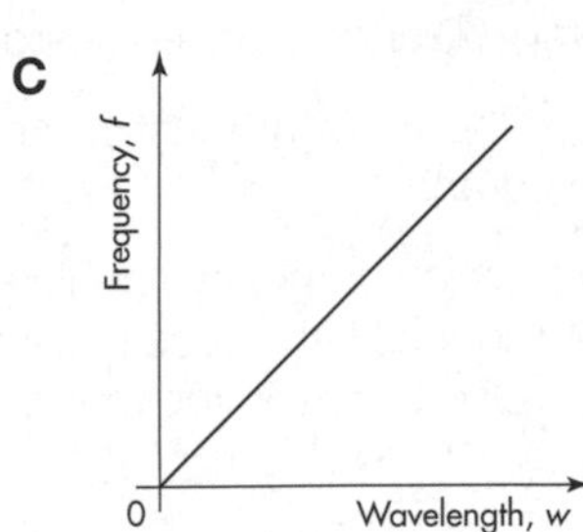

b What happens to the frequency as the wavelength increases?

10 The mass of a sphere is proportional to the cube of its radius. This graph shows the relationship between mass and radius for spheres made from the same material.

What is the mass of a sphere with a radius 20 cm made from the same material?

Worked Examination Questions

C **1** Jon invests £3000 in a saving account and then each month adds a further £500.

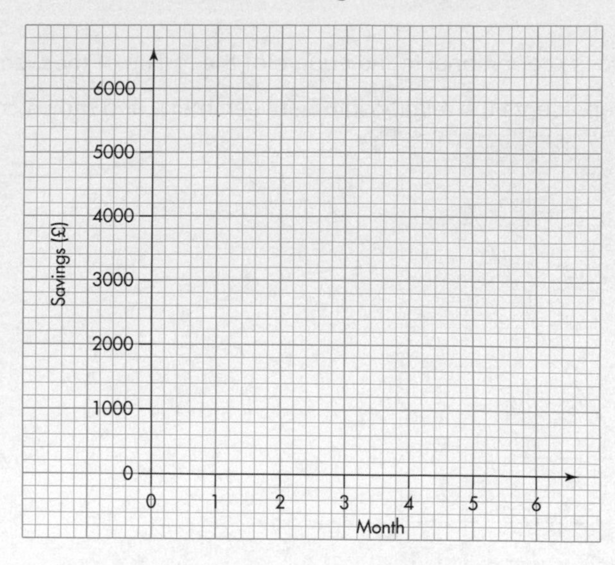

a Represent this information on the graph.

b Write down a formula for his total savings, T, in terms of his monthly savings, m.

a As the initial investment is £3000, the graph starts at (0, 3000).

Each month the amount rises £500 so the gradient is 500 and the graph passes through, for example, (6, 6000).

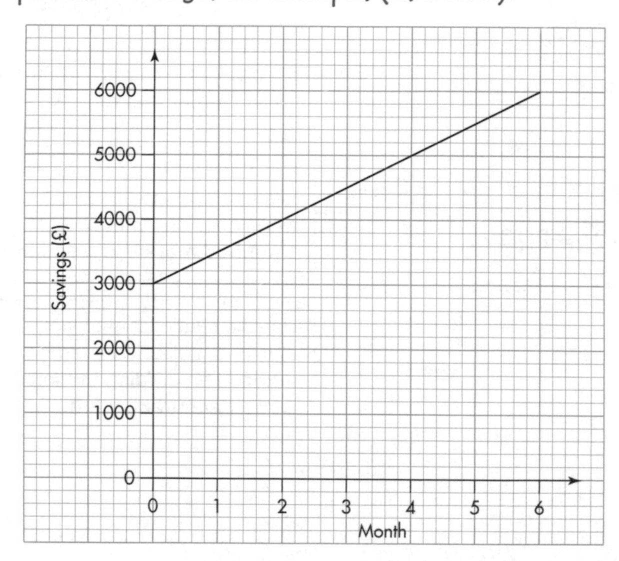

You gain **1 mark** for starting the graph at £3000 on the vertical scale. You gain **1 mark** for a line drawn with gradient 500.

b As the amount saved per month is £500, the monthly savings will be $500 \times m$ or $500m$. This is added to £3000 to work out the total saved, so the formula is: $T = 3000 + 500m$

You will score **2 marks** for a correct answer or 1 mark if part of the answer is correct.

Total: 4 marks

Worked Examination Questions

A* **2** A company makes circular plastic discs. The discs are the same thickness but have different radii.

The mass of each disc is proportional to the square of the radius of the disc.

The following graph represents the relationship between mass and radius.

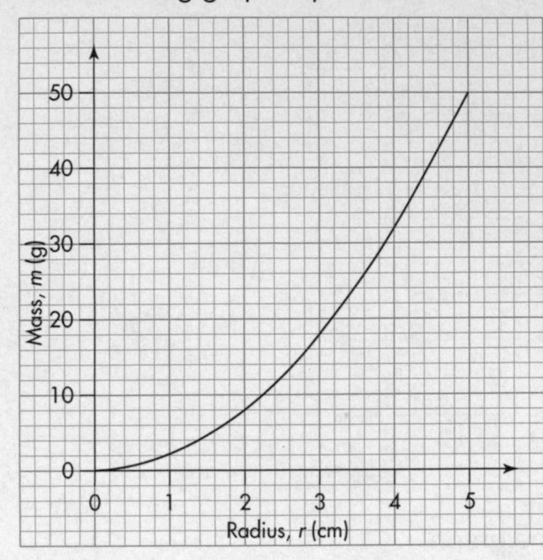

1000 discs with a radius of 4 cm are melted down and made into discs with a radius of 10 cm.

Use the graph to work out how many 10 cm discs were made.

Mass of a 4 cm disc = 32 g.

> This can be directly from the graph so earns **1 mark**.

Mass of 1000 of these = 32 000 g.

> Multiplying the mass by 1000 is just a mental skill so earns **1 mark**.

The relationship is $m = k \times r^2$

So $32 = k \times 4^2 \Rightarrow k = 2$

Check when $r = 5$, $m = 50$, $50 = 2 \times 5^2$ so $k = 2$

> Use the points on the graph to work out the relationship between r and m. It is always worth checking with two points. This earns **1 method mark** and **1 accuracy mark**.

Mass of a 10 cm disc = $2 \times 10^2 = 200$ g.

So 32 000 ÷ 200 = 160 discs.

> Use the relationship to work out the mass of a 10 cm disc. This earns **1 method mark**. Then dividing this into the total mass earns **1 accuracy mark**.

Total: 6 marks

Answers

Chapter 1 Probability: Set notation and Venn diagrams

Quick check

1 French and German German and Spanish
French and Spanish

2 **a** $\frac{1}{4}$ **b** $\frac{3}{10}$ **c** $\frac{3}{4}$

3 **a** $\frac{1}{2}$ **b** $\frac{5}{6}$ **c** $\frac{1}{3}$

Exercise 1A

1 **a** $\frac{5}{12}$ **b** $\frac{7}{12}$ **c** $\frac{1}{2}$
d $\frac{1}{2}$

2 **a** They are the same. **b** They are the same.

3 **a** $A = \{3, 6, 9\}$ **b** $\frac{3}{10}$ **c** $\frac{7}{10}$
d $B = \{2, 3, 5, 7\}$ **e** $\frac{2}{5}$ **f** $\frac{3}{5}$

4 **a** 0.9 **b** 0.7

5 **a** 0.75 **b** 0.45

Exercise 1B

1 **a**

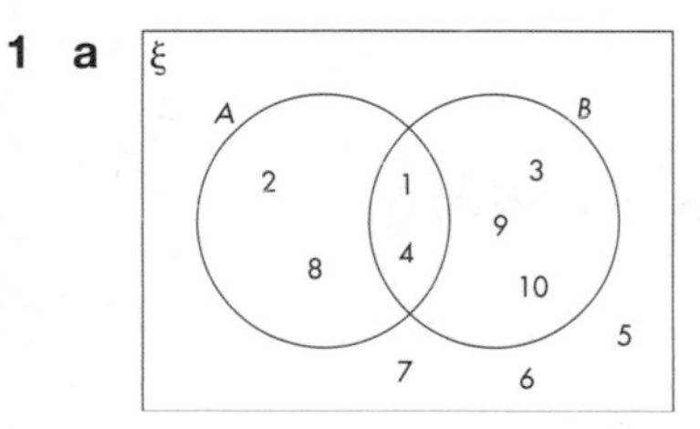

b **i** $\frac{2}{5}$ **ii** $\frac{3}{5}$ **iii** $\frac{1}{2}$
iv $\frac{1}{2}$ **v** $\frac{7}{10}$ **vi** $\frac{1}{5}$

2 **a** 130
b **i** $\frac{8}{13}$
ii The probability that a student chosen at random walks to and from school
c $\frac{5}{26}$

3 **a** **i** 0.52 **ii** 0.48 **iii** 0.65
iv 0.35 **v** 0.82 **vi** 0.35
b 0.3

4 **a**

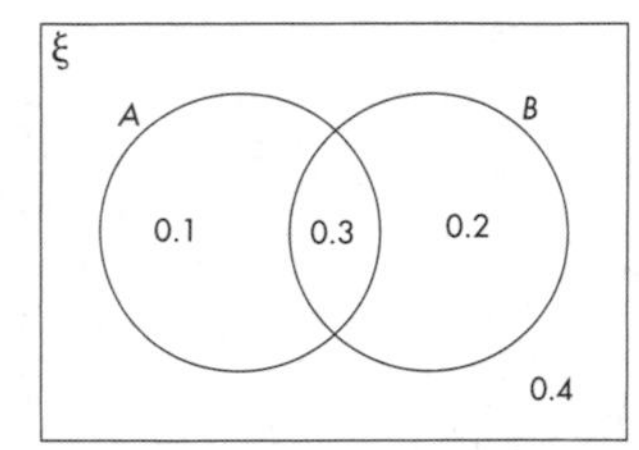

b **i** 0.5 **ii** 0.6 **iii** 0.3

5 **a**

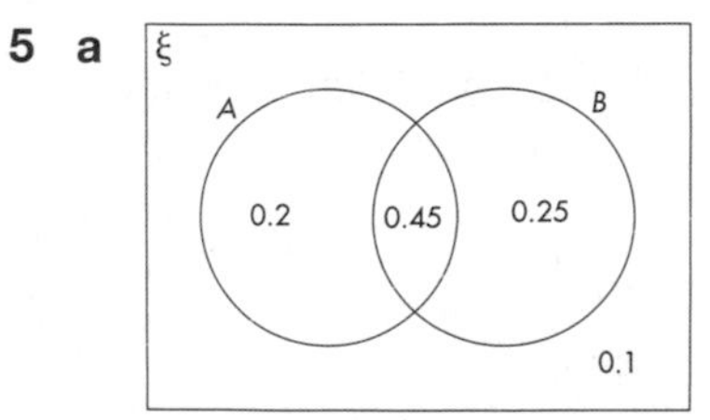

b **i** 0.65 **ii** 0.7 **iii** 0.9
iv 0.45

6 0.4

7 0.5

Exercise 1C

1 **a**

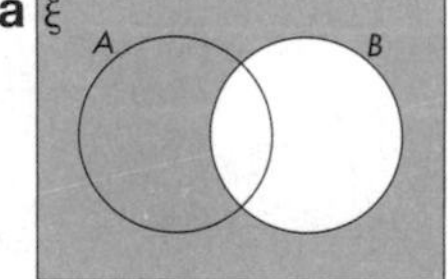

b

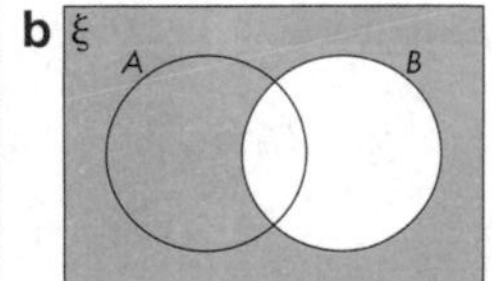

2 **a** $A' \cap B$
b $A \cap B$

3 $\frac{1}{2}$

4 a

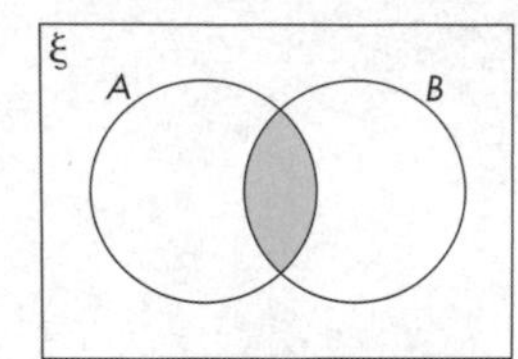

b

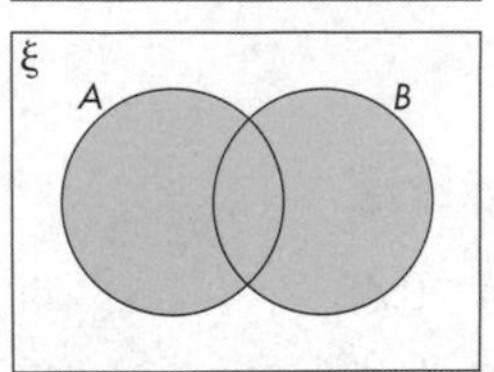

c

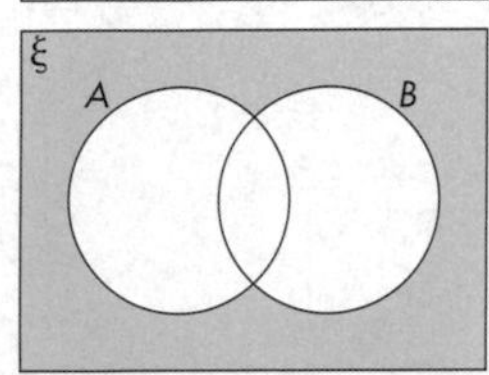

d

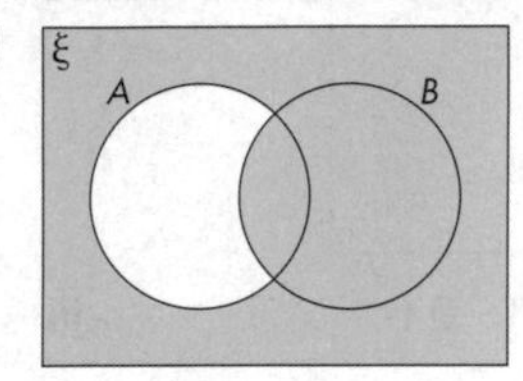

5 Various answers with at least 2 letters in the intersection

6 a $(A \cup B)'$ or $(A' \cap B)'$

b $A' \cup B'$ or $(A \cap B)$

7 a

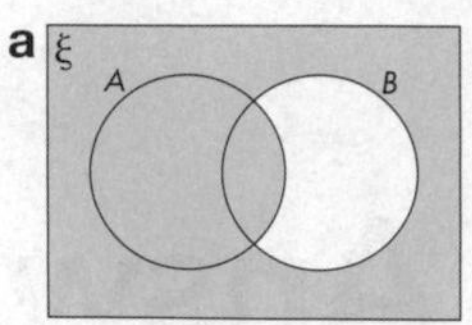

b

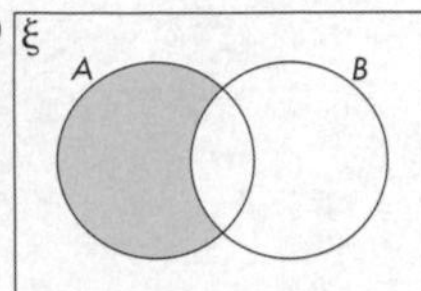

Examination questions

1 a P(*A*) + P(*B*) = 1.1 could never be true

b P(*A*) + P(*B*) = 0.7 could be true if all elements of *A* are also elements of *B*.

c P(*A*) + P(*B*) = 0.9 could be true if so, elements of *A* are also elements of *B*.

2 a $A = \{2, 4, 6, 8, 10, 12\}$

b $\frac{1}{2}$ **c** $\frac{1}{2}$

d $B = \{3, 6, 9, 12\}$ **e** $\frac{1}{3}$

f $\frac{2}{3}$ **g** They are multiples of 6.

3 a

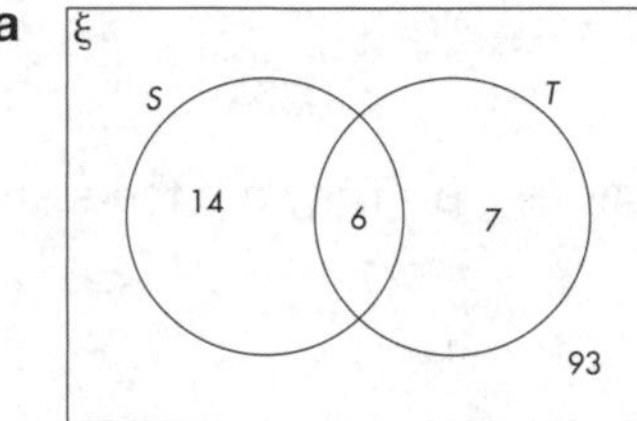

b $\frac{1}{6}$

4 0.3

5 a 110 **b** $\frac{19}{55}$ **c** $\frac{3}{22}$

d $\frac{36}{55}$

6 a $\frac{1}{2}$ **b** $\frac{7}{10}$ **c** $\frac{1}{5}$

Chapter 2 Algebra: The circle and simultaneous equations

Quick check

1 $x = 1.5$, $y = 2.25$

2 $x = \frac{2}{3}$ or $-\frac{1}{4}$

3 a $x = 8.53$ cm

b $y = 48.2°$

c $z = 14.2$ cm

Exercise 2A

1 a

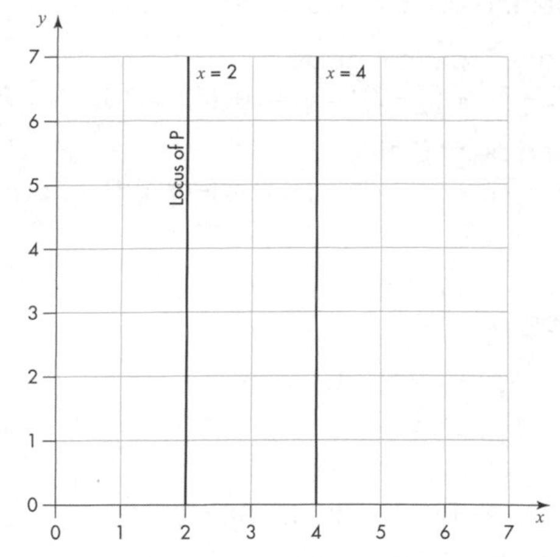

b $x = 2$

2 a

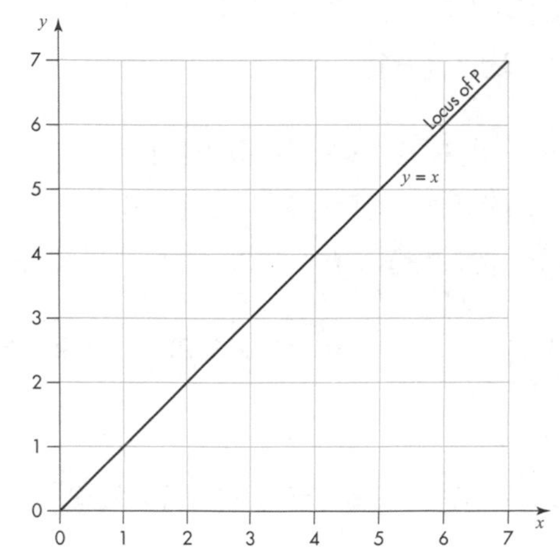

b $y = x$

3 a (3, 1) and (5, 3)

b

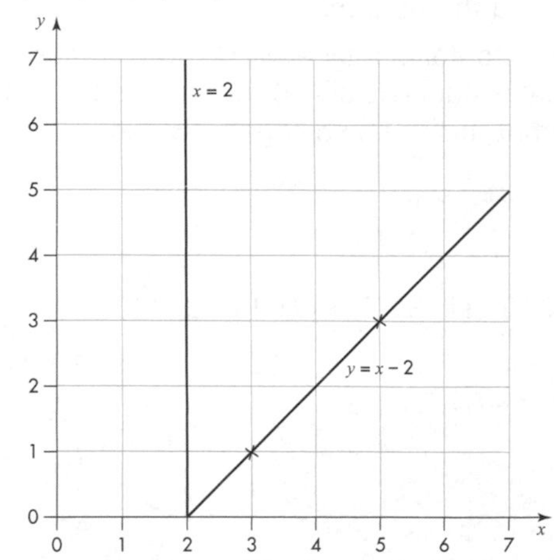

c $y = x - 2$

4 a (0, 1), (2, 3), (3, 4) and (5, 6)

b

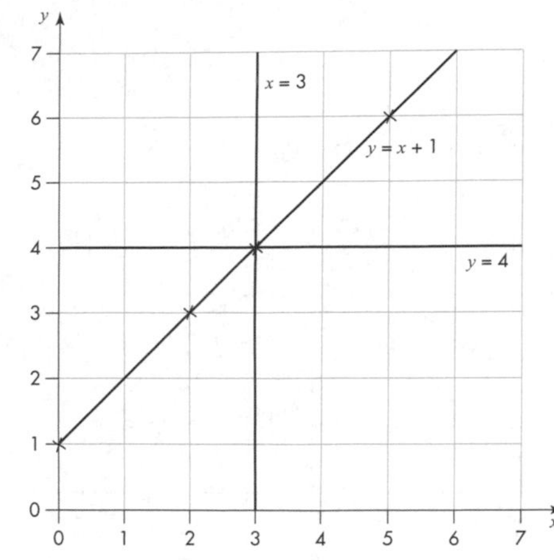

c $y = x + 1$

5 a

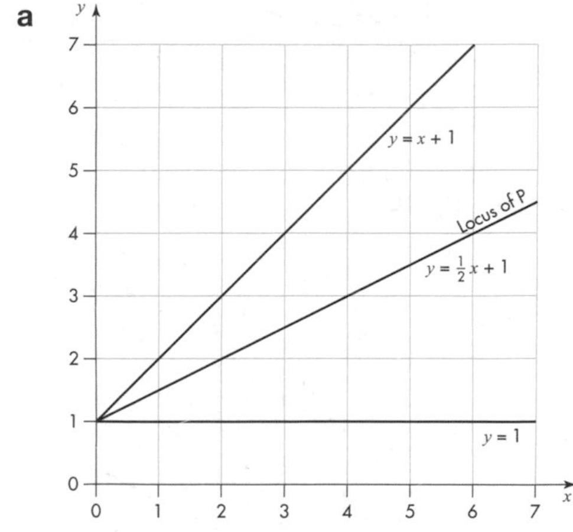

b $y = \frac{1}{2}x + 1$

6 a

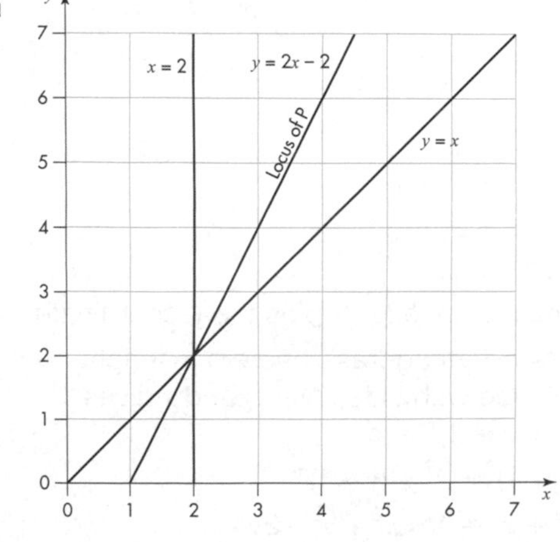

b $y = 2x - 2$

7 $y = 3x$

8 $y = \frac{1}{2}x + 1\frac{1}{2}$

9 **a** $x^2 + y^2 = 4$ **b** $x^2 + y^2 = 36$

10 **a**

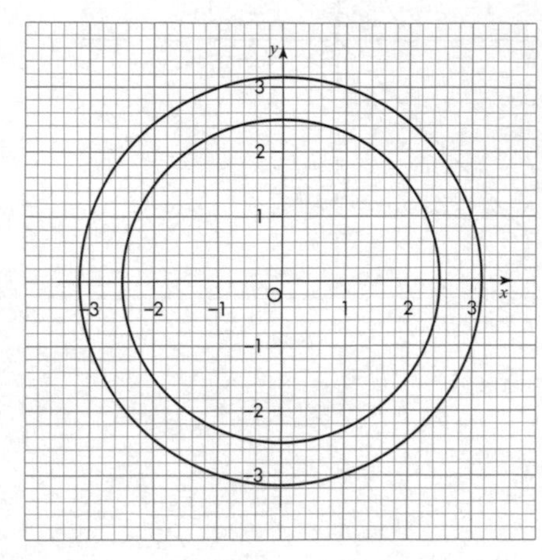

b 4π

11 **a** (–3, 1), (1, 3), (3, 4)

b

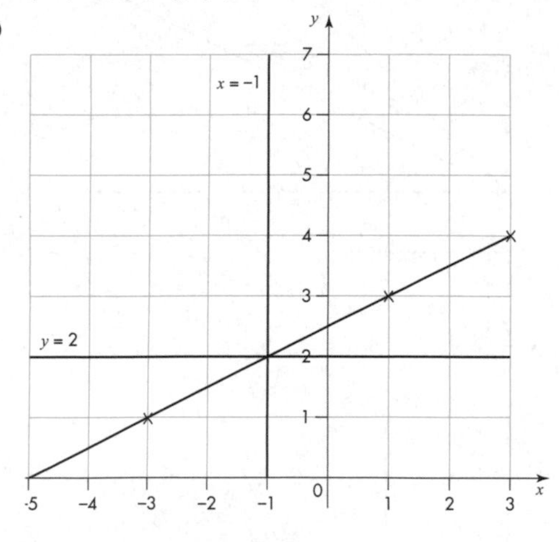

c $y = \frac{1}{2}x + 2\frac{1}{2}$

12 **a** The distance is –1 plus the y-coordinate

b Using Pythagoras' theorem in a right-angled triangle with sides $(y - 1)$ and x gives $d^2 = (y - 1)^2 + x^2$

c $(y + 1)^2 = x^2 + (y - 1)^2$

$y^2 + 2y + 1 = x^2 + y^2 - 2y + 1$

$x^2 = 4y$

$y = \frac{1}{4}x^2$

13 **a** For C_2, point A gives $3^2 + 4^2 = 9 + 16 = 25$ and point B gives $4^2 + 3^2 = 16 + 9 = 25$

For C_1, point A gives $(3 - 7)^2 + (4 - 7)^2 = (-4)^2 + (3)^2 = 16 + 9 = 25$

And point B gives $(4 - 7)^2 + (3 - 7)^2 = (-3)^2 + (-4)^2 = 9 + 16 = 25$

b $x + y = 7$

Exercise 2B

1 **a** $x = 1, y = -2$ **b** $x = 6, y = 5$
c $x = 6, y = -1$

2 **a** (–5, –2), (2, 5) **b** $(2, 3), (-9, -\frac{2}{3})$
c (–1, 6), (6, –1)

3 **a** $(\frac{1}{2}, 2\frac{1}{2}), (-3, -1)$ **b** $(\frac{1}{3}, 5\frac{2}{3}), (-\frac{1}{2}, 4)$
c $(\frac{1}{2}, -3\frac{1}{2}), (-\frac{1}{2}, -4\frac{1}{2})$

4 **a** $(-\frac{1}{5}, -\frac{7}{5}), (1, 1)$ **b** $(1, 0), (-\frac{3}{5}, \frac{4}{5})$
c $(-\frac{1}{3}, 1\frac{2}{3}), (-1, -1)$ **d** $(-\frac{4}{3}, -\frac{9}{5}), (2, 1)$
e $(-\frac{2}{3}, -\frac{1}{3}), (-\frac{2}{7}, \frac{3}{7})$ **f** (1, 3)

5 **a** (3, 2) **b** **i** (–3, 2) **ii** (3, –2)

Exercise 2C

1 $(-\sqrt{10}\cos 50°, \sqrt{10}\sin 50°) \approx (-2.03, 2.42)$

2 $(5\cos 70°, -5\sin 70°) \approx (1.71, 4.70)$

3 The angle between OP and the positive x-axis is 60 in the fourth quadrant.

The horizontal distance will be cos 60° and the vertical distance is sin 60° but as it is the fourth quadrant this will be a negative value.

4 150°

5 $(\cos 310°, \sin 310°) \approx (0.643, -0.766)$

6 (4.59, –6.55)

7 110°

8 $(\sqrt{20}\cos 240°, \sqrt{20}\sin 240°)$ or $(-\sqrt{5}, -\sqrt{15}) \approx (-2.24, -3.87)$

9 (4 cos 255°, 4 sin 255°) ≈ (–1.04, –3.86)

10 9π

Examination questions

1 a

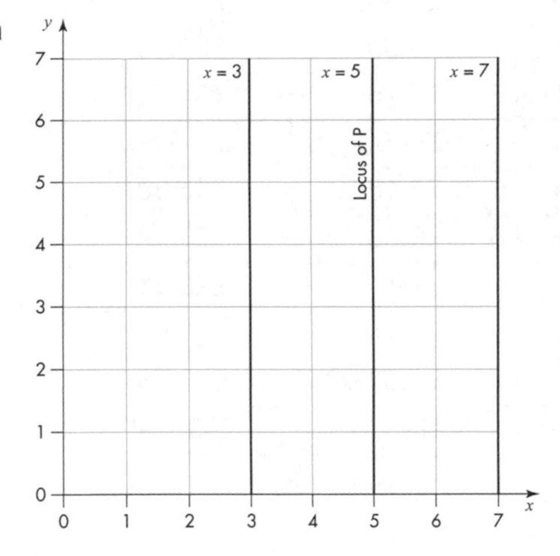

b $x = 5$

2 (0.423, –0.906)

3 (7 cos 40°, 7 sin 40°) and (–7 cos 40°, –7 sin 40°) or (7 cos 220°, 7 sin 220°)

4 a $x^2 + y^2 = 9$

b (3 cos 25°, 3 sin 25°) ≈ (2.72, 1.27)

5

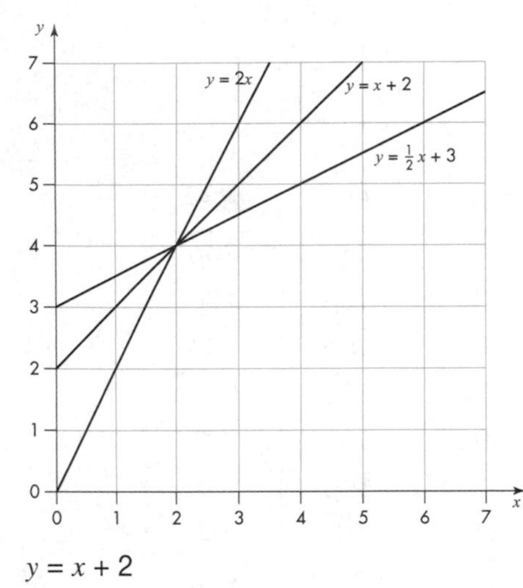

$y = x + 2$

6 $(-\frac{7}{5}, \frac{1}{5})$, (–1, 1)

7 (–3, –3), (5, 1)

Chapter 3 Finance: Financial and business applications

Quick check

1 a 5 **b** 30 **c** 2.5

2 a 18 **b** 20 **c** 3

3 a £2.20 **b** £6

Exercise 3A

1 a –£1.40 **b** +£1.80 **c** +£3.60
d –£2.20 **e** +£2.40 **f** +£2.67
g –86p **h** +£5.25 **i** –£1.25

2 £88 000

3 a £3.20 **b** 200

4

a	£6.70
b	£7.99
c	£1.80
d	£3.99
e	£17.99
f	£13.77
g	£12.00
h	£15.50
i	£15.80
j	£10.10

5 a £2.10 **b** £63 **c** 48

6 200

7 £1

8 Yes he does. They cost £2.40 each. 20% profit is 48p per shirt. So he sells for more than £2.88 so over 20%.

9 10p per metre

Exercise 3B

1

	10%	5%	20%
a	10p	5p	20p
b	20p	10p	40p
c	30p	15p	60p
d	40p	20p	80p
e	50p	25p	£1
f	£1	50p	£2
g	£2	£1	£4
h	£5	£2.50	£10
i	£10	£5	£20

2 **a** £1.60 **b** £2.25 **c** £3.75
d £4.95 **e** £6 **f** £8.20
g £9.30 **h** £10.75

3 Yes she can afford repairs as VAT = £64 and the total cost is £384 which is less than £400

4 **a** £411.25
b Less than £350 as VAT must actually be 20% of the price excluding VAT.
c Icebox is cheaper at £408

5

	Amount of VAT	Amount including VAT
a	£3.10	£18.60
b	£2.50	£15.00
c	£3.92	£23.52
d	£6.48	£38.88
e	£7.18	£43.08
f	£3.36	£20.16
g	63p	£13.23
h	77p	£16.17
i	99p	£20.79
j	88p	£18.48
k	£3.37	£70.77
l	£6.06	£127.26

6

	Amount of VAT	Amount including VAT
a	£2.68	£16.10
b	£4.83	£28.99
c	£2.70	£16.19
d	£3.36	£20.14
e	£3.89	£23.32
f	£5.76	£34.58
g	£24.93	£149.56
h	£2.16	£45.38
i	£3.58	£75.12
j	97p	£20.31
k	£1.14	£24.01
l	£3.66	£76.78
m	£19.22	£403.54

Exercise 3C

1

	Simple interest per annum	Total amount of simple interest	Total value of investment
a	£10	£30	£530
b	£15	£30	£530
c	£18	£36	£636
d	£16	£64	£864
e	£40	£80	£1080
f	£45	£90	£1590
g	£50	£200	£5200
h	£160	£320	£8320
i	£300	£900	£10 900

2

	Amount invested	Rate of interest per annum (R)	Number of years invested (I)	Total amount of simple interest
a	£500	3%	2	£30
b	£500	1%	3	£15
c	£700	2%	4	£56
d	£1000	2%	4	£80
e	£1000	5%	4	£200
f	£1500	10%	2	£300
g	£5000	1%	4	£200
h	£8000	3%	3	£720
i	£10 000	2%	3	£600

3 Reason 1: Could reinvest after 1 year and invest the interest so making more.

Reason 2: Can get the money back sooner if needed.

4 James is not correct as interest will be £420 and value £3920.

5

	Amount borrowed (*P*)	Rate of interest per annum (*R*)	Simple interest per annum	Number of years borrowed (*T*)	Total amount of simple interest (*I*)	Total cost of loan
a	£500	2.3%	£11.50	2	£23	£523
b	£500	3.3%	£16.50	2	£33	£533
c	£600	4.33%	£25.98	2	£51.96	£651.96
d	£800	2.75%	£22.00	4	£88.00	£888
e	£1000	4.8%	£48.00	2	£96.00	£1096
f	£1500	1.3%	£19.50	2	£39.00	£1539
g	£5000	0.2%	£10.00	4	£40.00	£5040
h	£8000	1.9%	£152.00	2	£304.00	£8304
i	£10 000	1.8%	£180.00	3	£540.00	£10 540

6 Loan costs £50 for setting up + £30 simple interest = £80

Investment pays out £40

Chloe does not make a profit; she makes a £40 loss.

xercise 3D

1

	Amount to be taxed	Amount of income tax
a	Nil	£0
b	£1000	£200
c	Nil	£0
d	£2000	£400
e	£1300	£260
f	£2800	£560
g	£9600	£1920
h	£10 380	£2076
i	£17 590	£3518
j	£20 700	£4140

2

	Salary	Personal allowance	Amount to be taxed	Rate of income tax	Amount of income tax
a	£6000	£7000	Nil	22%	£0
b	£8000	£7000	£1000	25%	£250
c	£5000	£7000	Nil	26%	£0
d	£9000	£7000	£2000	23%	£460
e	£7500	£6200	£1300	22%	£286
f	£9600	£6800	£2800	22%	£616
g	£15 000	£5400	£9600	25%	£2400
h	£16 350	£5970	£10 380	24%	£2491.20
i	£22 500	£4910	£17 590	28%	£4925.20
j	£26 000	£5300	£20 700	27%	£5589

3 £1600

4 £1210

5 Aby has a taxable income of £27 500.

Ben has a taxable income of £33 400.

So Ben will pay more income tax.

6 £540

7 Any two pairs of values with a difference of £17 000.

For example, salary £20 000 and personal allowance £3000 or salary £25 000 and personal allowance £8000.

8 They have different personal allowances.

9

	Total amount to be taxed	Amount to be taxed at 20%	Amount to be taxed at 40%
a	£48 000	£37 400	£10 600
b	£35 000	£35 000	Nil
c	£22 000	£22 000	Nil
d	£47 500	£37 400	£10 100
e	£67 800	£37 400	£30 400
f	£23 100	£23 100	Nil
g	Nil	Nil	Nil
h	£29 000	£29 000	Nil

10 **a** £11 720, **b** £7000, **c** £4400, **d** £11 520, **e** £19 640, **f** £4620, **g** Nil, **h** £5800

11

	Total amount to be taxed	Amount to be taxed at 20%	Amount to be taxed at 40%	Amount to be taxed at 50%
a	£98 000	£37 400	£60 600	Nil
b	£174 000	£37 400	£112 600	£24 000
c	£183 000	£37 400	£112 600	£33 000
d	£111 500	£37 400	£74 100	Nil
e	£292 800	£37 400	£112 600	£142 800
f	£88 700	£37 400	£51 300	Nil
g	£150 000	£37 400	£112 600	Nil
h	£152 500	£37 400	£112 600	£2500

12 **a** £31 720, **b** £64 520, **c** £69 020, **d** £37 120, **e** £123 920, **f** £28 000, **g** 52 520, **h** £53 770

13 **a** **i** £56 520 **ii** £58 020

b 2009 – £103 480

2010 – £102 980

So 2009.

14 £5000

15 Personal allowance may be more than £5000.

Exercise 3E

1 £6250

2 £4250

3 £3220

4 For example, starting with £100. After 10% increase this is £110, 10% of £110 is £11 so after this the value is £99.

5 30%

6 **a** $33\frac{1}{3}$% **b** £2133.33

7 $33\frac{1}{3}$%

8 20%

9 56%

Exercise 3F

1

a	5.1%	Monthly	5.22%
b	3.5%	Monthly	3.56%
c	2.8%	Monthly	2.84%
d	5.8%	Monthly	5.96%
e	3.9%	Quarterly	3.96%
f	2.4%	Quarterly	2.42%
g	1.2%	Quarterly	1.21%
h	0.2%	Quarterly	0.20%
i	4.0%	Every 6 months	4.04%
j	3.5%	Every 6 months	3.53%
k	7.1%	Every 6 months	7.23%
l	0.8%	Every 6 months	0.80%

2 Blue and White account is only option if saving less than £500 or likely to want to withdraw money before 2 years.

AER for Terrier account is 4%.

AER for Blue and White account is 3.56% which is less than Terrier account but more flexible.

Use Terrier account if planning to leave money in fo a long time otherwise use Blue and White account.

3 **a** AER 3.56% **b** AER 3.55%

c AER 3.53% **d** D: AER 3.5%

So **a** gives the better rate.

4

	a AER	b Value of investment after 1 year
i	2.73%	£1027.34
ii	6.27%	£1594.10
iii	3.25%	£2064.95
iv	2.83%	£1233.95
v	1.51%	£812.07
vi	0.60%	£3018.04
vii	4.55%	£2875.14
viii	3.33%	£7362.06
ix	6.09%	£8513.72

5 Monastry: AER is 4.35% with one withdrawal. So Monastry has a worse rate with one withdrawal allowed.

This means that she is better off with Britwide.

6

a	1.6%	Monthly	1.59%
b	5.0%	Monthly	4.89%
c	2.1%	Monthly	1.98%
d	1.7%	Quarterly	1.69%
e	0.6%	Quarterly	0.60%
f	5.5%	Quarterly	5.39%
g	3.4%	Every 6 months	3.37%
h	7.2%	Every 6 months	6.88%
i	9.0%	Every 6 months	8.81%

Examination questions

1 £225

2 £20

3 **a** £11.04 **b** 24

4 **a** £102

b 10% of £102 = £10.20 (or explanation that £8.50 is 10% of £85 which excludes the VAT)

5 **a** £1000 **b** £80

6 £5400

7 **a** £1660

b No change, taxable pay is the same

8 **a** £1680

b No change, taxable pay is the same

9 Last year, tax paid = £6842, this year tax paid = £6940, so she pays more this year.

10 **a** £1380

b 6 years although almost same value after 5 years.

11 Increased, for example, £100 → £110 → £121 and £100 → £90 → £81

Up £21 and down £19 so a net increase.

12 **a** £51 720 **b** £58 920

13 He pays £7480 (basic rate) + £45 040 (higher rate) + £50 750 (additional rate) = £103 270

One third of his pay is £86 666.67 so he does pay more.

14 **a** AER 4.96% **b** AER 4.99%

c AER 5.01% **d** AER 5%

So **c** is the best rate.

15 Ruby will have £1040. Emmy will have an AER of 3.97% giving her £1039.70

So Ruby will have 40p more.

Chapter 4 Finance: Personal and domestic finance

Quick check

1 **a** £130

b £60

c £64.50

2 **a** £61.60

b £72

c £105

3 **a** 5

b 8

c 11

Exercise 4A

1 £283.40

2 £16 000 as a weekly wage of £300 is £300 × 52 = £15 600 per year.

3

a	10	£6.40	£64.00
b	20	£7.50	£150.00
c	30	£6.90	£207.00
d	40	£10.20	£408.00
e	35	£11.30	£395.50
f	37	£9.90	£366.30
g	15	£12.80	£192.00

4 Current job pays £414.40 × 52 = £21 548.80 per year
5% of £3000 = £150 per month
So commission pays £1800 per year.
So new job pays £21 800 per year.
Very little to choose between the jobs (so probably stay in current job). However the new job does pay marginally more, so a student could say that he should change jobs – that wouldn't be incorrect.

5 £385.40

6 a 4% of £50 000 = £2000
10% of £50 000 = £5000
£18 000 + £5000 (=£23 000) is more than £20 000 + £2000 (£22 000)
b Less of the pay is guaranteed.

7 4 hours

8

	Commission paid	Total salary
a	£2000	£11 200
b	£3000	£11 500
c	£2000	£10 000
d	£600	£12 100
e	£1500	£14 000
f	£750	£11 350
g	£1200	£16 000

9 a Vic's basic salary is £1000 per month.
5% of £2000 = £100
So Vic is paid £1000 + £100 = £1100, the same as Bob.
b £300

Exercise 4B

1 **a** €575, **b** €805, **c** $1050, **d** $1125, **e** ¥117 000, **f** ¥156 000, **g** $2100, **h** $2520, **i** 18 865 kr, **j** 21 560 kr

2 **a** £173.91, **b** £434.78, **c** £533.33, **d** £633.33, **e** £153.85, **f** £269.23, **g** £744.05, **h** £952.38, **i** £927.64, **j** £1113.17

3 Changing £500 to US dollars at an exchange rate of £1 = $1.50 is better as it gives $750
Changing £480 at an exchange rate of £1 = $1.53 with a £20 commission also costs £500 but gives $734.40

4 a 6300 kr b 150 kr c £22.22

5 Josh: £211.27
Olivia: £206.90 + £4.14 commission = £211.04
Josh has the better deal.

6 a £1 = $1.88
b The greatest drop was from June to July.
c There is no trend in the data so you cannot tell if it will go up or down.

Exercise 4C

1 £5764.80

2 £1193.40

3

	Total amount of repayment	Total interest paid
a	£1163.04	£163.04
b	£1744.32	£244.32
c	£2493.72	£493.72
d	£3116.88	£616.88
e	£5814.72	£814.72
f	£10 690.20	£3190.20
g	£12 942.72	£4942.72
h	£19 311.60	£9311.60

4 £10 684.32

5 £479.52

6 A loan taken out for a longer period will mean that there is more interest to repay so the monthly repayments will include the cost of the extra interest payment.

7 Fast Loan costs £644.88 more. (Quick Loan £3116.88, Fast Loan £3761.76)

8 Comfort furniture is more expensive. (Seats For You: 24 months @ £50 = £1200, Comfort furniture 36 months @ £36.85 = 1326.60)

Exercise 4D

1 150

2 78p, 80.3p, 84.2p, 85p, 87.4p, 93.6p

3 £74.73

4 **a** May and June
b Prices were below the base year prices in January, February and March but then went above the base year prices.

5 20% of 30p is 6p. So a rise of 20% would only take the price to 36p, not 41p. The newspaper was not correct as the rise is more than 20%.

6 £9.45

7 350

8 **a** September 2008
b Inflation falls but prices are still higher than 12 months earlier.

9 The general cost of living in 2009 dropped to 98% of that in 2008.

10 £51.50

11 **a** It was a holiday month.
b **i** 138–144 thousand **ii** 200–210 thousand

Exercise 4E

1

a	=A1+3
b	=A2*C5
c	=E2–5
d	=B9/A9
e	=A6+A7
f	=SUM(B2:B12)
g	=C2^3
h	=(A3+A4)/12
i	=(B3/100+1)^4
j	=F7/E7–20

a =A3*60/B3 **b** 120, 60

3 **a** =B2*C2
b

	A	B	C	D
1	Item	Number required	Cost per item	Total cost per item
2	Adults bus fare	4	£5.50	£22.00
3	Child bus fare	8	£1.80	£14.40
4	Adult cinema ticket	4	£5.60	£22.40
5	Child cinema ticket	8	£3.50	£28.00
6	Early bird discount	12	£1	£12.00

c The total amount to pay after deducting the discount.

4 **a** 9 **b** 8

5 **a** =A2*5 – 13 **b** =A5 + 1
c

	A	B	C
1	Term number n	nth term $4n + 2$	nth term $5n - 13$
2	1	6	–8
3	2	10	–3
4	3	14	2
5	4	18	7
6	5	22	12

d 15

6 **a** =B3 * 20 / 100
b Multiplying by 1.2 is equivalent to adding on 20%

	A	B	C	D
1	Item	Price excluding VAT	VAT	Price including VAT
2	Shirt	£20	£4	£24
3	Dress	£35	£7	£42
4	Coat	£84	£16.80	£100.80
5	Shoes	£45	£9	£54
6	Tie	£11	£2.20	£13.20

7 **a** =B3–C3 **b** =D2*0.2 or =D2*20/100

c =D3*0.2 or = D3*20/100

d

	A	B	C	D	E
1	**Name**	**Salary**	**Taxable allowance**	**Taxable pay**	**Amount of income tax to pay**
2	A Smithies	£15 000	£8000	£7000	£1400
3	L Peltier	£11 500	£6000	£5500	£1100
4	S Black	£24 000	£7500	£16 500	£3300
5	U Gul	£18 500	£3450	£15 050	£3010
6	A Chopra	£16 000	£8750	£7250	£1450

8 **a** = A2*1.49 **b** =A2*1.15

c

	A	B	C
1	Amount in pounds	Amount in US dollars	Amount in euros
2	£10	$14.90	€11.50
3	£25	$37.25	€28.75
4	£50	$74.50	€57.50
5	£100	$149	€115

d An exchange rate for 1 euro into US dollars (€1 = $1.30).

9 **a** =A2*B2*C2/100

b

	D
1	Amount of interest (*I*)
2	£4.20
3	£21.60
4	£13.00
5	£100.00

10 **a** **i** =A3*24 **ii** =E3–C3

b **i** =A4*36 **ii** =E4–C4

c

	E	F
1	**Total amount repayable**	**Interest payable**
2	£1095.96	£95.96
3	£1189.68	£189.68
4	£1288.44	£288.44

Exercise 4F

1 2, 5, 8, 11, 14

2 **a** £1500

b It is less than or equal to £6500

3 For example

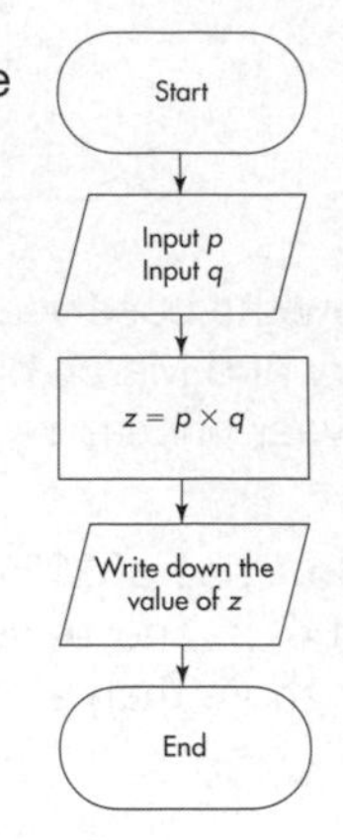

4 **a** €1062

b €468.46 (note £3 deducted before money converted)

c £3 for converting £500 or less, free for converting more than £500

5 **a** £168 **b** £131.25 **c** £45

d For example: Change 1.2 to 0.2 and 1.05 to 0.05. Then "Write down the amount of VAT".

6

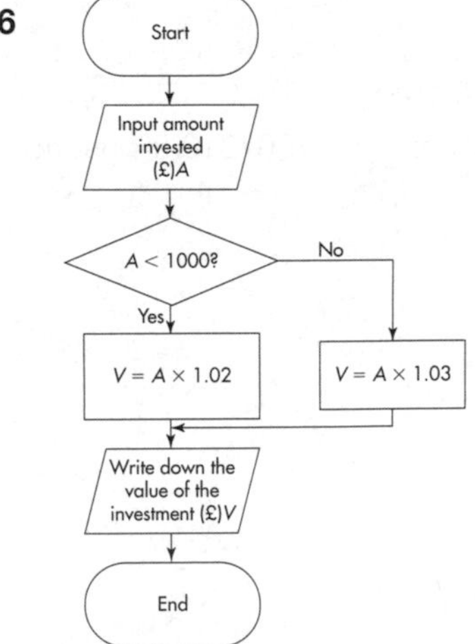

7 For example:

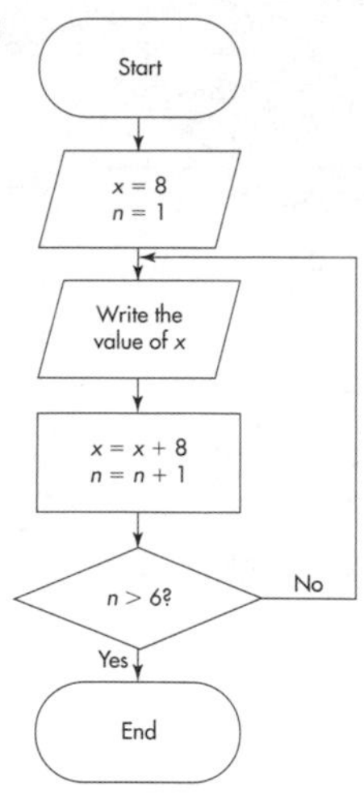

Examination questions

1 £1300 per month better, as it is equivalent to £1300 × 12 = £15 600 per year.

2 **a** £4150

b Under the old rate of pay she would receive £1000 + £300 = £1300

Under the new rate she will receive £1100 + £150 = £1250 so she is worse off.

3 **a** $725 **b** £862.07

4 **a** £613.60 **b** £60

5 **a** €2.50 **b** Over €100

c €307.50 converts to £267.53

6 **a** The CPI has risen by 12.4% **b** £30.91

c Prices fell from December 2009 to January 2010.

7 **a** 2 years **b** £480

8 **a** £31.55 **b** 86p **c** 128.7

9 **a** £784.35

b Prices for these goods and services may not have changed in line with the RPI.

10 **a** =B2

b **i** =B2*2 **ii** £14.60

c =B2*C2 **d** =SUM(D2:D8)

11 **a** =B3*1.14

b

	A	B	C
1	Item	2005 price	2010 price
2	Loaf	£1	£1
3	Skirt	£20	£22.80
4	Towel	£12	£13.68
5	Vacuum cleaner	£230	£262.20

12 **a** $V = P \times C \div 100$ **b** £103.50

Chapter 5 Finance: Linear programming

Quick check

1 **a** 3.5 **b** 4

2 **a** $x > 4.5$ **b** $x \leqslant 3.5$

3

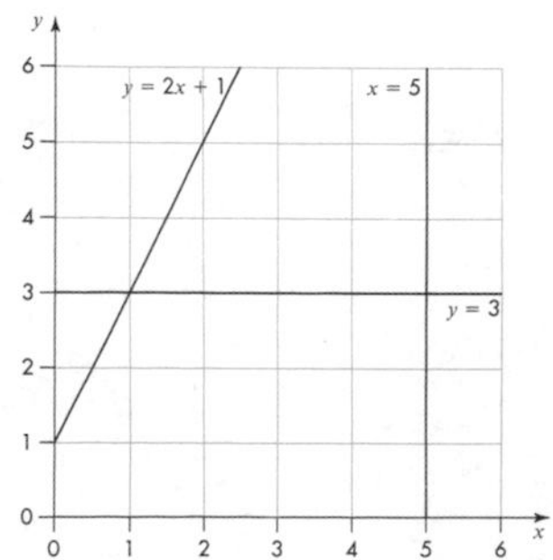

Exercise 5A

1 **a–b**

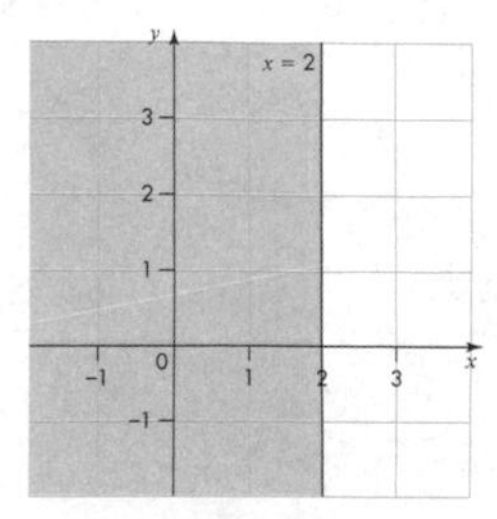

2 **a–b**

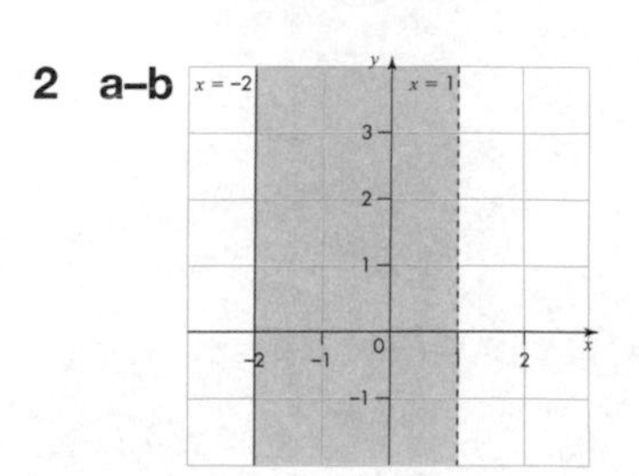

3 a–b

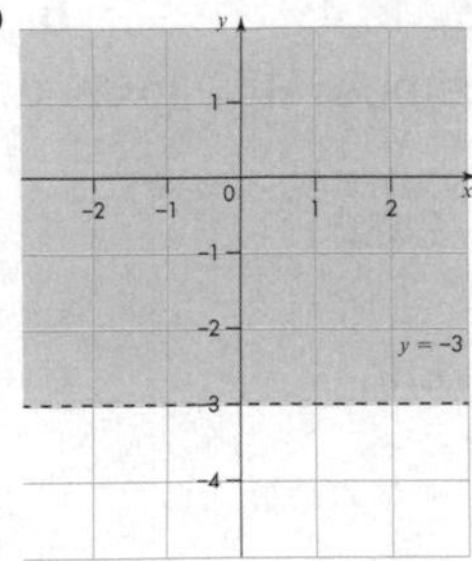

4 a–c

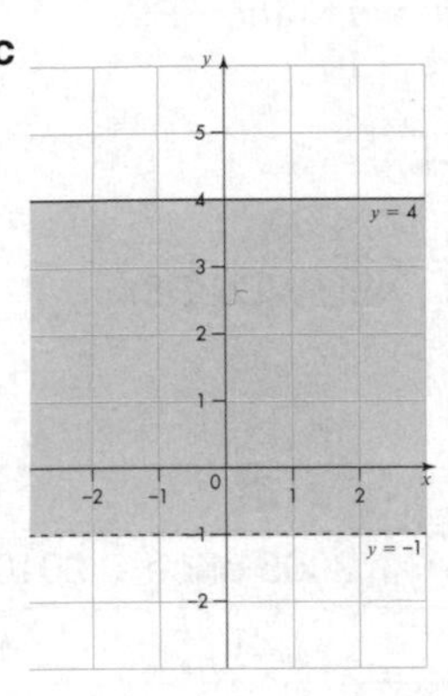

5 a

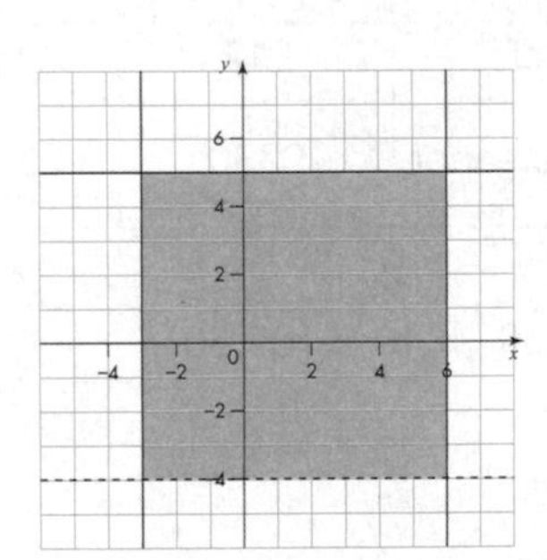

b i Yes **ii** Yes **iii** No

6 a–b

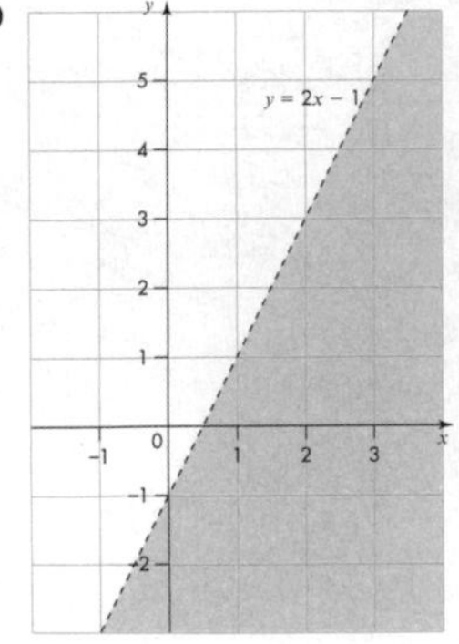

7 a–b

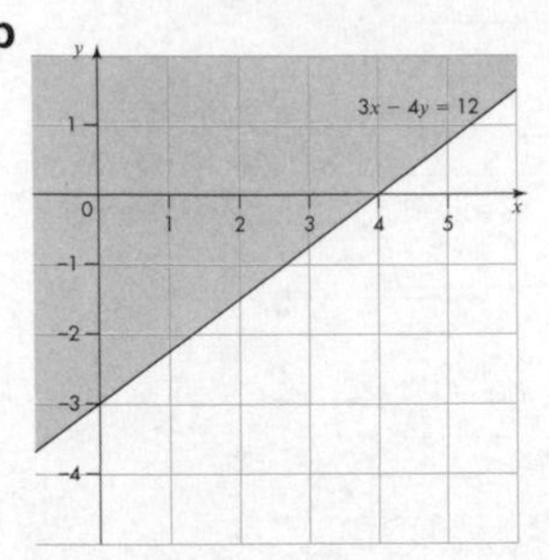

8 a–b

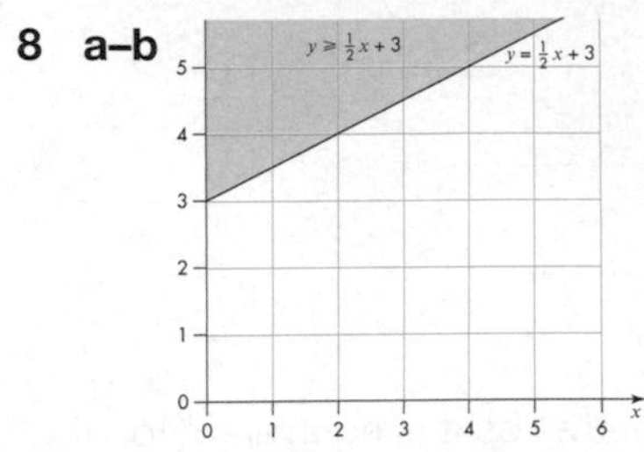

9 a–d

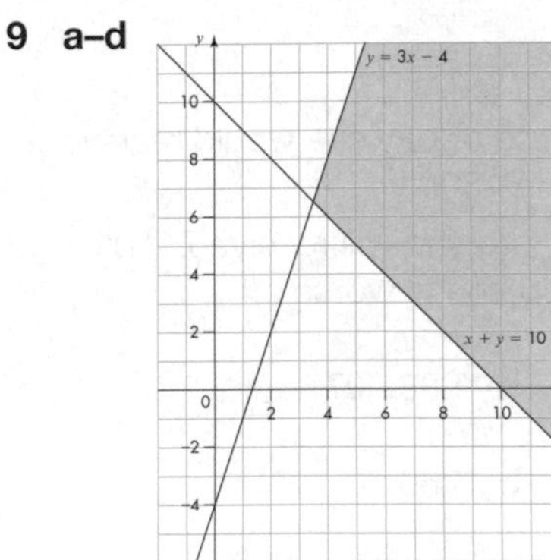

e i No **ii** No **iii** Yes

10 a–f

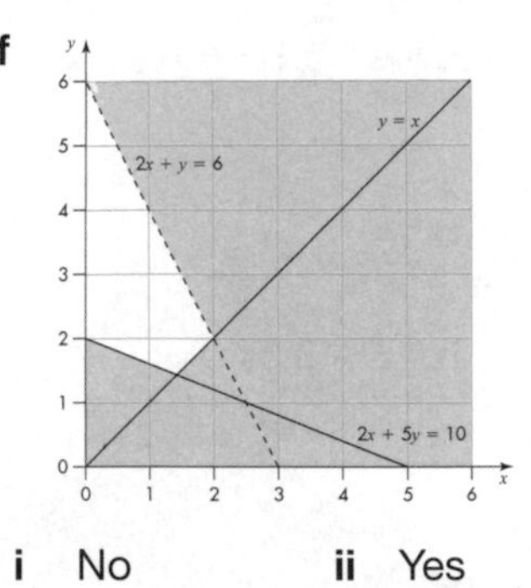

g i No **ii** Yes **iii** Yes

11 a

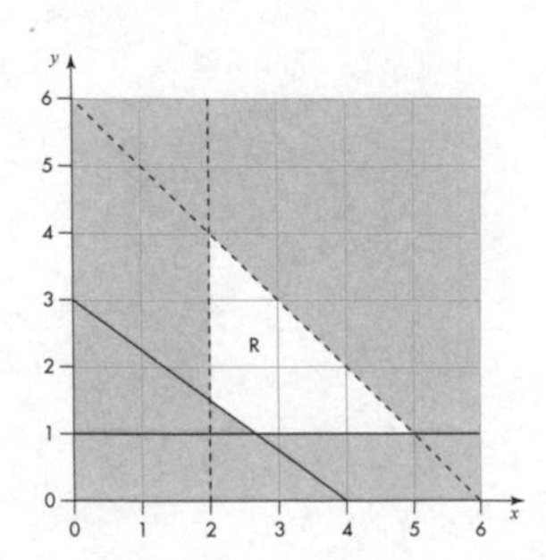

b There are only 3 points for which the coordinates are integers. These are (4, 1), (3, 1) and (3, 2). Working out $2x + y$ for each of these is 9, 7 and 8 respectively so (4, 1) is the point with maximum value for $2x + y$.

c Working out $3x + 2y$ for the points above gives 14, 11 and 13 respectively so (3, 1) is the point with minimum value for $3x + 2y$.

12 a

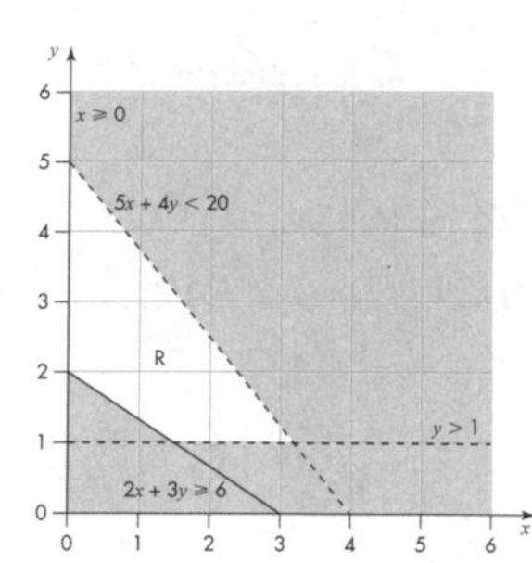

b (2, 2) **c** (1, 2), (2, 2), (0,2)

xercise 5B

1 a £$148x$ **b** £$125y$

c £$(148x + 125y)$

2 a 53p **b** $(19x + 15y)$p

c $(19x + 15y + 3z)$p

d $(19d + 15f + 21)$p

3 a £$(Ax + By)$ **b** £$(Ax + 2Bx)$

c £$(9A + (9 + y)B)$

4 a May be true. **b** May be true.

c Must be false.

5 E Excels hold $40E$; S Storms hold 50S. There must be more than 1500 seats, $40E + 50S \geqslant 1500$. Dividing through by 10 gives $4E + 5S \geqslant 150$.

6 a i W rides cost £$1.50W$. This cannot exceed £6.00, so $1.50W \leqslant 6.00$. Dividing through by 1.5 gives $W \leqslant 4$.

ii Likewise $2D \leqslant 6$, giving $D \leqslant 3$.

iii Total cost is $1.50W + 2D \leq 6.00$. Multiplying through by 2 gives $3W + 4D \leq 12$.

b $D \leqslant 2$

c i Yes **ii** No **iii** No **iv** Yes

a $45x + 25y \leqslant 200 \Rightarrow 9x + 5y \leqslant 40$

$y \geqslant x + 2$

Exercise 5C

1 a i Cost $30x + 40y \leqslant 300 \Rightarrow 3x + 4y \leqslant 30$

ii At least 2 apples, so $x \geqslant 2$

iii At least 3 pears, so $y \geqslant 3$

iv At least 7 fruits, so $x + y \geqslant 7$

b i No **ii** No **iii** No **iv** Yes

2 a i Space: $4x + 3y \leqslant 48$

ii Cost: $300x + 500y \leqslant 6000 \Rightarrow 3x + 5y \leqslant 60$

b i Yes **ii** No **iii** Yes **iv** Yes

3 a i Number of seat required is $40x + 50y \geqslant 300 \Rightarrow 4x + 5y \geqslant 30$

ii Number of 40-seaters $x \leqslant 6$

iii Number of 50-seaters $y \leqslant 5$

b i Yes **ii** Yes **iii** Yes **iv** Yes

c Combination **iii**, which costs £760

d Five 40-seaters and two 50-seaters cost £740

4 $2x + 3y \leqslant 48$, $1.25x + 2.5y \leqslant 37.5 \Rightarrow x + 2y \leqslant 30$

5 $2x + 1.5y \leqslant 300 \Rightarrow 4x + 3y \leqslant 600$;
$50x + 40y \geqslant 6000 \Rightarrow 5x + 4y \geqslant 600$; $x \leqslant 2y$

Exercise 5D

1 a i Time: $30x + 45y \geqslant 180 \Rightarrow 2x + 3y \geqslant 12$

ii Cost: $75x + 100y \leqslant 300 \Rightarrow 3x + 4y \leqslant 24$

b $x \geqslant 2$, $y \geqslant 2$

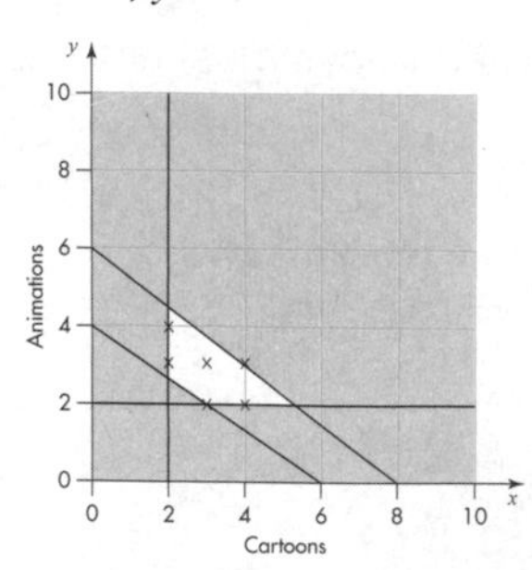

c Longest time: 4 cartoons and 3 animations

d Least cost: 3 cartoons and 2 animations

2 a i Water: $x + 2y \leqslant 30$

ii Cost: $250x + 150y \leqslant 3750 \Rightarrow 5x + 3y \leqslant 75$

iii Wants more rainbow than fantails: $x < y$

b $x \geqslant 1$, $y \geqslant 1$,

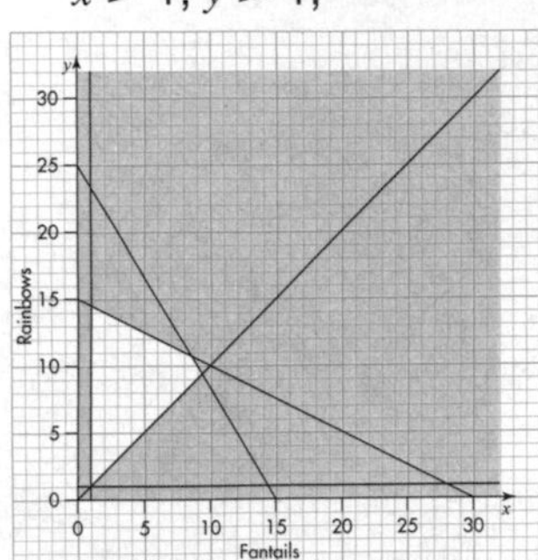

c Greatest number = 19 fish: 8 fantails and 11 rainbows

d Cheapest option = 3 fish: 1 fantail and 2 rainbows

3 a $2x + 4y \geqslant 40$, $x \leqslant 10$, $y \leqslant 7$,
$100x + 150y \leqslant 1800 \Rightarrow 2x + 3y \leqslant 36$

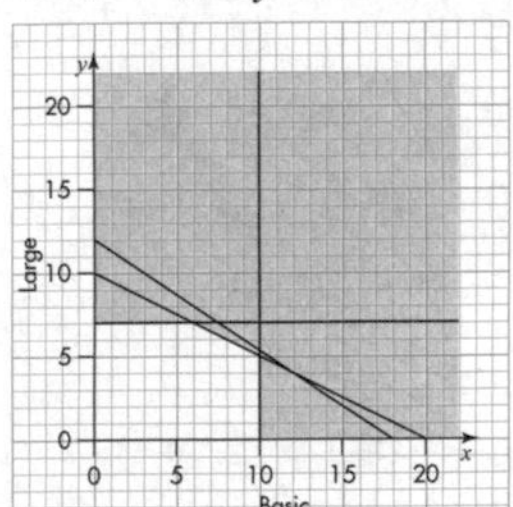

b Cheapest combination: 5 basi and 7 large

c He had forgotten that boys must not share rooms with girls.

4 a $8x + 20y \geqslant 100 \Rightarrow 2x + 5y \geqslant 25$,
$15x + 30y \leqslant 180 \Rightarrow x + 2y \leqslant 12$, $x \geqslant y$

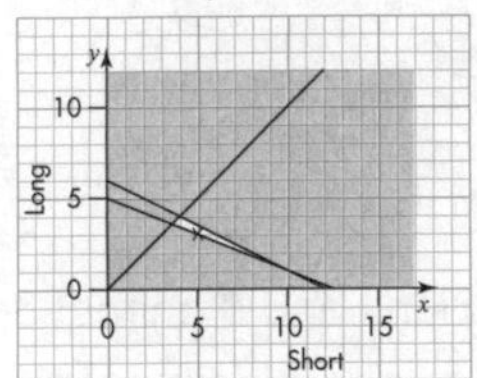

b Smallest number of questions is 8: (4, 4) and (5, 3)

c (4, 4) gives 112 marks, (5, 3) gives 100 marks

d (4, 4) takes 3 hours, (5, 3) takes 2 hours 45 minutes.

5 a $4x + 6y \leqslant 24 \Rightarrow 2x + 3y \leqslant 12$,
$300x + 500y > 1500 \Rightarrow 3x + 5y > 15$,
$x \leqslant 4$, $y \leqslant 3$

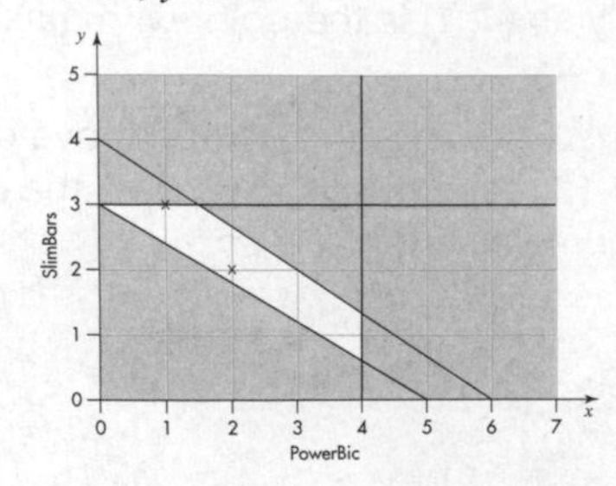

b 4, comprising (2, 2) or (1, 3)

c 1 PowerBic and 3 SlimBars at £13.

Examination questions

1 a $45x + 90y$

b $45x + 90y \leqslant 360 \Rightarrow x + 2y \leqslant 8$

2 a May be true **b** Must be false

c May be true, e.g. $x = 30$, $y = 20$

d May be true, e.g. $x = 3$, $y = 40$

3 a–c

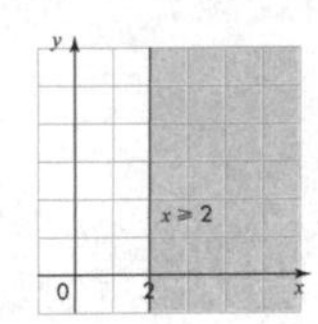

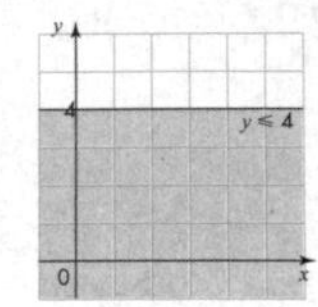

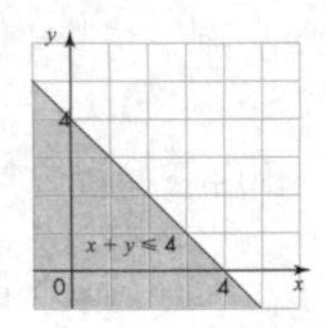

4 a–b

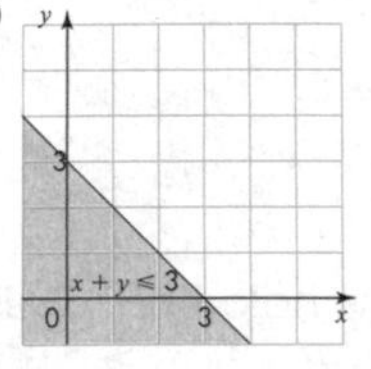

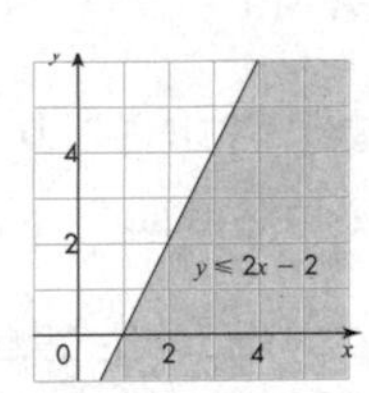

5 a $x + y \leqslant 10$, $x + 3y \geqslant 15$, $x + y \geqslant 0$

b

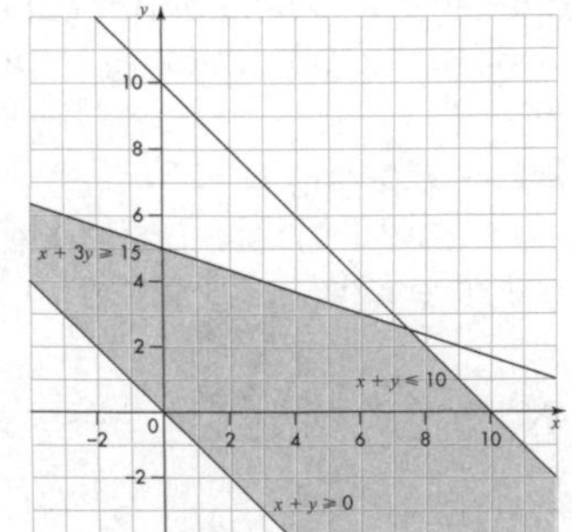

6 a Land: $1500x + 2000y \leqslant 12\,000 \Rightarrow 3x + 4y \leqslant 24$;
Workers: $5x + 3y \leqslant 30$

b

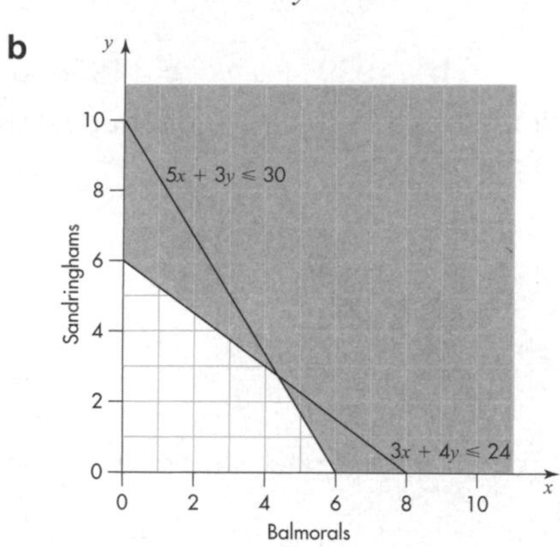

c 4 Balmorals and 3 Sandringhams

7 a

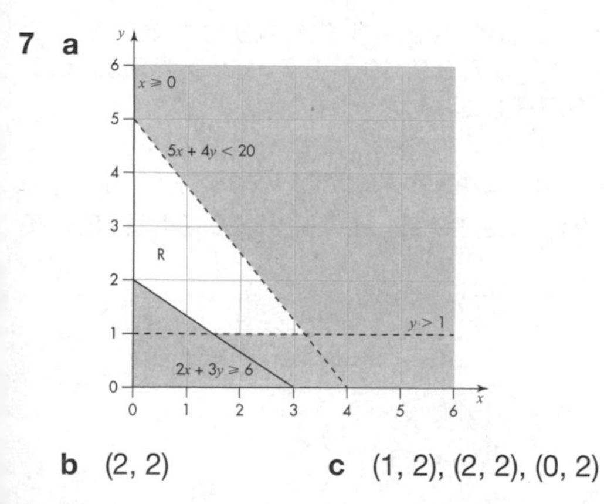

b (2, 2) **c** (1, 2), (2, 2), (0, 2)

8 a $4x + 5y \geqslant 20$, $5x + 6y \leqslant 30$, $x \leqslant 4$, $y \leqslant 3$

b

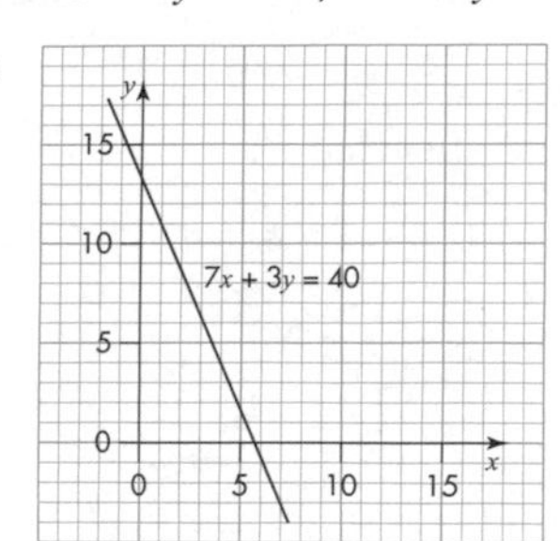

c 4 Canardleys and 1 Rapture

9 a $175x + 75y \leqslant 1000$ (divide by 25)

b

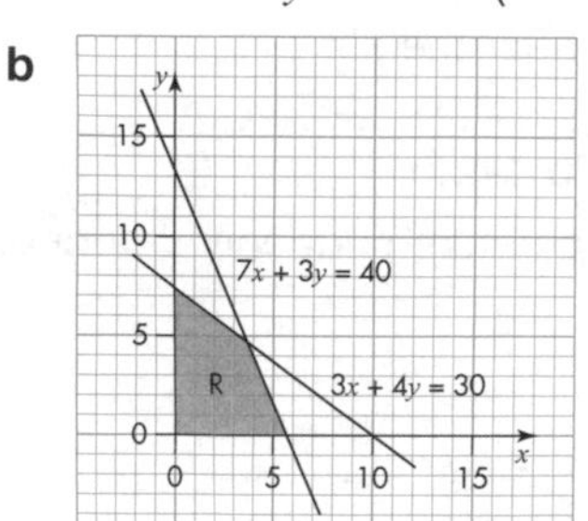

c 5

Chapter 6 Statistics: Time series and moving averages

Quick check

1 a 7 **b** 16 **c** 104 **d** 35.1

2 a–b

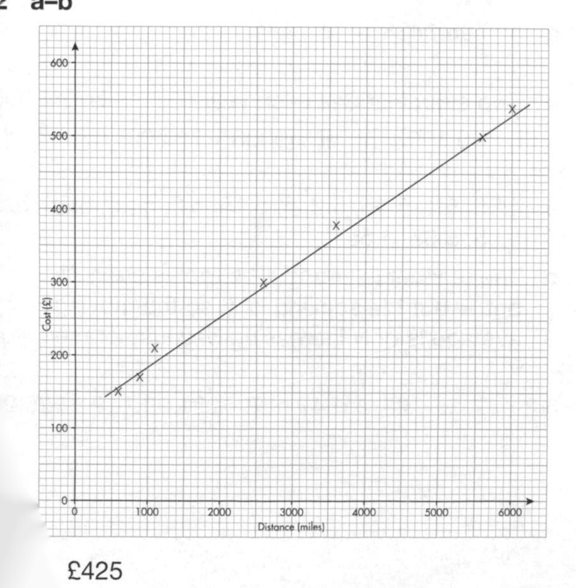

£425

Exercise 6A

1 a

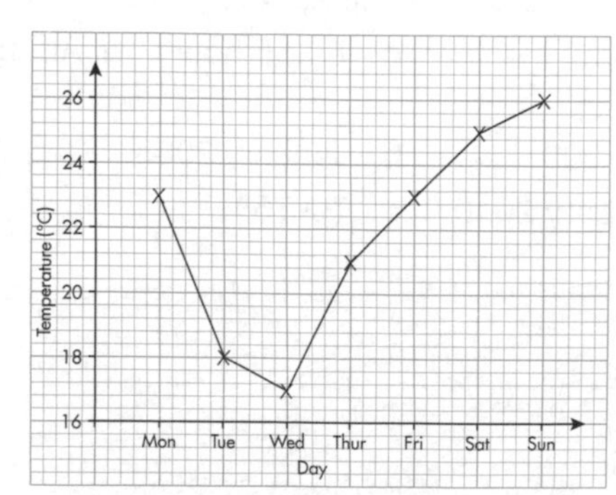

b No, as the data only shows a trend and the intermediate vales have no meaning.

2 a

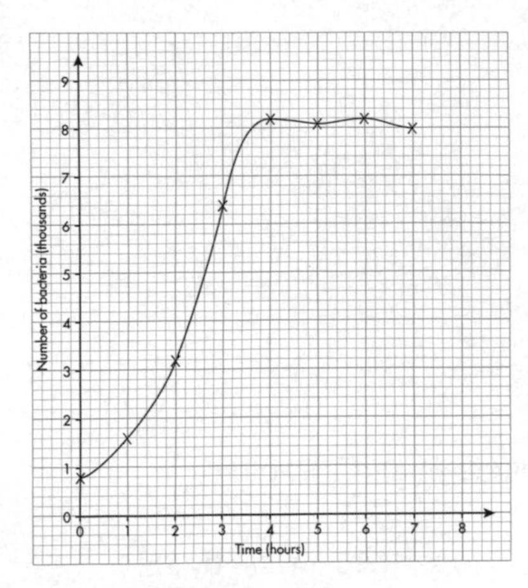

b About 3.5 hours after the start.

3 a 12 °C **b** 12.5 °C

c 1300 as the temperature suddenly dropped.

d 9 °C

4 a

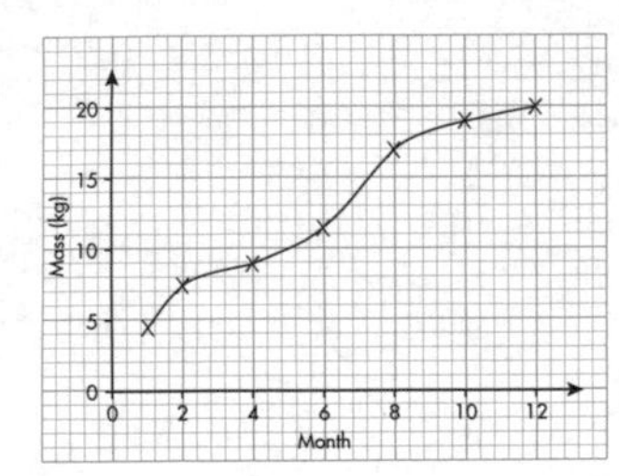

b 8.25 kg **c** 18 kg

d The puppy seems to have stopped growing so extending the graph may not have any relevance.

5 a 1 am to 5.30 am, 1:30 pm to 6 pm

b 1 pm **c i** 12.5 hours **ii** About 4 am

6 a £10 000 **b** January 2008

c January 2010 **d** 70–75%

7 a

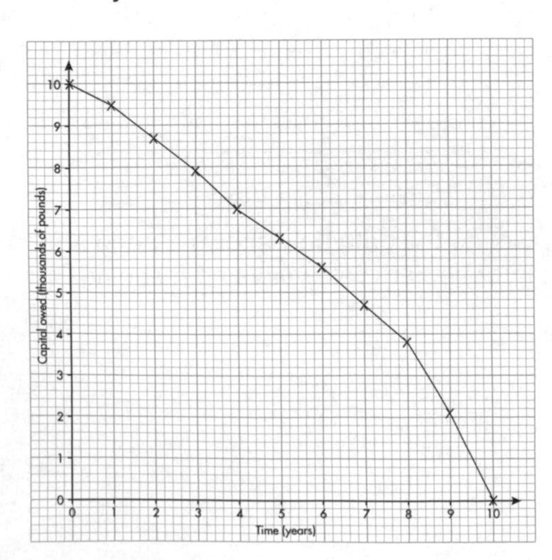

b 8 years. The amount owing dropped a lot after 8 years.

8 a 1917 **b** 1929 **c** 1958 to 1968

d Yes, as if the trend from 1991 continues it would be over £5 in 2010.

Exercise 6B

1

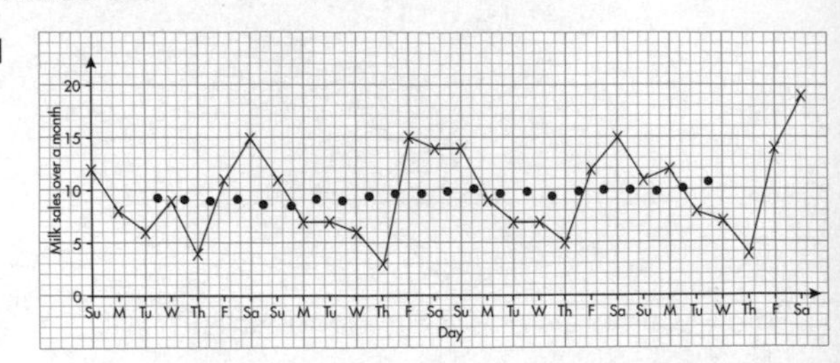

Moving averages are: 9.3, 9.1, 9.0, 9.1, 8.7, 8.6, 9.1 9.0, 9.4, 9.7, 9.7, 9.9, 10.1, 9.7, 9.9, 9.4, 9.9, 10.0, 10.0, 9.9, 10.1, 10.7

2 a The shop is open 4 days a week.

b 20, 21, 21.5, 22.25, 22.75, 23.25, 24.25, 25, 26

c

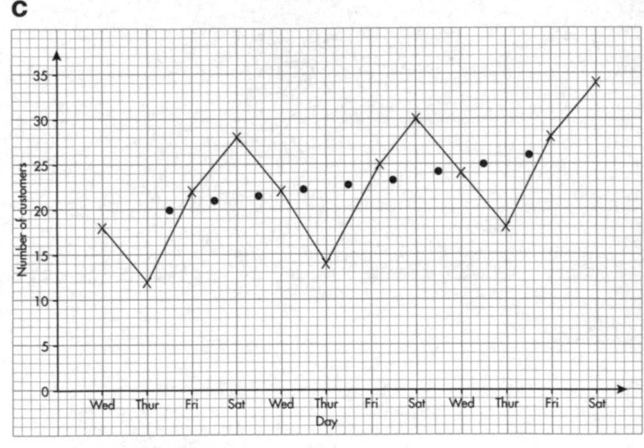

3 a 430 and 438 **b** 431

4 a Two examinations each year **b** 44 300

c i 40 100 **ii** (9000 + 71 200) ÷ 2 = 40 10

5 a 4.95, 5.1, 5.15, 5.2, 5, 5.15, 5.2, 5.25, 5.35, 5.5, 5.55, 5.65, 5.55

b The teacher is wrong. The tables are getting better but the spelling isn't so the moving average is only increasing because of the table

6 6.2 **7** $a = 25$, $b = 47$ **8** £21.90 **9** £25.2

Exercise 6C

1 a

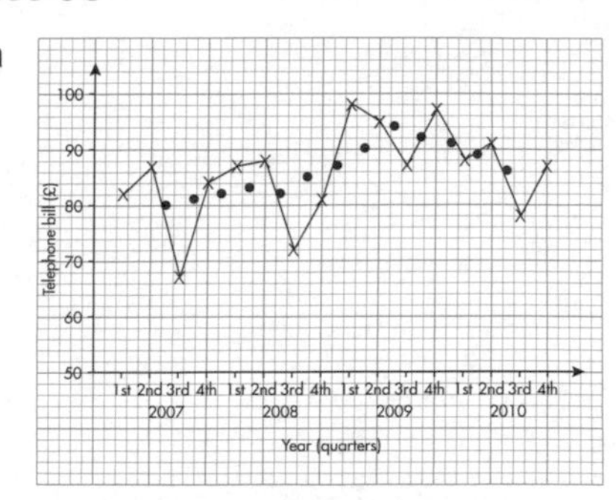

b Moving averages are: 80, 81, 82, 83, 82, 85, 87, 90, 94, 92, 91, 89, 86

c Recent fall may be due to moving to cheaper provider or using e-mail rather than making calls.

d Trend suggests next moving average about 83.5, so first quarter 2011 bill is £78.

2 a

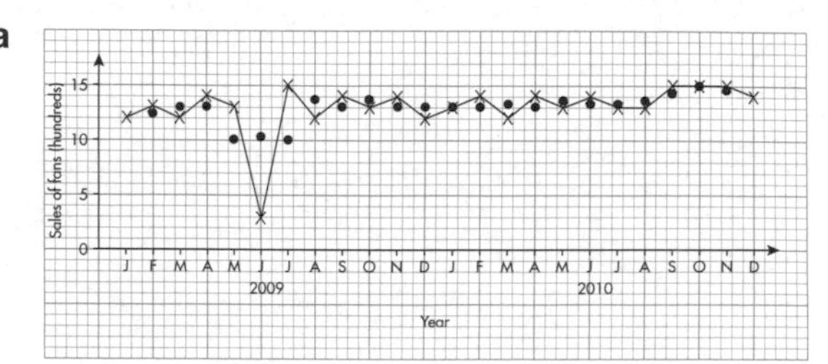

Moving averages are: 12.3, 13.0, 13.0, 10.0, 10.3, 10.0, 13.7, 13.0, 13.7, 13.0, 13.0, 13.0, 13.0, 13.3, 13.0, 13.7, 13.3, 13.3, 13.7, 14.3, 15.0, 14.7

b Apart from a blip in June 2009, sales showing slight improvement.

c Trend suggests next moving average about 14.6, so January 2011 sales estimate is 15.

3 a

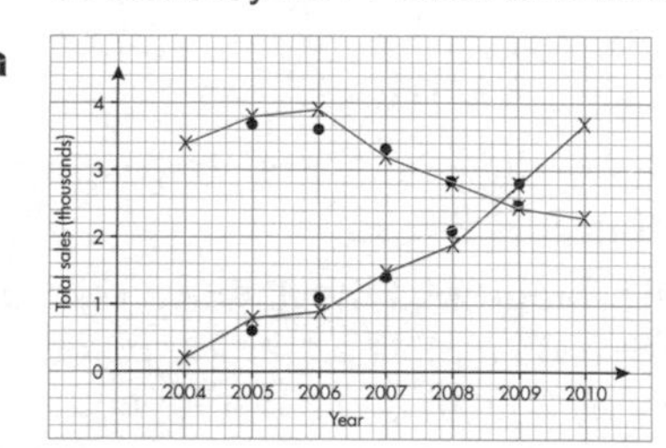

b Moving averages are: (videos) 3.7, 3.6, 3.3, 2.8, 2.5; (DVDs) 0.6, 1.1, 1.4, 2.1, 2.8

c Sales of DVD players increasing strongly, video recorder sales falling.

d Trend suggests next moving averages about 2.3 (videos) and 3.5 (DVDs), so 2011 sales estimates are 2.1 (videos) and 4.0 (DVDs).

4 a 60 and 64 **b** 64

c i

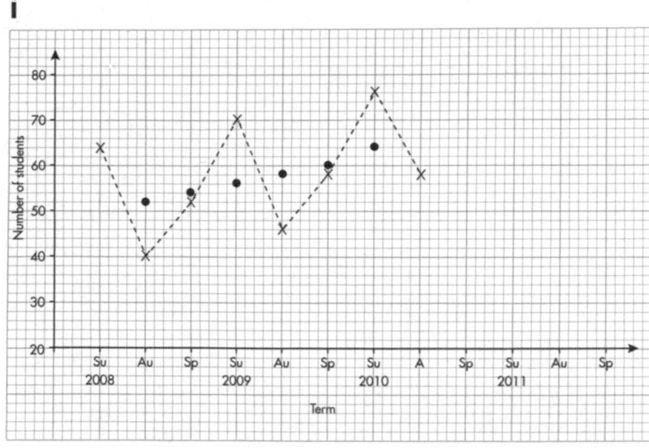

ii Estimated value for next moving average is 68, which gives a value for summer 2011 of $3 \times 68 - 64 - 58 = 82$

5 a 31 600, 29 400 and 28 100.

b Plotting the moving averages and extending them gives a value of 27 500 for the next moving average. This gives an entry for June 2012 of $2 \times 27\,500 - 12\,200 = 42\,800$.

6 Moving averages are 150, 153, 156.5, 158.5, 161, 163, 167

Plotting these gives a value of about 169 for the next moving average.

This gives a value for winter 2010 of $2 \times 169 - 188 = 150$

7 Extending the trend of the moving averages gives a value for the next moving average of about 35. This gives the value for January 2012 of $3 \times 35 - 36 - 28 = 41$

8 a £46.10

b Because the next value is higher than any of the previous 4

c The next moving average will be £41.50 which gives a value for December 2011 of £33.40

9 Moving averages are 108.10, 107.24, 108.39, 105.89, 109.54, 111.40, 112.55, 118.70, 118.60, 119.90, 120.40, 123.28, 124.33.

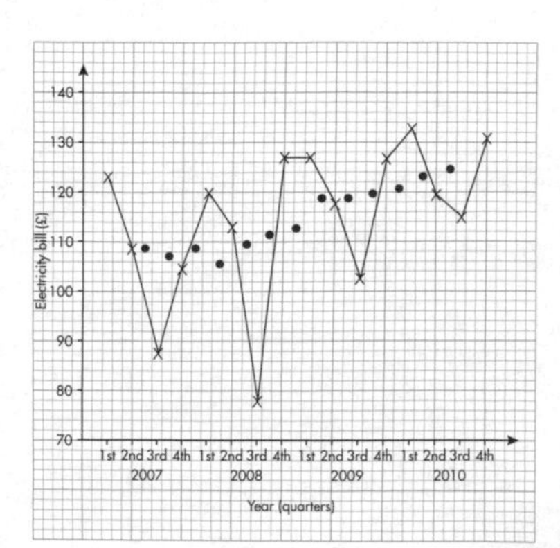

10

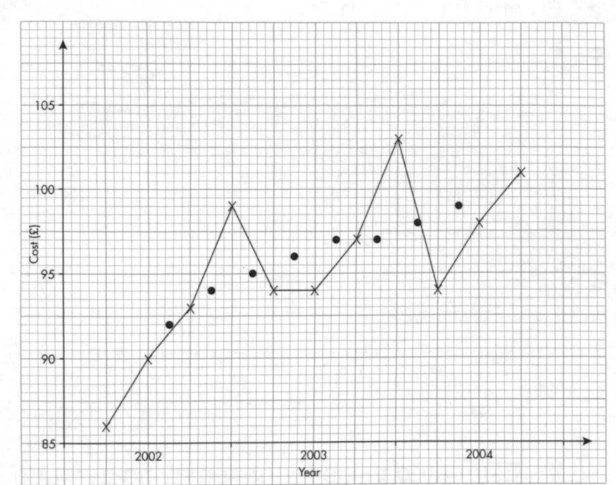

11

Raw data	12	8	22	15	11	25	18	17	28	21
3-point MA		14	15	16	17	18	20	21	22	

Examination questions

1 **a** 202 to 203 **b** 92%

c $197 \div 170 \times 100 = 115.88$ so 16% increase to 2 sf.

2 **a**

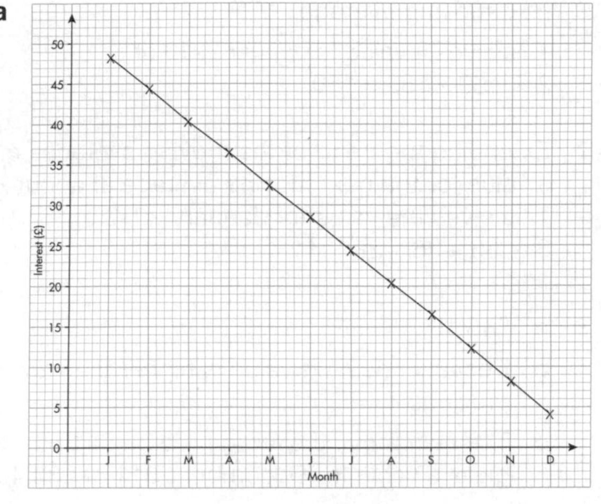

b

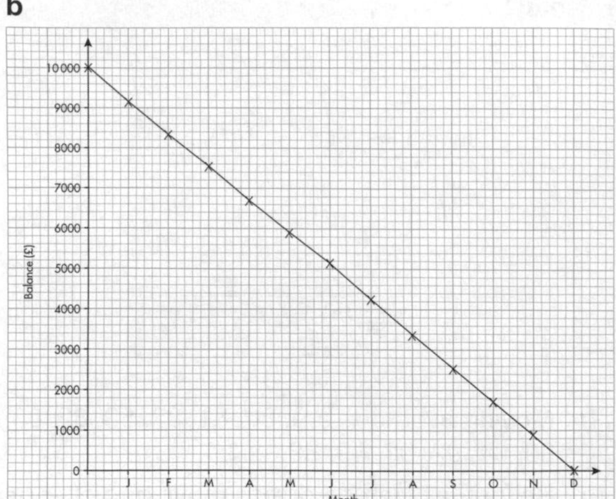

3 **a** There are two examinations a year **b** 43 500

4 **a** $(42 + 37 + 62) \div 3 = 141 \div 3 = 47$ **b** 52

5 **a** 51

b

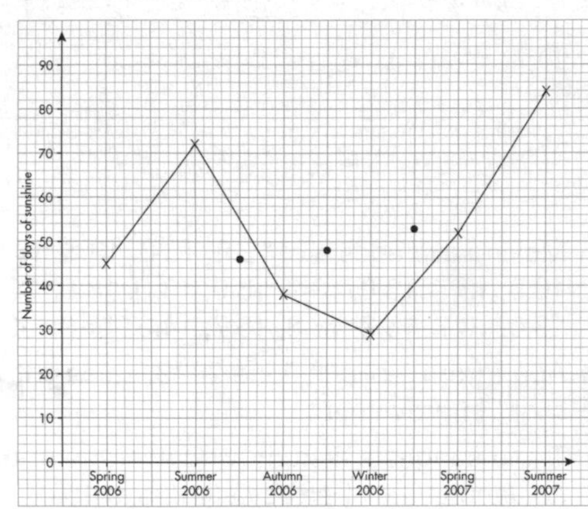

6 **a** $(48 + 30 + 81) \div 3 = 159 \div 3 = 53$

b **i** The pattern goes up 3, 4, 5, 6, 7 so the next number is $80 + 8 = 88$

ii Take the three-point moving average for summer 2005 (88) as a starting point, multiply by 3 (264) and subtract actual student numbers for spring and summer 2005, $264 - 57 - 114 = 93$

Chapter 7 Statistics: Risk

Quick check

1 **a** $\frac{1}{2}$ **b** $\frac{1}{3}$

c $\frac{2}{3}$ **d** 0

2 **a** M **b** F

c T **d** T

3 **a** 0.15, 0.24, 0.28, 0.33

b Yes as 4 occurs significantly more times than it should and 1 occurs significantly fewer times than it should.

Exercise 7A

1

Less than 50 m	0
50 m	32
500 m	20
1 km	2
2 km	1.5
15 km	0.5

2 Note that the hazard factor is very subjective and open to argument.

a i Higher than 90 feet.

ii Hazard factor 7–10 as it could cause a fatal accident.

b i Higher than 250.

ii Hazard factor 1–2 as can easily go out and buy more tiles.

c i Slightly smaller.

ii Hazard factor if wrong size 10 (in context) as it will need a completely new pane.

d i Save £3 or slightly more in case the licence goes up.

ii Hazard factor 0–1 as it will only need a few pounds adding if it does go up.

e i Buy three rolls, as can't buy part rolls.

ii Hazard factor 0 as 3 rolls is plenty.

f i Just over 150 g.

ii Hazard factor 0 as there is likely to be plenty of flour in the bag, but 10 (in context) if there is just 150 g and it is all used in the recipe. This will mean the dough will stick to the hands and the surface and could be ruined.

g i 200 at most. Some runners may not turn up.

ii Hazard factor 1–2 as there may be a few shirts left over, which might be a small waste of money.

h i Very hard to estimate but more than 150. The club will have to take a chance and guess how many will be needed. This will depend on previous year's entry numbers, the weather on the day, etc. Say 200 are ordered and not all used it is a waste. If more than 200 turn up there will be some upset runners although more shirts could be ordered and sent out later. There is a risk for those entering on the day, of course that they may not get a T-shirt.

ii Hazard factor 4–6.

3 a 3.37 b 3.37 : 1.63 ≈ 2.1 : 1

c 432 : 1

4 1 ⇒ B, 2 ⇒ E, 3 ⇒ D, 4 ⇒ C, 5 ⇒ A

5 India 0.033, Russia 0.20, USA 0.053, UK 0.02[illegible] Germany 0.011, Colombia 0.61, Japan 0.005, Ireland 0.0086

In order, Colombia, Russia, USA, India, UK, Germany, Ireland, Japan

6 a For example, men take more risks; they engage in more dangerous occupations.

b Age 55–64 = 1 : 0.62, Age 35–44 = 1 : 0.65, Age 5–14 = 1 : 0.75

c Decreases in comparison. Weaker males have already died off, or they take less risks, no longer engage in dangerous occupations.

7

	a Crimes per 1000 population		**b Crimes per 1000 km^2**	
Year 2008	**New York**	**North Dakota**	**New York**	**North Dakota**
Violent	3.98	1.66	554	5.90
Property	19.9	18.9	2775	67.1
Robbery	1.63	0.112	227	0.398
Aggravated assault	2.16	1.19	301	4.20
Burglary	3.37	3.28	470	11.6
Larceny theft	15.3	14.30	2126	50.6
Vehicle theft	1.29	1.37	179	4.87

c Most crimes are have a lower incidence per 1000 population in North Dakota and some are about half those of New York but many such as burglary are about the same and vehicle theft is higher in North Dakota. The claim may not be true.

8 a 120 feet b 2 seconds

c i 5 mph ii 60 feet iii 0.75–1 second

d 0.25 seconds

9 a School results better, higher average earnings, more businesses.

b Housing costs more, more crime, more cars, more accidents, weekly bills cost more.

10 a Dot com $(-2000 \times 0.1) + (-1000 \times 0.22) + (200 \times 0.34) + (1000 \times 0.34) = £164$
Blue chip $(-1000 \times 0.08 + (200 \times 0.9) + (1000 \times 0.02) = £120$

b Short-term losses are higher but long-term gains are higher.

c Very unlikely to lose any money.

Examination questions

1 a Low probability but a catastrophic effect as most plane crashes are totally fatal.

b 1 ⇒ A, 2 ⇒ C , 3 ⇒ D, 4 ⇒ B

2 a Risk of death in any one week is horse riding 1.7×10^{-6}, swimming 6×10^{-7}, running 4×10^{-5}, cycling 7.5×10^{-5} so in order of risk of fatality cycling, running, horse riding, swimming.

b Risk of an injury in any one week: horse-riding 5.1×10^{-4}, swimming 1×10^{-4}, running 5×10^{-4}, cycling 1.6×10^{-3} so in order of risk of injury cycling, horse riding, running, swimming.

c Risk of fatality: running goes up to 6×10^{-5} and cycling goes down to 3.75×10^{-5} so running becomes more risky than cycling.

Risk of injury: running goes up to 7.5×10^{-4} and cycling goes down to 8×10^{-4} so running becomes more risky than horse riding.

3 a Risk factor of the tsunami $= 0.003 \times 700 = 2.1$
Risk factor of the typhoon $= 0.07 \times 20 = 1.4$

b $2.1 \div 1.4 = 1.5$ so the risk factor of the tsunami is 1.5 times the risk factor of the typhoon.

4 a Bus 4, train 2.4, car 6, motorcycle 1.5

b The motorcycle gives the lowest risk factor for being late but has the highest chance of an accident. The train has the next lowest lateness risk factor and is the safest, so she is probably better off taking the train each day.

5 Indiana 6.5×10^{-7}, Florida 6.9×10^{-7}, Texas 6.8×10^{-7}, Wisconsin 3.7×10^{-6}, Colorado 7×10^{-6}, Maine 3.1×10^{-5}, Alaska 7.1×10^{-5}

So the rank order of danger is Alaska, Maine, Colorado, Wisconsin, Florida, Texas, Indiana.

6 $150 \times 0.03 = 4.5$, $50 \times 0.12 = 6$, $6 \div 4.5 = 1.33$ times as likely.

Chapter 8 Number: Understand and use Venn diagrams to solve problems

1 a $P(A) = \frac{7}{10}$ **b** $P(B') = \frac{3}{10}$

2 a

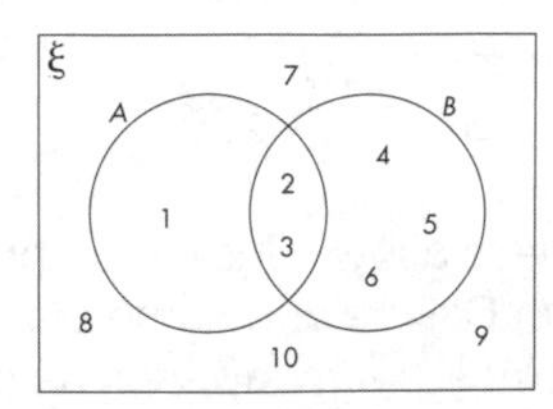

b i $P(A) = \frac{3}{10}$ **ii** $P(A \cup B) = \frac{6}{10}$

iii $P(A \cap B) = \frac{2}{10} = \frac{1}{5}$

Exercise 8A

1 a ξ F R 50 30 100 **b** 30

2 3 **3** 6 **4** 41 **5** None **6** 17 **7** 10

8 a

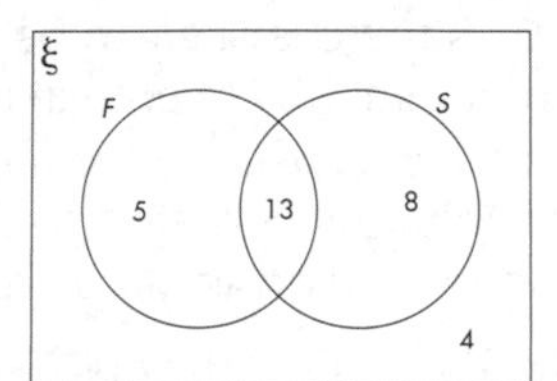

b 13

9 a

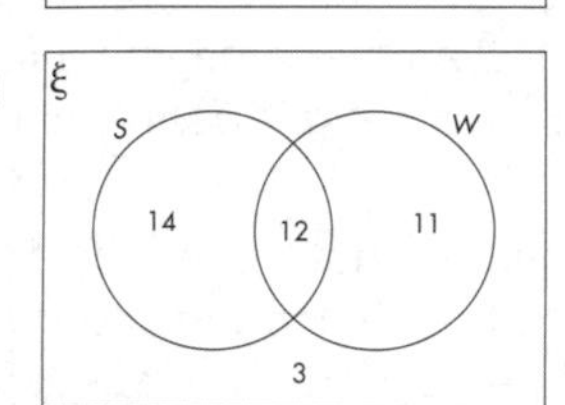

b 23

10 a 220 **b** 940 **11 a** 60 **b** 0

12 48 **13** 72

14 ξ P Q 18 10 18 16

Examination questions

1 a

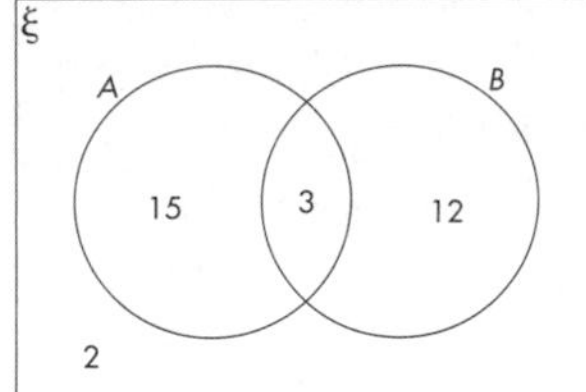

b 27

2 7

3 a

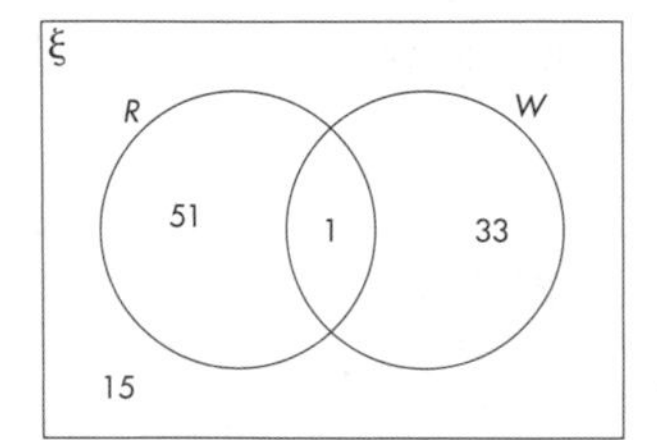

b 1

4 a 46

ξ, T, F: 175, 64, 46; 15

b

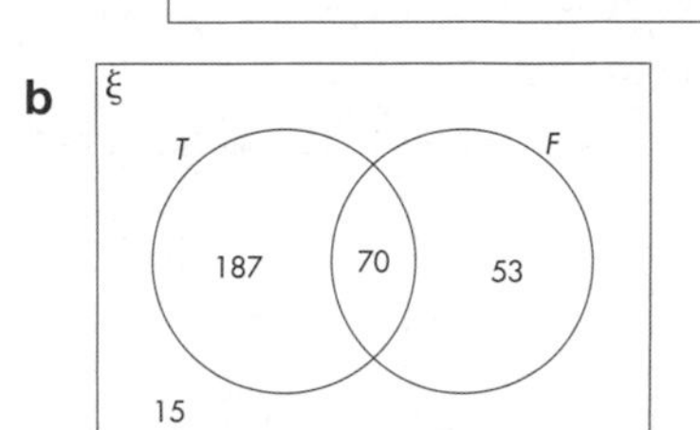

5 110 **6** 27

7

ξ, P, Q: 34, 4, 17; 13

Chapter 9 Algebra: Algebraic equivalence and quadratic sequences

Quick check

1 a 2 **b** –4

2 a 2, 5, 8, 11, 14
b 7, 11, 15, 19, 23
c 1, 7, 13, 19, 25

3 a $3n + 2$ **b** $24 - 4n$ **c** $5n - 1$

Exercise 9A

1 a Line 2, $y = 2x - 1$ **b** 4

2 a $-2 \times 3 + 4 = -6 + 4 = -2$ **b** $x = 3$

3 a $18 - 3 = 15$, $15 \div 2 = 7.5$, $18 \div 4 = 4.5$, $4.5 + 3 = 7.5$
b $y = x - \frac{3}{2}$, $y = 18 - \frac{3}{2} = 16\frac{1}{2}$
c $y = \frac{x-3}{2}$ **d** 209

4 a 13 **b** 3
c i 22 **ii** Yes, $3 \times 25 - 2 = 73$

5 a $x + y = 6$ **b** $x + y = 9$
c No as the y-coordinate is 2 less than the x-coordinate not 2 more; the line is $y = x + 2$.

6 a $2 \times -1 + 1 = -2 + 1 = -1$, so no.
b $y = \frac{x-1}{2}$
c This is the inverse flow diagram:

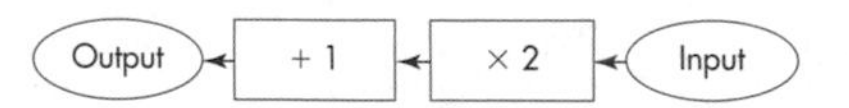

7 a Take $x = 5$, $y = 2 \times 5 + 4 = 14$; take $x = 14$, $y = 14 \div 2 - 2 = 7 - 2 = 5$
b Input → × 2 → + 4 → Output

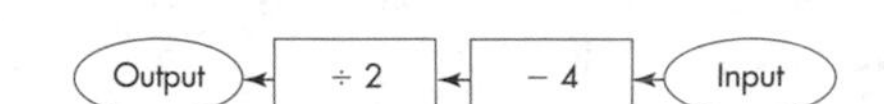

c $\frac{x-4}{2} = \frac{x}{2} - 2$

8 a –2 Output ← × –1 ← – 6 ← Input
b –4
c Let $x = 2$, $y = 6 - 2 = 4$; let $y = 4$, $y = 6 - 4 = 2$; so self-inverse

9 Let $x = 3$, $y = \frac{1}{3}$; let $x = \frac{1}{3}$, $y = 1 \div \frac{1}{3} = 3$; let $x = \frac{2}{5}$, $y = \frac{5}{2}$; let $x = \frac{5}{2}$, $y = \frac{2}{5}$; so self-inverse

10 $(2x)^2 = 4x^2 \Rightarrow x^2 \times 4 = 4x^2$

11 **a** $2x - 1 = x + 2$, $2x - x = 2 + 1$, $x = 3$, $y = 3 + 2 = 5$ so the point is (3, 5).
b Input $x = 3$ to each diagram. The output is 5 in each case.

Exercise 9B

1 **a** **i** 34, 43 **ii** goes up 3, 4, 5, 6, etc.
b **i** 24, 31 **ii** goes up 1, 2, 3, 4, etc.
c **i** 54, 65 **ii** goes up 5, 6, 7, 8, etc.
d **i** 57, 53 **ii** goes down 10, 9, 8, 7, etc.

2 **a** 4, 7, 12, 19, 28 **b** 2, 8, 18, 32, 50
c 2, 6, 12, 20, 30 **d** 4, 9, 16, 25, 36
e 2, 8, 16, 26, 38 **f** 4, 7, 14, 25, 40

3 **a** $2n + 1$ **b** n
c $n(2n + 1) = 2n^2 + n$
d $2n^2 + n + 1$

4 **a** n **b** $n + 1$
c $n(n + 1)$ **d** 9900 square units

5 **a** Yes, constant difference is 1 **b** No
c Yes, constant difference is 2 **d** No
e Yes, constant difference is 1 **f** No

6 **a** $4n + 4$ **b** n^2
c $n^2 + 4n + 4$ **d** $n^2 + 4n + 4$
e The sides of the large squares are of length $n + 2$ so the total number of squares is $(n + 2)^2$ which is the same answer as **c**.

7 **a** Table 10, 15, 21; 6, 10, 15; 16, 25, 36
b **i** 45 **ii** 100

Exercise 9C

1 **a** **i** 36, 49 **ii** n^2
b **i** 35, 48 **ii** $n^2 - 1$
c **i** 38, 51 **ii** $n^2 + 2$
d **i** 39, 52 **ii** $n^2 + 3$
e **i** 34, 47 **ii** $n^2 - 2$
f **i** 35, 46 **ii** $n^2 + 10$

2 **a** **i** 37, 50 **ii** $(n + 1)^2 + 1$
b **i** 35, 48 **ii** $(n + 1)^2 - 1$
c **i** 41, 54 **ii** $(n + 1)^2 + 5$
d **i** 50, 65 **ii** $(n + 2)^2 + 1$
e **i** 48, 63 **ii** $(n + 2)^2 - 1$

3 **a** **i** $n^2 + 4$ **ii** 2504
b **i** $3n + 2$ **ii** 152
c **i** $(n + 1)^2 - 1$ **ii** 2600
d **i** $n(n + 4)$ **ii** 2700
e **i** $n^2 + 2$ **ii** 2502
f **i** $5n - 4$ **ii** 246

4 **a** $2n^2 - 3n + 2$ **b** $3n^2 + 2n - 3$
c $\frac{1}{2}n^2 + \frac{5}{2}n + 1$ **d** $\frac{1}{2}n^2 + 4\frac{1}{2}n - 2$
e $\frac{1}{2}n^2 + 1\frac{1}{2}n + 6$ **f** $\frac{1}{2}n^2 + 1\frac{1}{2}n + 2$

5 $6n^2$ **6** **a** 26 **b** $1\frac{1}{2}n^2 + \frac{1}{2}n$ **c** 8475

7 **a** 45
b nth term is $\frac{1}{2}n^2 + \frac{1}{2}n$ so $\frac{1}{2} \times 15 \times 15 + \frac{1}{2} \times 15 =$ 120, so no.

8 Front face is n^2, sides faces are $n \times (n + 1) = n^2 + n$ so total surface area is $2 \times n^2 + 4 \times (n^2 + n) = 6n^2 + 4n$.

9 Sequence is 1, 7, 19, 37. nth term is $3n^2 - 3n + 1$ so the 100th hexagonal number is 29 701.

Examination questions

1 **a** −10
b Input = 0.5, output = 0.5. Draw the line $y = x$ and find the intersection.

2 **a** 5 → 10 → 18 → 9 → 4
b $n \to 2n \to 2n + 8 \to n + 4 \to 4$

3 4, 9, 18

4 First term is $(2 \times 0 \times -1) \div 5 = 0$, second term is $(2 \times 1 \times 0) \div 5 = 0$, third term is $(2 \times 2 \times 1) \div 5 = 0$

5 $2n^2 - 2n + 3$

6 **a** nth term is $n^2 + 2n$, so if $n = 12$, $12^2 + 2 \times 12 =$ 168, so yes. **b** 1680

7 nth term is $2n^2 - n$, so if $n = 20$, $2 \times 20^2 - 20 = 3$

Chapter 10 Geometry: Tessellations and tiling patterns

Quick check

1 a 155° **b** 54°, 216°

2 108°

Pentomino activities

Activity D

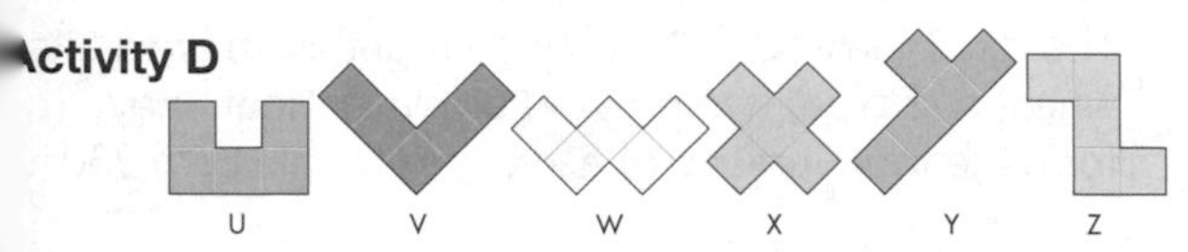

Activity E

The pentominoes with line symmetry are I, T, U, V, W, X.

The pentominoes with rotational symmetry are I, X, Z

Activity F

There are different ways of tessellating the pentominoes. These are just one way.

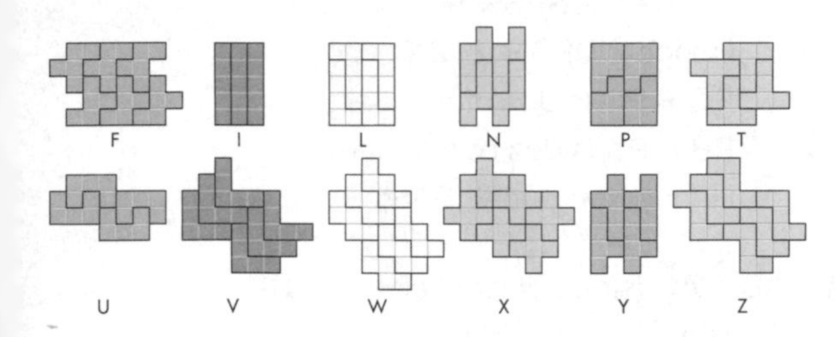

Activity G

These are just one of the possible answers in each case.

a

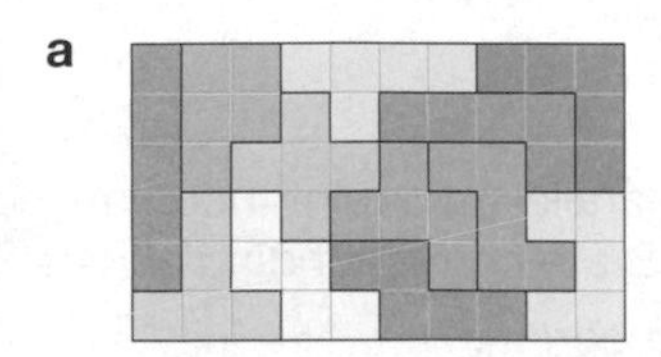

b

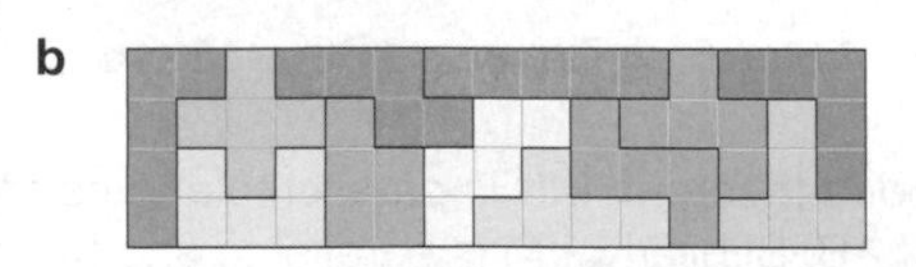

c

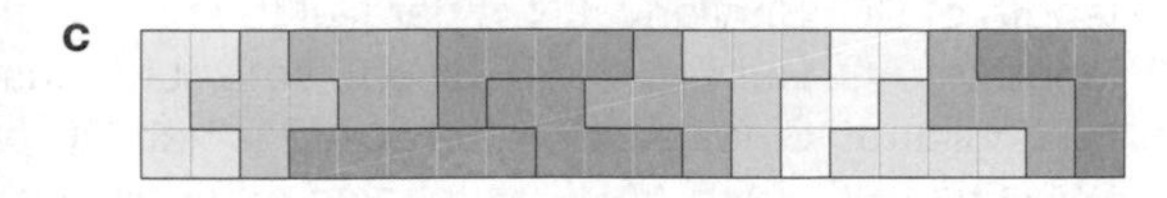

Exercise 10A

1 E, C, D, A, B

a 105°

b Because 105° does not fit with either the octagon or the hexagon.

3 a Square and equilateral triangle

b At a vertex, two squares and an equilateral triangle and a regular hexagon meet so the angle in the hexagon will be 360° – 90° – 90° – 60° = 120°

4 a

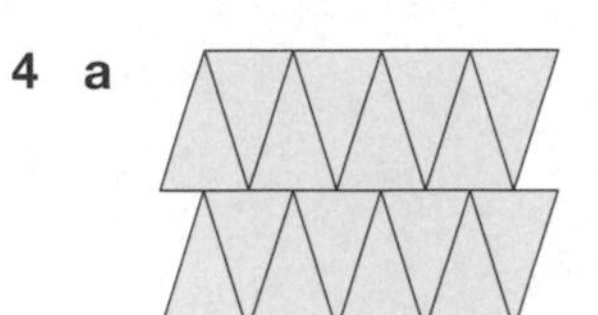

b

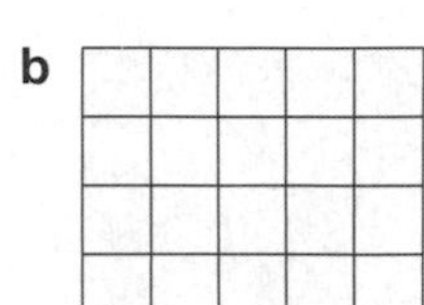

c The angles where two squares and three equilateral triangles meet will be 90° + 90° + 60° + 60° + 60° = 360° so they tessellate.

5

6 a The interior angle of a regular pentagon is 108°. This does not divide into 360° exactly.

b The pentagons meet regular hexagons with internal angles of 120°. Therefore the sum of two of the angles of the pentagon is 240°. A regular pentagon has interior angles of 108°, which does not divide into 240°. In fact, the angles of the pentagon are 90°, 90°, 120°, 120°, 120°.

7 $4 \times (180 - 3x) + 2x = 360$, $720 - 10x = 360$, $x = 36$

8 a Two of the shapes will fit together to form a parallelogram.

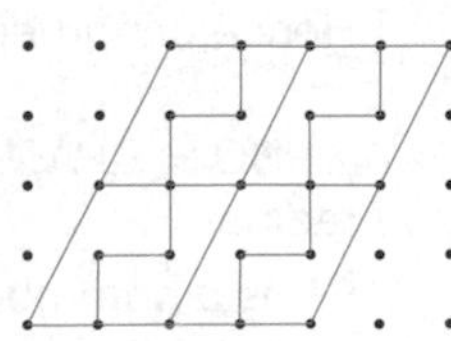

b Because there is no part of the shape that will fit into the angle of 45°.

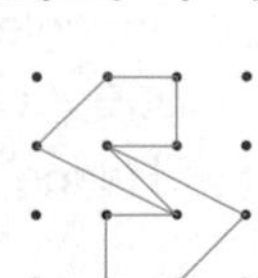

Examination questions

1 E, C, D, A, B

2 **a** A regular hexagon and a square

b The interior angle of a hexagon is 120°, the square has an angle of 90°, so the decagon has an interior angle of 360° – 120° – 90° = 150°.

3

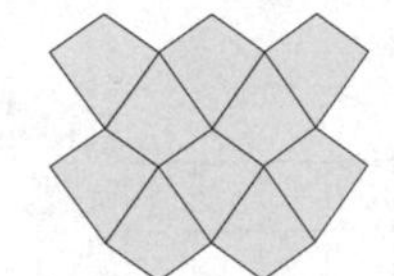

4

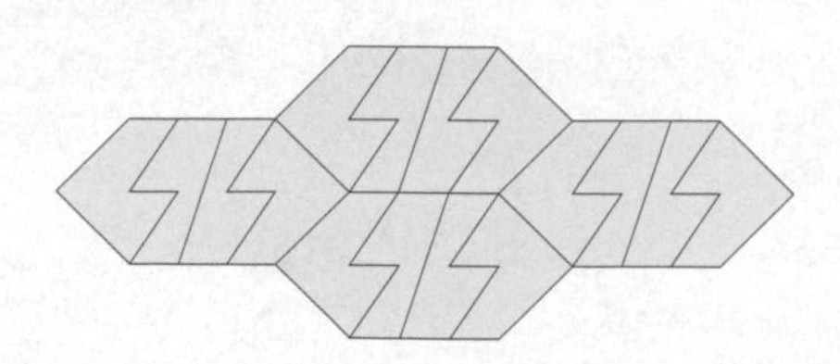

5 230°

6 The red, blue and yellow shape together make a kite; where two blue, two red and a yellow meet, they form the four vertices of a kite, with angle sum 360°.

Chapter 11 Geometry

Quick check

1 2.4 : 9.6

2 **a** 90° **b** 90° **c** 30°

Exercise 11A

1 **a** No as AAA is not a condition for congruency.

b No as the lengths could be different. Only one angle is guaranteed to be the same.

c No as even though one angle and one side are the same the other angles could be anything.

d Yes as the angles must be the same and one side is the same so ASA would apply.

2 **a** No SSA **b** Yes ASA **c** No AAA

d Yes SAS **e** Yes SSS

3 BC = AD (isosceles trapezium)

DC is a common side

∠D = ∠C (isosceles trapezium)

Hence congruent (SAS)

4 ∠PBO = ∠PAO = 90° (angle between tangent and radius)

OP is a common hypotenuse of a right-angled triangle

OB = OA (radius of circle)

Hence congruent (RHS).

5 ∠B = ∠E (angles in a regular pentagon)

AB = AE (sides of a regular pentagon)

BC = ED (sides of a regular pentagon)

Hence congruent (SAS).

6 **a** ∠PBN = ∠PSN = 90° (angle of square)

PN is a common hypotenuse of two right-angled triangles

PS = PB (sides of a square)

Hence congruent (RHS)

b ∠QPB + ∠BPS = 90°

∠DPS + ∠BPS = 90°

Hence ∠DPS = ∠QPB

PQ = PD (sides of a square)

PB = PS (sides of a square)

Hence congruent (SAS).

7 AB = AC (sides of isosceles ΔABC)

AP is a common side

PB = PC (tangents from a common point are equal

Hence congruent (SSS).

Exercise 11B

1 **a** PQ and BC are parallel (midpoint theorem), so ∠AQP and ∠ACB are corresponding angles.

b By Pythagoras' theorem,

$PQ = \sqrt{4.5^2 + 6^2} = 7.5$ cm so BC = 15 cm

2 By midpoint theorem, MN is parallel to and half the length of AB. Similarly, LM is parallel to and half th length of BC. Similarly ,LN is parallel to and half th length of AC. But LM = MC (given), so ΔALM must be equilatral. Since AL is half AB, AM is half QC ar LM is half BC, then ΔABC must also be equilateral

3 **a** 9 cm **b** 3 cm

4 By midpoint theorem, PQ is parallel to and half th length of BC and QR is parallel to and half the length of AB. Hence opposite sides are equal in length and parallel, so PQRB is a parallelogram

5 AQ = BQ (Q is midpoint of AB)
$\angle AQP = \angle PQB = 90°$ (parallel to CB, by midpoint theorem)
PQ is a common side in ΔAQP and ΔPQB
Hence ΔAQP and ΔBQP are congruent (SAS) so
$AP = PB = \frac{1}{2}AC$

6 AB = AC (given)
AD is a common side
BD = DC (D is midpoint of BC)
Hence ΔABD and ΔADC are congruent (SSS).
$\therefore \angle BDA = \angle CDA = 90°$ (angles on straight line)
EF is parallel to BC (midpoint theorem)
Let G be the point where AD intersects EF. Hence EF is perpendicular to AD.
$EG = \frac{1}{2}BD$ and $GF = \frac{1}{2}DC$ (midpoint theorem)
Hence EF bisects AD.

7 **a** $\frac{1}{2} \times 9 \times 10 \times \sin B = 31.82 \Rightarrow B = 45°$
b $AC^2 = 10^2 + 9^2 - 2 \times 9 \times 10 \times \cos 45° = 53.72$
$AC = 7.33$ cm

8 $LP = \frac{1}{2}AC$ (midpoint theorem) = AL (L is midpoint AB, given)
$\angle BLM = \angle ALP$ (opposite angles)
Hence ΔAPL is congruent to ΔMLB (SAS)
$CQ = \frac{1}{2}AB$ (midpoint theorem) = LB (L is midpoint AB, given)
CM = MB (M is midpoint of CB, given)
$\angle CMQ = \angle LMB$ (opposite angles)
Hence ΔCMQ is congruent to ΔMLB (SAS)
Both ΔCMQ and ΔAPL are congruent to ΔMLB so they must be congruent to each other.

xercise 11C

1 **a** 7.5 cm **b** 7.2 cm

2 No, as $5 : 9 \neq 4 : 7$

3 **a** 8 cm **b** 5.625 cm

4 5.25 cm **5** 14.4 cm **6** 8 cm

7 **a** 6.8 cm **b** 9.2 cm

8 **a** 1 : 3 **b** 2 : 3 **c** 2 : 9

Exercise 11D

1 6.75 cm **2** 7.5 cm

3 **a** 6 cm **b** 4.8 cm

4 2.6 cm

5 **a** 3.5 cm **b** 5.25 cm

6 14.5 cm **7** Yes radius is 10 cm in each case

8 **a** 8 cm **b** 5 cm

Examination questions

1 5.25 cm

2 **a** 4.5 cm **b** 3.75 cm

3 11 cm **4** 6.75 cm

5 SR is parallel to and half the length of AC. (midpoint theorem)
Similarly PQ is parallel to and half the length of AC.
Similarly, taking diagonal DB, RQ and SP are parallel to and half the length of DB.
As the diagonals of a rectangle are the same length, SRPQ has two pairs of opposite sides that are equal and parallel. Hence it is a rhombus.

6 $MP = \frac{1}{2}CB = NC$ (midpoint theorem on ΔCBD)
$NP = \frac{1}{2}DC = MC$ (midpoint theorem on ΔCPD)
MN is a common side
Hence ΔMNP and ΔMNC are congruent (SSS).

hapter 12 Algebra: Graphs

uick check

1 35 mph

a There is no movement. The vehicle is stationary.
b Where the gradient is steepest

Exercise 12A

1 **a** 20 m/s **b** 100 metres
c 150 metres **d** 750 metres

2 **a** 15 km **b** 5 km/h

3 **a** AB, greatest area **b** 45 miles **c** 135 miles

4 15 m/s

5 a

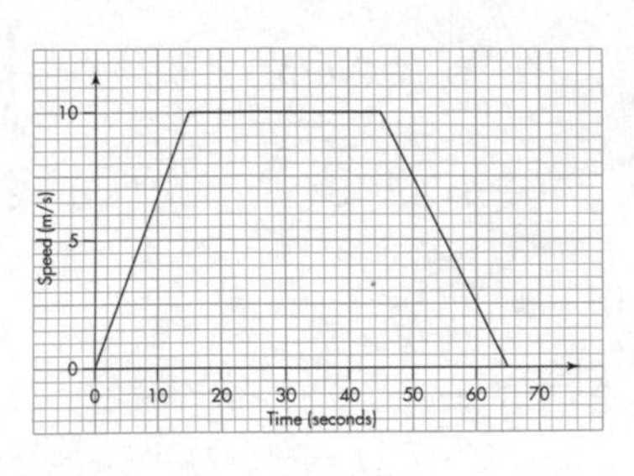

b 475 metres

6 a Could be true or false
b Must be true
c Must be true
d Could be true or false
e Could be true or false

Exercise 12B

1 80 miles, underestimate

2 250 metres, overestimate

3 900 metres, underestimate

4 75 metres, underestimate

5 180 metres, underestimate

6 8 miles, underestimate

7 20 miles, overestimate

8 a i Car starts from rest and speeds up to 10 m/s after 20 seconds. It then travels at a constant speed of 10 m/s for 30 seconds, and then speeds up again to reach 20m/s in the next 10 seconds.
ii 120 metres
b i The lorry increases speed at a steady rate whereas the car speeds up quickly at first but then levels off to a constant speed and then speeds up at an increasing rate to reach 20 m/s.
ii Lorry travels further (600 metres as against car, approximately 550 metres) as area under graph is greater.

Exercise 12C

1 a It costs £2.50 for the first 600 metres and then 20p for every extra 200 metres.
b Charges go up in steps, for example 650 metres would cost £2.70 and so would 700 metres. It is not charged as 1p for every 10 metres.

2 a

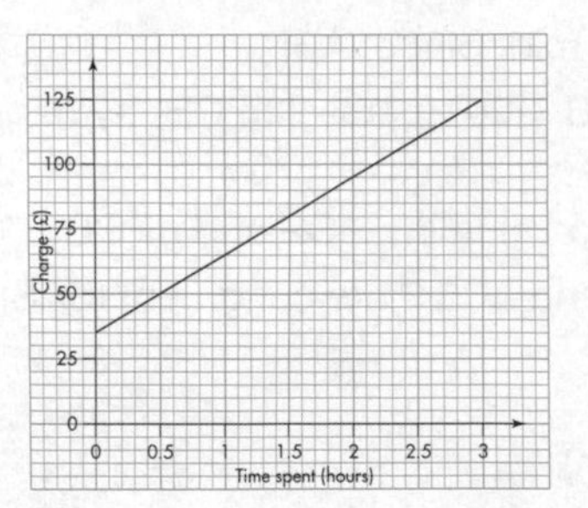

b £95

3 a £15 **b** 300 minutes **c** 5p

4 a £20 **b** £140 **c** £10
d $T = d + 12m$

5 a £80 **b** 30 **c** £2.40

6 a 10 mph
b Faster as gradient more or line steeper.

7 5 m/s^2, 0 m/s^2, 2.5 m/s^2

8 40 km/h^2, 30 km/h^2, 100 km/h^2

9 a $\frac{3v}{10}$ m/s^2 **b** 337.5 m

Exercise 12D

1 a tangent drawn **b** 10 m/s **c** 0 m/s

2 a i 10–12 km/h **ii** 20–22 km/h
b 2 hours

3 0.8 m/s^2 **4** 1.4 m/s^2

5 a 7.3 m/s^2 **b** –7.3 m/s^2

6 –(0.8–0.9) m/h^2

7 Any two points where the gradient of one is the negative of the other, e.g. at 1s and 3s.

Exercise 12E

1

2 Equation 1 is graph C, Equation 2 is graph B, Equation 3 is graph A

3

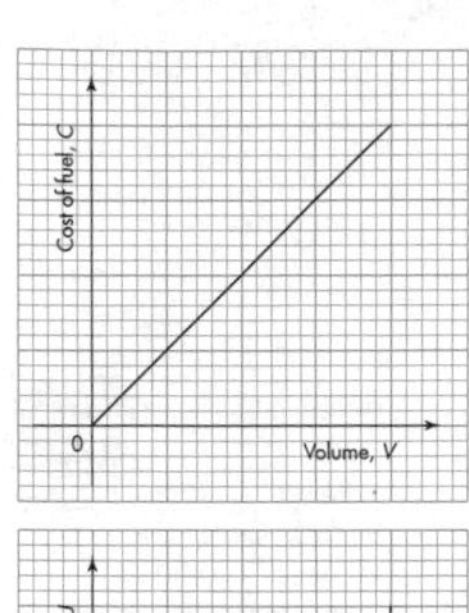

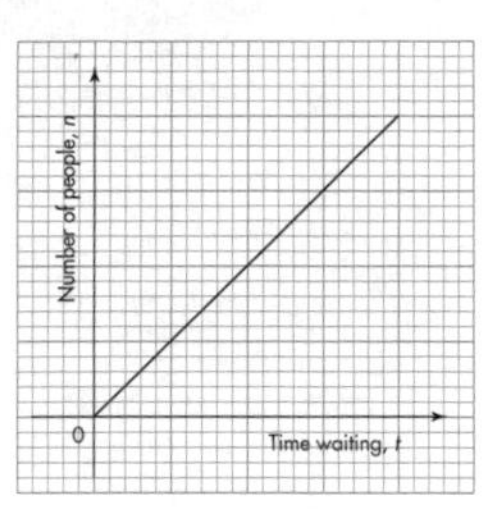

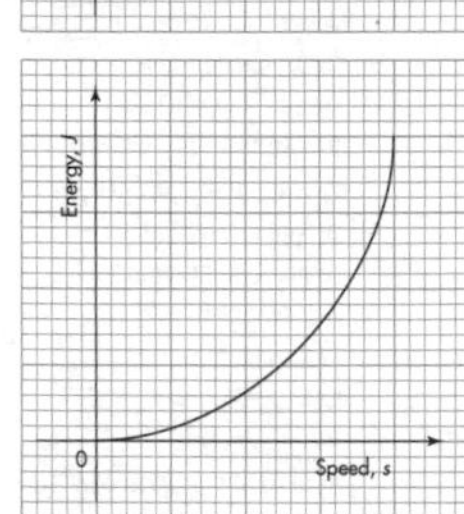

4 **a** $k = 2$ **b** $k = 20$

c $k = 36$ **d** $k = 250$

5 **a**

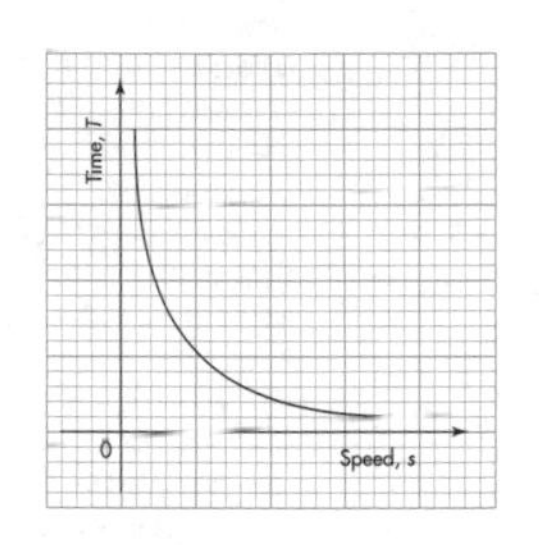

b

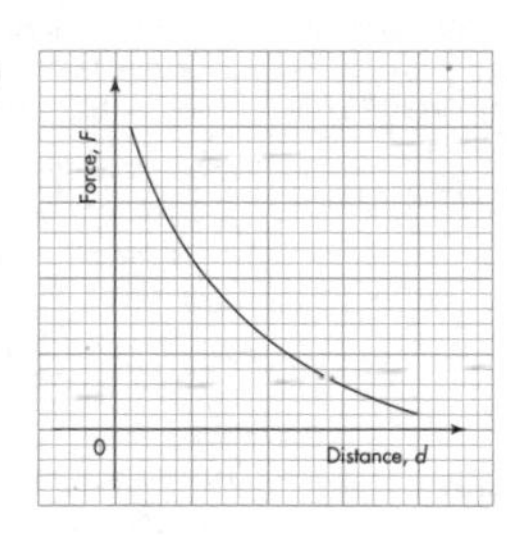

c

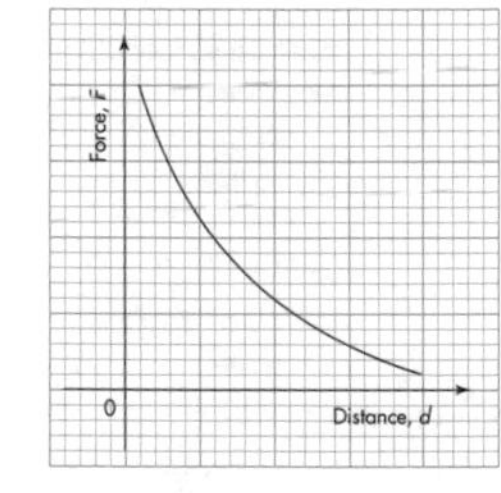

6 **a** Graph B

b reading from the graph, when $K = 40$, $v = 4$, so $40 = 0.5 \times m\ 3\ 4^2$, $m = 5$.

7 **a** Graph A

b Reading from the graph, when $V = 20$, $P \approx 72$, so $20 \times 72 = k \times 20$ so $k \approx 72$. Check with a second point, when $V = 30$, $P \approx 49$, so $30 \times 49 = k \times 20$, so $k \approx 73.5$

Examination questions

1 **a** £30 **b** 5p

c **i**

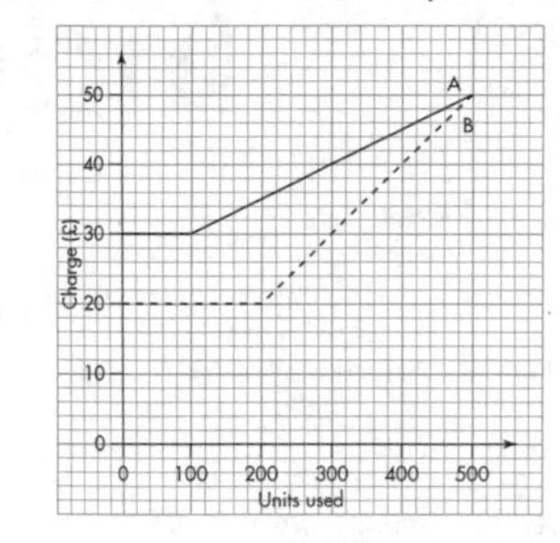

ii Company A, since company B line is going over Company A line.

2 **a** BC and DE **b** AB and CD

c 62.5 km **d** 10.4 km/h

3 **a**

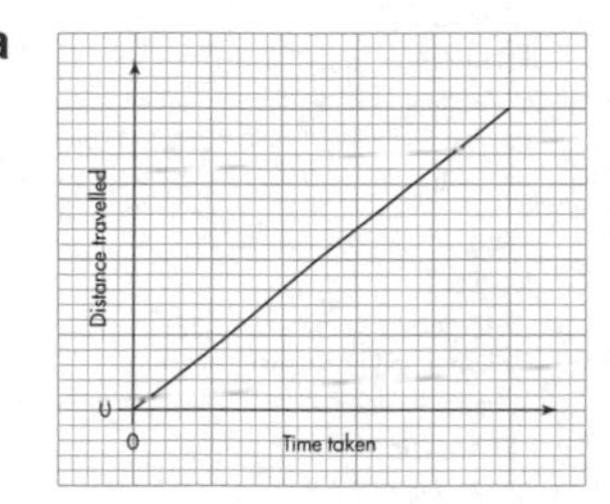

b Speed is constant

4 1.8 km **5** **a** 550 metres **b** $0.\dot{3}$ m/s^2

6 **a** 600 metres

b **i** 300 metres

ii Not worked out the exact area, used a triangle as approximate area.

7 **a** 0.56 m/s^2

b Stops accelerating before starting decelerating

8 **a** 16.7 km/h^2

b When $t = 0$ as steepest

9 **a** B **b** It decreases

10 200 g

Index